T0269044

The aim of this book is to explain the shape of Greek mathematical thinking. It can be read on three levels: first as a description of the practices of Greek mathematics; second as a theory of the emergence of the deductive method; and third as a case-study for a general view on the history of science. The starting point for the enquiry is geometry and the lettered diagram. Reviel Netz exploits the mathematicians' practices in the construction and lettering of their diagrams, and the continuing interaction between text and diagram in their proofs, to illuminate the underlying cognitive processes. A close examination of the mathematical use of language follows, especially mathematicans' use of repeated formulae. Two crucial chapters set out to show how mathematical proofs are structured and explain why Greek mathematical practice manages to be so satisfactory. A final chapter looks into the broader historical setting of Greek mathematical practice.

Reviel Netz is a Research Fellow at Gonville and Caius College, Cambridge, and an Affiliated Lecturer in the Faculty of Classics.

IDEAS IN CONTEXT 51

THE SHAPING OF DEDUCTION IN GREEK MATHEMATICS

IDEAS IN CONTEXT

The books in this series will discuss the emergence of intellectual traditions and of related new disciplines. The procedures, aims and vocabularies that were generated will be set in the context of the alternatives available within the contemporary frameworks of ideas and institutions. Through detailed studies of the evolution of such traditions, and their modification by different audiences, it is hoped that a new picture will form of the development of ideas in their concrete contexts. By this means, artificial distinctions between the history of philosophy, of the various sciences, of society and politics, and of literature may be seen to dissolve.

The series is published with the support of the Exxon Foundation.

A list of books in the series will be found at the end of the volume.

THE SHAPING OF DEDUCTION IN GREEK MATHEMATICS

A Study in Cognitive History

REVIEL NETZ

PUBLISHED BY THE PRESS SYNDICATE OF THE UNIVERSITY OF CAMBRIDGE
The Pitt Building, Trumpington Street, Cambridge, United Kingdom

CAMBRIDGE UNIVERSITY PRESS
The Edinburgh Building, Cambridge CB2 2RU, UK
40 West 20th Street, New York NY 10011–4211, USA
477 Williamstown Road, Port Melbourne, VIC 3207, Australia
Ruiz de Alarcón 13, 28014 Madrid, Spain
Dock House, The Waterfront, Cape Town 8001, South Africa

http://www.cambridge.org

First published 1999
First paperback edition 2003

Typeset in 11/12½pt Baskerville No. 2 [GC]

A catalogue record for this book is available from the British Library

Library of Congress cataloguing in publication data

Netz, Reviel.
The shaping of deduction in Greek mathematics: a study in
cognitive history / Reviel Netz.
p. cm. – (Ideas in context; 51)
Includes bibliographical references and index.
ISBN 0 521 62279 4 (hardback)
1. Mathematics, Greek. 2. Logic. I. Title. II. Series.
QA27.G8N47 1999
510′.938–dc21 98–20463 CIP

ISBN 0 521 62279 4 hardback
ISBN 0 521 54120 4 paperback

Transferred to digital printing 2003

To Maya

Contents

Preface

This book was conceived in Tel Aviv University and written in the University of Cambridge. I enjoyed the difference between the two, and am grateful to both.

The question one is most often asked about Greek mathematics is: 'Is there anything left to say?' Indeed, much has been written. In the late nineteenth century, great scholars did a stupendous work in editing the texts and setting up the basic historical and mathematical framework. But although the materials for a historical understanding were there, almost all the interpretations of Greek mathematics offered before about 1975 were either wildly speculative or ahistorical. In the last two decades or so, the material has finally come to life. A small but highly productive international community of scholars has set up new standards of precision. The study of Greek mathematics today can be rigorous as well as exciting. I will not name here the individual scholars to whom I am indebted. But I can – I hope – name this small community of scholars as a third institution to which I belong, just as I belong to Tel Aviv and to Cambridge. Again I can only express my gratitude.

So I have had many teachers. Some were mathematicians, most were not. I am not a mathematician, and this book demands no knowledge of mathematics (and only rarely does it demand some knowledge of Greek). Readers may feel I do not stress sufficiently the value of Greek mathematics in terms of mathematical content. I must apologise – I owe this apology to the Greek mathematicians themselves. I study form rather than content, partly because I see the study of form as a way into understanding the content. But this content – those discoveries and proofs made by Greek mathematicians – are both beautiful and seminal. If I say less about these achievements, it is because I have looked elsewhere, not because my appreciation of them is not as keen as it should be. I have stood on the shoulders of giants –

to get a good look, from close quarters, at the giants themselves. And if I saw some things which others before me did not see, this may be because I am more short-sighted.

I will soon plunge into the alphabetical list. Three names must stand out – and they happen to represent the three communities mentioned above. Sabetai Unguru first made me read and understand Greek mathematics. Geoffrey Lloyd, my Ph.D. supervisor, shaped my view of Greek intellectual life, indeed of intellectual life in general. David Fowler gave innumerable suggestions on the various drafts leading up to this book – as well as giving his inspiration.

A British Council Scholarship made it possible to reach Cambridge prior to my Ph.D., as a visiting member at Darwin College. Awards granted by the ORS, by the Lessing Institute for European History and Civilization, by AVI and, most crucially, by the Harold Hyam Wingate Foundation made it possible to complete graduate studies at Christ's College, Cambridge. The book is a much extended and revised version of the Ph.D. thesis, prepared while I was a Research Fellow at Gonville and Caius College. It is a fact, not just a platitude, that without the generosity of all these bodies this book would have been impossible. My three Cambridge colleges, in particular, offered much more than can be measured.

I owe a lot to Cambridge University Press. Here, as elsewhere, I find it difficult to disentangle 'form' from 'content'. The Press has contributed greatly to both, and I wish to thank, in particular, Pauline Hire and Margaret Deith for their perseverance and their patience.

The following is the list – probably incomplete – of those whose comments influenced directly the text you now read (besides the three mentioned already). My gratitude is extended to them, as well as to many others: R. E. Aschcroft, Z. Bechler, M. F. Burnyeat, K. Chemla, S. Cuomo, A. E. L. Davis, G. Deutscher, R. P. Duncan-Jones, P. E. Easterling, M. Finkelberg, G. Freudental, C. Goldstein, I. Grattan-Guinness, S. J. Harrison, A. Herreman, J. Hoyrup, E. Hussey, P. Lipton, I. Malkin, J. Mansfeld, I. Mueller, J. Ritter, K. Saito, J. Saxl, D. N. Sedley, B. Sharples, L. Taub, K. Tybjerg, B. Vitrac, L. Wischik.[1]

[1] I have mentioned above the leap made in the study of Greek mathematics over the last two decades. This owes everything to the work of Wilbur Knorr, who died on 18 March 1997, at the age of 51. Sadly, he did not read this book – yet the book would have been impossible without him.

Abbreviations

GREEK AUTHORS

Abbreviation	Work (standard title)	Author
de Aedific.	*de Aedificiis*	Procopius
Amat.	*Amatores*	[Plato]
APo.	*Analytica Posteriora*	Aristotle
APr.	*Analytica Priora*	Aristotle
Av.	*Aves*	Aristophanes
Cat.	*Categoriae*	Aristotle
CF	*On Floating Bodies*	Archimedes
CS	*On Conoids and Spheroids*	Archimedes
DC	*Measurement of Circle*	Archimedes
D.L.	Lives of Philosophers	Diogenes Laertius
EE	*Ethica Eudemia*	Aristotle
El. Harm.	*Elementa Harmonica*	Aristoxenus
de Eloc.	*Demetrius on Style*	Demetrius
EN	*Ethica Nicomachea*	Aristotle
Epin.	*Epinomis*	[Plato] (Plato?)
Euthd.	*Euthydemus*	Plato
Euthyph.	*Euthyphro*	Plato
Grg.	*Gorgias*	Plato
HA	*Historia Animalium*	Aristotle
Hip. Mai.	*Hippias Maior*	Plato
Hip. Min.	*Hippias Minor*	Plato
In de Cael.	*In Aristotelis de Caelo Commentaria*	Simplicius
In Eucl.	*In Euclidem*	Proclus
In Phys.	*In Aristotelis Physica Commentaria*	Simplicius

Abbreviation	Work (standard title)	Author
de Int.	*de Interpretatione*	Aristotle
In SC	*In Archimedes' SC*	Eutocius
In Theaetet.	*Anonymi Commentarius In Platonis Theaetetum*	Anonymous
Lgs.	*de Legibus*	Plato
Mech.	*Mechanica*	[Aristotle]
Mem.	*Memorabilia*	Xenophon
Metaph.	*Metaphysica*	Aristotle
Meteor.	*Meteorologica*	Aristotle
Meth.	*The Method*	Archimedes
Nu.	*Nubes*	Aristophanes
Ort.	*Risings and Settings*	Autolycus
Parm.	*Parmenides*	Plato
de Part.	*de Partibus Animalium*	Aristotle
PE	*Plane Equilibria*	Archimedes
Phaedr.	*Phaedrus*	Plato
Phys.	*Physica*	Aristotle
QP	*Quadrature of the Parabola*	Archimedes
Rep.	*Republica*	Plato
SC	*On Sphere and Cylinder*	Archimedes
SE	*Sophistici Elenchi*	Aristotle
SL	*Spiral Lines*	Archimedes
Theaetet.	*Theaetetus*	Plato
Tim.	*Timaeus*	Plato
Vit. Alc.	*Vita Alcibiadis*	Plutarch
Vita Marc.	*Aristotelis Vita Marciana*	Anonymous
Vita Pyth.	*de Vita Pythagorica*	Iamblichus

ROMAN AUTHORS

Abbreviation	Work (standard title)	Author
Ann.	*Annales*	Tacitus
Nat. Hist.	*Naturalis Historia*	Pliny the Elder
ND	*de Natura Deorum*	Cicero
de Rep.	*de Republica*	Cicero
Tusc.	*Tusculanae Disputationes*	Cicero

DOCUMENTARY SOURCES

Abbreviation	Standard title
BGU	*Berliner griechische Urkunden*
FD	*Fouilles de Delphes*
ID	*Inscriptions Délos*
IG	*Inscriptionae Graecae*
IGChEg.	*Inscriptionae Graecae* (Christian Egypt)
IK	*Inschriften aus Kleinasien*
Ostras	*Ostraka* (Strasburg)
P. Berol.	*Berlin Papyri*
PCair.Zen.	*Zenon Papyri*
PFay.	*Fayum Papyri*
P. Herc.	*Herculaneum Papyri*
PHerm Landl.	*Landlisten aus Hermupolis*
POxy.	*Oxyrhynchus Papyri*
YBC	*Yale Babylonian Collection*

OTHER ABBREVIATIONS

Abbreviation	Reference (in bibliography)
CPF	*Corpus dei Papiri Filosofici*
DK	Diels–Kranz, *Fragmente der Vorsokratiker*
KRS	Kirk, Raven and Schofield (1983)
L&S	Long and Sedley (1987)
LSJ	Liddell, Scott and Jones (1968)
Lewis and Short	Lewis and Short (1966)
TLG	*Thesaurus Linguae Graecae*
Usener	Usener (1887)

NOTE ON GENDER

When an indefinite reference is made to ancient scholars – who were predominantly male – I use the masculine pronoun. The sexism was theirs, not mine.

The Greek alphabet

Capital *approximately the form used in ancient writing*	Lower case *a form used in modern texts*	Name of letter
Α	α	Alpha
Β	β	Bēta
Γ	γ	Gamma
Δ	δ	Delta
Ε	ε	Epsilon
Ζ	ζ	Zēta
Η	η	Ēta
Θ	θ	Thēta
Ι	ι	Iōta
Κ	κ	Kappa
Λ	λ	Lambda
Μ	μ	Mu
Ν	ν	Nu
Ξ	ξ	Xi
Ο	ο	Omicron
Π	π	Pi
Ρ	ρ	Rhō
Σ	σ ς[1]	Sigma
Τ	τ	Tau
Υ	υ	Upsilon
Φ	φ	Phi
Χ	χ	Chi
Ψ	ψ	Psi
Ω	ω	Ōmega

[1] A modern form for the letter in final position.

Note on the figures

As is explained in chapter 1, most of the diagrams in Greek mathematical works have not yet been edited from manuscripts. The figures in modern editions are reconstructions made by modern editors, based on their modern understanding of what a diagram should look like. However, as will be argued below, such an understanding is culturally variable. It is therefore better to keep, as far as possible, to the diagrams as they are found in Greek manuscripts (that is, generally speaking, in Byzantine manuscripts). While no attempt has been made to prepare a critical edition of the Greek mathematical diagrams produced here, almost all the figures have been based upon an inspection of at least some early manuscripts in which their originals appear, and I have tried to keep as close as possible to the visual code of those early diagrams. In particular, the reader should forgo any assumptions about the lengths of lines or the sizes of angles: unequal lines and angles may appear equal in the diagrams and vice versa.

In addition to the ancient diagrams (which are labelled with the original Greek letters), a few illustrative diagrams have been prepared for this book. These are distinguished from the ancient diagrams by being labelled with Latin letters or with numerals.

While avoiding painterly effects, ancient diagrams possess considerable aesthetic value in their austere systems of interconnected, labelled lines. I wish to take this opportunity to thank Cambridge University Press for their beautiful execution of the diagrams.

Introduction

This book can be read on three levels: first, as a description of the practices of Greek mathematics; second, as a theory of the emergence of the deductive method; third, as a case-study for a general view on the history of science. The book speaks clearly enough, I hope, on behalf of the first two levels: they are the explicit content of the book. In this introduction, I give a key for translating these first two levels into the third (which is implicit in the book). Such keys are perhaps best understood when both sides of the equation are known, but it is advisable to read this introduction before reading the book, so as to have some expectations concerning the general issues involved.

My purpose is to help the reader relate the specific argument concerning the shaping of deduction to a larger framework; to map the position of the book in the space of possible theoretical approaches. I have chosen two well-known landmarks, Kuhn's *The Structure of Scientific Revolutions* and Fodor's *The Modularity of Mind*. I beg the reader to excuse me for being dogmatic in this introduction, and for ignoring almost all the massive literature which exists on such subjects. My purpose here is not to argue, but just to explain.

THE STRUCTURE OF SCIENTIFIC REVOLUTIONS

The argument of Kuhn (1962, 1970) is well known. Still, a brief résumé may be useful.

The two main conceptual tools of Kuhn's theory are, on the one hand, the distinction between 'normal science' and 'scientific revolutions' and, on the other hand, the concept of 'paradigms'. Stated very crudely, the theory is that a scientific discipline reaches an important threshold – one can almost say it begins – by attaining a paradigm. It then becomes normal science, solving very specific questions within the framework of the paradigm. Finally, paradigms may change and, with them, the entire position of the discipline. Such changes constitute scientific revolutions.

What Kuhn meant by paradigms is notoriously unclear. One sense of a 'paradigm' is a set of metaphysical assumptions, such as Einstein's concept of time. This sense is what has been most often discussed in the literature following Kuhn. The focus of interest has been the nature of the break involved in a scientific revolution. Does it make theories from two sides of the break 'incommensurable', i.e. no longer capable of being judged one against the other?

I think this is a misguided debate: it starts from the least useful sense of 'paradigm' (as metaphysical assumptions) – least useful because much too propositional. To explain: Kuhn has much of interest to say about normal science, about the way in which a scientific community is united by a set of practices. But what Kuhn failed to articulate is that practices are just that – practices. They need not be, in general, statements in which scientists (implicitly or explicitly) believe, and this for two main reasons.

First, what unites a scientific community need not be a set of *beliefs*. Shared beliefs are much less common than shared practices. This will tend to be the case in general, because shared beliefs require shared practices, but not vice versa. And this must be the case in cultural settings such as the Greek, where polemic is the rule, and consensus is the exception. Whatever is an object of belief, whatever is verbalisable, will become visible to the practitioners. What you believe, you will sooner or later discuss; and what you discuss, especially in a cultural setting similar to the Greek, you will sooner or later debate. But the real undebated, and in a sense undebatable, aspect of any scientific enterprise is its non-verbal practices.

Second, beliefs, in themselves, cannot *explain* the scientific process. Statements lead on to statements only in the logical plane. Historically, people must intervene to get one statement from the other. No belief is possible without a practice leading to it and surrounding it. As a correlate to this, it is impossible to give an account of the scientific process without describing the practices, over and above the beliefs.

This book is an extended argument for this thesis in the particular case of Greek mathematics. It brings out the set of practices common to Greek practitioners, but argues that these practices were generally 'invisible' to the practitioners. And it shows how these practices functioned as a glue, uniting the scientific community, and making the production of 'normal science' possible. The study is therefore an empirical confirmation of my general view. But the claim that 'paradigms' need not be propositional in nature should require no empirical confirmation. The propositional bias of Kuhn is a mark of his times. *The Structure of Scientific Revolutions* may have signalled the end of positivism in the history and philosophy of science, but it is itself essentially a positivist study, belonging (albeit critically) to the tradition of the *International Encyclopedia of Unified Science*, its original place of publication. It is a theory about the production of propositions from other propositions. To us, however, it should be clear that the stuff from which propositions are made need not itself be propositional. The process leading to a propositional attitude – the process leading to a person's believing that a statement is true – consists of many events, and most of them, of course, are not propositional. Kuhn's mistake was assimilating the process to the result: 'if the result is propositional, then so should the process be'. But this is an invalid inference.

Much has happened since Kuhn, and some of the literature in the history of science goes beyond Kuhn in the direction of non-propositional practices. This is done mainly by the sociologists of science. I respect this tradition very highly, but I do not belong to it. This book should not be read as if it were 'The Shapin of Deduction', an attempt to do for mathematics what has been so impressively done for the natural sciences.[1] My debt to the sociology of science is obvious, but my approach is different. I do not ask just what made science the way it was. I ask what made science successful, and successful in a real intellectual sense. In particular, I do not see 'deduction' as a sociological construct. I see it as an objectively valid form, whose discovery was a positive achievement. This aspect of the question tends to be sidelined in the sociology of science. Just as Kuhn assimilated the process to the result, making them both propositional, so the sociologists of science (in line with contemporary pragmatist or post-modern philosophers) assimilate the result to the process. They stress the non-propositional (or, more important for them, the non-objective or arbitrary) aspects of the process leading to scientific results. They do so in order to relativise science, to make it seem less propositional, or less ideology-free, or less objective.

But I ask: what sort of a process is it, which makes possible a positive achievement such as deduction? And by asking such a question, I am

[1] E.g. (to continue with the distinguished name required by the pun) in Shapin (1994).

led to look at aspects of the practice which the sociologist of science may overlook.[2]

To return to Kuhn, then, what I study can be seen, in his terms, as a study of the paradigms governing normal science. However, this must be qualified. As regards my paradigms, they are sets of practices and are unverbalised (I will immediately define them in more precise terms). As regards normal science, there are several differences between my approach and that of Kuhn. First, unlike – perhaps – Kuhn, and certainly unlike most of his followers, the aim of my study is explicitly to explain what makes this normal science successful in its own terms. Further: since my view is that what binds together practitioners in normal science is a set of practices, and not a set of beliefs, I see revolutions as far less central. Development takes the form of evolution rather than revolution. Sets of practices are long-lived, in science as elsewhere. The historians of the *Annales* have stressed the conservatism of practice in the material domain – the way in which specific agricultural techniques, for instance, are perpetuated. We intellectuals may prefer to think of ourselves as perpetually original. But the truth is that the originality is usually at the level of contents, while the forms of presentation are transmitted from generation to generation unreflectively and with only minor modifications. We clear new fields, but we till them as we always did. It is a simple historical observation that intellectual practices are enduring. Perhaps the most enduring of them all has been the Greek mathematical practice. Arguably – while modified by many evolutions – this practice can be said to dominate even present-day science.[3]

THE MODULARITY OF MIND

It is still necessary to specify what sort of practices I look at. The simple answer is that I look at those practices which may help to explain the success of science. In other words, I look at practices which may have an influence on the cognitive possibilities of science. To

[2] While such an approach is relatively uncommon in the literature, I am not the first to take it; see, for instance, Gooding (1990), on Faraday's experimental practices.

[3] There is a question concerning the relation between mathematics and other types of science. I do not think they are fundamentally distinct. The question most often raised in the literature, concerning the applicability of Kuhn to mathematics, is whether or not there are 'mathematical scientific revolutions' (in the sense of deep metaphysical shifts. See e.g. Gillies (1992)). But what I apply to mathematics is not the concept of scientific revolution, but that of normal science, and in this context the distinction between mathematics and other types of science seems much less obvious.

clarify what these may be, a detour is necessary, and I start, again from a well-known study, Fodor's *The Modularity of Mind* (1983).

Fodor distinguishes two types of cognitive processes: 'input/output mechanisms' (especially language and vision), on the one hand, and 'central processes' (for which a key example is the fixation of belief – the process leading to a person's believing in the truth of a statement) on the other. He then argues that some functions in the mind are 'modules'. By 'modules' are meant task-specific capacities (according to this view syntax, for instance, is a module; that is, we have a faculty which does syntactic computations and nothing else). Modules are automatic: to continue the same example, we do syntactic computations without thinking, without even wishing to do so. Syntactic parsing of sentences is forced upon us. And modules are isolated (when we do such computations in this modular way, we do not bring to bear any other knowledge). Modules thus function very much as if they were computer programs designed for doing a specified job. The assumption is that modules are innate – they are part of our biological make-up. And, so Fodor argues, modules are coextensive with input/output mechanisms: whatever is an input/output mechanism is a module, while nothing else is a module. The only things which are modular are processes such as vision and language, and nothing else in our mind is modular. Most importantly, central processes such as the fixation of belief are not modular. They are not task-specific (there is nothing in our brain whose function is just to reach beliefs), they are not automatic (we do not reach beliefs without conscious thoughts and volitions), and, especially, they are not isolated (there are a great many diverse processes related to any fixation of belief). Since central processes appeal to a wide range of capacities, without any apparent rules, it is much more difficult to study central processes.

Most importantly for the cognitive scientist, this difference between modules and central processes entails that modules will be the natural subject matter of cognitive science. By being relatively simple (especially in the sense of being isolated from each other), modules can be described in detail, modelled, experimented on, meaningfully analysed in universal, cross-cultural terms. Central processes, on the other hand, interact with each other in complicated, unpredictable ways, and are thus unanalysable. Hence Fodor's famous 'First Law of the Nonexistence of Cognitive Science': 'The more global . . . a cognitive process is, the less anybody understands it.'[4]

[4] Fodor (1983) 107.

I am not a cognitive scientist (and this study is not an 'application' of some cognitive theory). I do not profess to pass any judgement on Fodor's thesis. But the facts of the development of cognitive science are clear. It has made most progress with Fodorean modules, especially with language. It has been able to say less on questions concerning Fodorean central processes. Clearly, it is very difficult to develop a cognitive science of central processes. But this of course does not mean that central processes are beyond study. It simply means that, instead of a cognitive science of such aspects of the mind, we should have a cognitive *history*. 'The Existence of Cognitive History' is the direct corollary to Fodor's first law. Fodor shows why we can never have a neat universal model of such functions as the fixation of belief. This is registered with a pessimistic note, as if the end of universality is the end of study. But for the historian, study starts where universality ends.

It is clear why cognitive history is possible. While there are no general, universal rules concerning, for example, reasoning, such rules do exist *historically*, in specific contexts. Reasoning, in general, can be done in an open way, appealing to whatever tools suggest themselves – linguistic, visual, for example – using those tools in any order, moving freely from one to the other. In Greek mathematics, however, reasoning is done in a very specific way. There is a method in its use of cognitive resources. And it must be so – had it not been selective, simplified, intentionally blind to some possibilities, it would have been unmanageable. Through the evolution of specific cognitive methods, science has been made possible. Specific cognitive methods are specific ways of 'doing the cognitive thing' – of using, for instance, visual information or language. To illustrate this: in this book, I will argue that the two main tools for the shaping of deduction were the diagram, on the one hand, and the mathematical language on the other hand. Diagrams – in the specific way they are used in Greek mathematics – are the Greek mathematical way of tapping human visual cognitive resources. Greek mathematical language is a way of tapping human linguistic cognitive resources. These tools are then combined in specific ways. The tools, and their modes of combination, are the cognitive method.

But note that there is nothing universal about the precise shape of such cognitive methods. They are not neural; they are a historical construct. They change slowly, and over relatively long periods they may seem to be constant. But they are still not a biological constant. On the one hand, therefore, central processes can be studied (and this is because they are, in practice, in given periods and places, performed methodically, i.e. not completely unlike modules). On the other hand, they cannot be studied by cognitive science, i.e. through experimental methods and universalist assumptions. They can only be studied as historical phenomena, valid for their period and place. One needs studies in *cognitive history*, and I offer here one such study.[5]

I have promised I would locate this book with the aid of two landmarks, one starting from Kuhn, the other starting from Fodor. These two landmarks can be visualised as occupying two positions in a table (see below), where cognitive history can be located as well.

		Sources of knowledge	
		Cultural	**Biological**
Status of knowledge	**Propositional knowledge**	Kuhnian history of science	
	Practices of knowledge	Cognitive history	Fodorean cognitive science

Cognitive history lies at the intersection of history of science and the cognitive sciences. Like the history of science, it studies a cultural artefact. Like the cognitive sciences, it approaches knowledge not through its specific propositional contents but through its forms and practices.

An intersection is an interesting but dangerous place to be in. I fear cognitive scientists may see this study as too 'impressionistic' while historians may see it as over-theoretical and too eager to generalise. Perhaps both are right; I beg both to remember I am trying to do what is neither cognitive science nor the history of ideas. Whether I have succeeded, or whether this is worth trying, I leave for the reader to judge.

[5] It remains to argue that the subject of my study is a central process and not a module. Whether 'deduction' as such is a module or not is a contested question. Rips (1994), for instance, thinks it is a module; Johnson-Laird (1983) disagrees. I cannot discuss here the detail of the debate (though I will say that much of my study may be seen as contributing to Johnson-Laird's approach), but in fact I need not take any stance in this debate. What I study is not 'deduction' as such; what I study is a specific form, namely the way in which Greek mathematicians argued for their results. It will be seen that the mechanisms involved are very complex, and very different from anything offered by those who argue that deduction is a module. If indeed there is some module corresponding to deduction, then it is no more than a first-level stepping stone *used* in mathematical deduction (in much the same way as the modules of vision are necessary for the perception of mathematical diagrams, but yet we will not try to reduce mathematical cognition into the modules of vision).

PLAN OF THE BOOK

The first four chapters of the study describe the tools of the Greek mathematical method. The first two chapters deal with the use of the diagram, and chapters 3 and 4 deal with the mathematical language.

How is deduction shaped from these tools? I do not try to define 'deduction' in this study (and I doubt how useful such a definition would be). I concentrate instead on two relatively simpler questions: first, what makes the arguments seem *necessary*? (That is, I am looking for the origins of the compelling power of arguments.) Second, what makes the arguments seem *general*? (That is, I am looking for the origins of the conviction that a particular argument proves the general claim.) These questions are dealt with in chapters 5 and 6, respectively. In these chapters I show how the elements of the style combine in large-scale units, and how this mode of combination explains the necessity and generality of the results.

The final chapter discusses the possible origins of this cognitive mode: what made the Greek mathematicians proceed in the way they did? I try to explain the practices of Greek mathematics through the cultural context of mathematics in antiquity, and, in this way, to put deduction in a historical context.

A specimen of Greek mathematics

Readers with no acquaintance with Greek mathematics may wish to see a sample of it before reading a description of its style. Others may wish to refresh their memory. I therefore put here a literal translation of Euclid's *Elements* II.5, with a reconstruction of its diagram.[1]

In this translation, I intervene in the text in several ways, including the following:

* I add the established titles of the six parts of the proposition. These six parts do not always occur in the same simple way as here, but they are very typical of Euclid's geometrical theorems. They will be especially important in chapter 6.
* I mark the sequence of assertions in both construction (with roman letters) and proof (with numerals). This is meant mainly as an aid for the reader. The sequence of assertions in the proof will interest us in chapter 5.
* Text in angle-brackets is my addition. The original Greek is extremely elliptic – a fact which will interest us especially in chapter 4.

Note also the following:

* Letters are used in diagram and text to represent the objects of the proposition in the middle four parts. These letters will interest us greatly in chapters 1–2.
* Relatively few words are used. There is a limited 'lexicon': this is the subject of chapter 3.
* These few words are usually used within the same phrases, which vary little. These are 'formulae', the subject of chapter 4.

[1] Note also that I offer a very brief description of the *dramatis personae* – the main Greek mathematicians referred to in this book – before the bibliography (pp. 316–22).

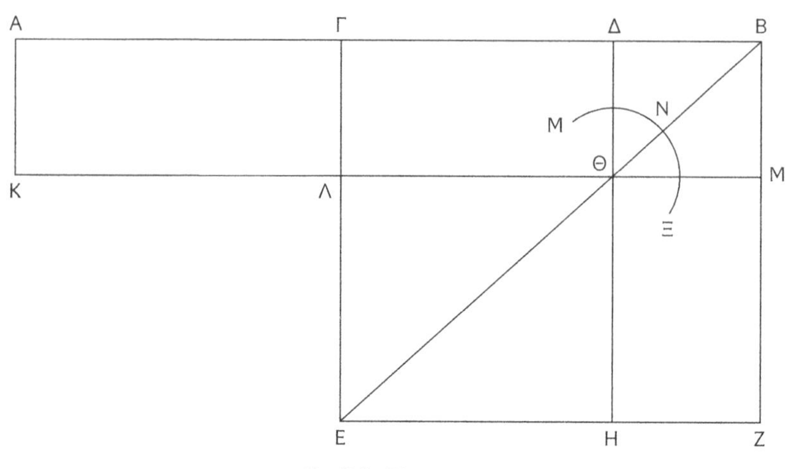

Euclid's Elements II.5.

[*protasis* (enunciation)]

If a straight line is cut into equal and unequal <segments>, the rectangle contained by the unequal segments of the whole, with the square on the <line> between the cuts, is equal to the square on the half.

[*ekthesis* (setting out)]

For let some line, <namely> the <line> AB, be cut into equal <segments> at the <point> Γ, and into unequal <segments> at the <point> Δ;

[*diorismos* (definition of goal)]

I say that the rectangle contained by the <lines> AΔ, ΔB, with the square on the <line> ΓΔ, is equal to the square on the <line> ΓB.

[*kataskeuē* (construction)]

(a) For, on the <line> ΓB, let a square be set up, <namely> the <square> ΓEZB,
(b) and let the <line> BE be joined,
(c) and, through the <point> Δ, let the <line> ΔH be drawn parallel to either of the <lines> ΓE, BZ,
(d) and, through the <point> Θ, again let the <line> KM be drawn parallel to either of the <lines> AB, EZ,
(e) and again, through the <point> A, let the <line> AK be drawn parallel to either of the <lines> ΓΛ, BM.

[*apodeixis* (proof)]

(1) And since the complement ΓΘ is equal to the complement ΘZ;
(2) let the <square> ΔM be added <as> common;
(3) therefore the whole ΓM is equal to the whole ΔZ.
(4) But the <area> ΓM is equal to the <area> AΛ,
(5) since the <line> AΓ, too, is equal to the <line> ΓB;

(6) therefore the <area> ΑΛ, too, is equal to the <area> ΔΖ.
(7) Let the <area> ΓΘ be added <as> common;
(8) therefore the whole ΑΘ is equal to the gnomon ΜΝΞ.
(9) But the <area> ΑΘ is the <rectangle contained> by the <lines> ΑΔ, ΔΒ;
(10) for the <line> ΔΘ is equal to the <line> ΔΒ;
(11) therefore the gnomon ΜΝΞ, too, is equal to the <rectangle contained> by the <lines> ΑΔ, ΔΒ.
(12) Let the <area> ΛΗ be added <as> common
(13) (which is equal to the <square> on the <line> ΓΔ);
(14) therefore the gnomon ΜΝΞ and the <area> ΛΗ are equal to the rectangle contained by the <lines> ΑΔ, ΔΒ and the square on the <line> ΓΔ;
(15) but the gnomon ΜΝΞ and the <area> ΛΗ, <as a> whole, is the square ΓΕΖΒ,
(16) which is <the square> on the <line> ΓΒ;
(17) therefore the rectangle contained by the <lines> ΑΔ, ΔΒ, with the square on the <line> ΓΔ, is equal to the square on the <line> ΓΒ.

[*sumperasma* (conclusion)]

Therefore if a straight line is cut into equal and unequal <segments>, the rectangle contained by the unequal segments of the whole, with the square on the <line> between the cuts, is equal to the square on the half; which it was required to prove.

CHAPTER I

The lettered diagram

PLAN OF THE CHAPTER

That diagrams play a crucial role in Greek mathematics is a fact often alluded to in the modern literature, but little discussed. The focus of the literature is on the verbal aspect of mathematics. What this has to do with the relative roles of the verbal and the visual in our culture, I do not claim to know. A description of the practices related to Greek mathematical diagrams is therefore called for. It will prove useful for our main task, the shaping of deduction.

The plan is: first, a brief discussion of the material implementation of diagrams, in section 1. Some practices will be described in section 2. My main claims will be that (a) the diagram is a necessary element in the reading of the text and (b) the diagram is the metonym of mathematics. I will conclude this section with a discussion of the semiotics of lettered diagrams. Section 3 will describe some of the historical contexts of the lettered diagram. Section 4 is a very brief summary.

This chapter performs a trick: I talk about a void, an absent object, for the diagrams of antiquity are not extant, and the medieval diagrams have never been studied as such.[1] However, not all hope is lost. The texts – whose transmission is relatively well understood – refer to diagrams in various ways. On the basis of these references, observations concerning the practices of diagrams can be made. I thus start from the text, and from that base study the diagrams.

[1] The critical edition most useful from the point of view of the ancient diagrams is Mogenet (1950). Some information is available elsewhere: the Teubner edition of the *Data*, for instance, is very complete on lettering; Jones's edition of Pappus and Clagett's edition of the Latin Archimedes are both exemplary, and Janus, in *Musici Graeci*, is brief but helpful. Generally, however, critical apparatuses do not offer substantial clues as to the state of diagrams in manuscripts.

1 THE MATERIAL IMPLEMENTATION OF DIAGRAMS

There are three questions related to the material implementation of diagrams: first, the contexts in which diagrams were used; second, the media available for drawing; finally, there is the question of the technique used for drawing diagrams – and, conversely, the technique required for looking at diagrams (for this is a technique which must be learned in its own right).

One should appreciate the distance lying between the original moment of inspiration, when a mathematician may simply have imagined a diagram, and our earliest extensive form of evidence, parchment codices. In between, moments of communication have occurred. What audience did they involve?

First, the 'solitaire' audience, the mathematician at work, like someone playing patience. Ancient images pictured him working with a diagram.[2] We shall see how diagrams were the hallmark of mathematical activity and, of course, a mathematician would prefer to have a diagram in front of him rather than playing the game out in his mind. It is very probable, then, that the process of discovery was aided by diagrams.

The contexts for communicating mathematical results must have been very variable, but a constant feature would have been the small numbers of people involved.[3] This entails that, very often, the written form of communication would be predominant, simply because fellow mathematicians were not close at hand. Many Greek mathematical works were originally set down within letters. This may be a trivial point concerning communicative styles, or, again, it may be significant. After all, the addressees of mathematical works, leaving aside the *Arenarius*,[4] are not the standard recipients of letters, like kings, friends or relations. They seem to have been genuinely interested mathematicians, and the inclusion of mathematics within a letter could therefore be an indication that works were first circulated as letters.[5]

[2] This is the kernel of the myth of Archimedes' death in its various forms (see Dijksterhuis (1938) 30ff.). Cicero's evocation of Archimedes 'from the dust and drawing-stick' (*Tusc.* v.64) is also relevant. Especially revealing is Archimedes' tomb, mentioned in the same context. What is Einstein's symbol? Probably '$E = MC^2$'. Archimedes' symbol was a diagram: '*sphaerae figura et cylindri*' (ibid. v.65).

[3] See the discussion in chapter 7, subsection 2.2 below (pp. 282–92).

[4] As well as Eratosthenes' fragment in Eutocius.

[5] Pappus' dedicatees are less easy to identify, but Pandrosion, dedicatee of book III, for instance, seems to have been a teacher of mathematics; see Cuomo (1994) for discussion.

Not much more is known, but the following observation may help to form some a priori conclusions. The lettered diagram is not only a feature of Greek mathematics; it is a predominant feature. Alternatives such as a non-lettered diagram are not hinted at in the manuscripts.[6] There is one exception to the use of diagrams – the *di' arithmōn*, 'the method using numbers'. While in general arithmetical problems are proved in Greek mathematics by geometrical means, using a diagram, sometimes arithmetical problems are tackled as arithmetical. Significantly, even this is explicitly set up as an exception to a well-defined rule, the *dia grammōn*, 'the method using lines'.[7] The diagram is seen as the rule from which deviations may (very rarely) occur.

It is therefore safe to conclude that Greek mathematical exchanges, as a rule, were accompanied by something like the lettered diagram. Thus an exclusively oral presentation (excluding, that is, even a diagram) is practically ruled out. Two methods of communication must have been used: the fully written form, for addressing mathematicians abroad, and (hypothetically) a semi-oral form, with some diagram, for presentation to a small group of fellow mathematicians in one's own city.

1.1 The media available for diagrams

It might be helpful to start by considering the media available to us. The most important are the pencil/paper, the chalk/blackboard and (gaining in importance) the computer/printer. All share these characteristics: simple manipulation, fine resolution, and ease of erasing and rewriting. Most of the media available to Greeks had none of these, and none had ease of erasing and rewriting.

The story often told about Greek mathematicians is that they drew their diagrams in sand.[8] A variation upon this theme is the dusted

[6] I exclude the fragment of Hippocrates of Chios, which may of course reflect a very early, formative stage. I also ignore for the moment the papyrological evidence. I shall return to it in n. 31 below.

[7] I shall return to this distinction below, n. 61.

[8] Sand may be implied by the situation of the geometry lesson in the *Meno*, though nothing explicit is said; if the divided line in the *Republic* was drawn in sand, then Cephalus' house must have been fairly decrepit. Aristotle refers to drawing in γῆ – e.g. *Metaph.* 1078a20; it may well be that he has the *Meno* in mind. Cicero, *de Rep.* I.28–9 and Vitruvius VI.I.I, have the following tale: a shipwrecked philosopher deduces the existence of life on the island on whose shores he finds himself by (Vitruvius' phrase) *geometrica schemata descripta* – one can imagine the wet sand on the shore as a likely medium. The frontispiece to Halley's edition of the *Conics*, reproduced as the cover of Lloyd (1991), is a brilliant *reductio ad absurdum* of the story.

surface. This is documented very early, namely, in Aristophanes' *Clouds*;[9] Demetrius, a much later author, misremembered the joke and thought it was about a wax tablet[10] – a sign of what the typical writing medium was. Indeed, the sand or dusted surface is an extremely awkward solution. The ostrakon or wax tablet would be sufficient for the likely size of audience; a larger group would be limited by the horizontality of the sand surfaces. And one should not think of sand as directly usable. Sand must be wetted and tamped before use, a process involving some exertion (and mess).[11] Probably the hard work was done by Euclid's slaves, but still it is important to bear in mind the need for *preparation* before each drawing. Sand is a very cheap substitute for a drawing on wood (on which see below), but it is not essentially different. It requires a similar amount of preparation. It is nothing like the immediately usable, erasable blackboard.

The possibility of large-scale communication should be considered – and will shed more light on the more common small-scale communication. There is one set of evidence concerning forms of presentation to a relatively large audience: the evidence from Aristotle and his followers in the peripatetic school.

Aristotle used the lettered diagram in his lectures. The letters in the text would make sense if they refer to diagrams – which is asserted in a few places.[12] Further, Theophrastus' will mentions maps on *pinakes* (for which see below) as part of the school's property.[13] Finally, Aristotle refers to *anatomai*, books containing anatomical drawings, which students were supposed to consult as a necessary complement to the lecture.[14]

What medium did Aristotle use for his mathematical and semi-mathematical diagrams? He might have used some kind of prepared tablets whose medium is nowhere specified.[15] As such tablets were,

[9] Ashes, sprinkled upon a table: Aristophanes, *Nu.* 177. To this may be added later texts, e.g. Cicero, *Tusc.* V.64; *ND* II.48.

[10] Demetrius, *de Eloc.* 152.

[11] I owe the technical detail to T. Riehl. My own experiments with sand and ashes, wetted or not, were unmitigated disasters – this again shows that these surfaces are not as immediately usable as are most modern alternatives.

[12] E.g. *Meteor.* 363a25–6, *APr.* 41b14. Einarson (1936) offers the general thesis that the syllogism was cast in a mathematical form, diagrams included; while many of his individual arguments need revision, the hypothesis is sound.

[13] D.L. V.51–2. [14] See Heitz (1865) 70–6.

[15] Jackson (1920) 193 supplies the evidence, and a guess that Aristotle used a *leukoma*, which is indeed probable; but Jackson's authority should not obscure the fact that this is no more than a guess.

presumably, portable, they could not be just graffiti on the Lyceum's walls. Some kind of special surface is necessary, and the only practical option was wood, which is the natural implication of the word *pinax*. To make such writing more readable, the surface would be painted white, hence the name *leukōma*, 'whiteboard' – a misleading translation. Writings on the 'whiteboard', unlike the blackboard, were difficult to erase.[16]

Two centuries later than Aristotle, a set of mathematical – in this case astronomical – *leukōmata* were put up as a dedication in a temple in Delos.[17] This adds another tiny drop of probability to the thesis that wide communication of mathematical diagrams was mediated by these whiteboards.[18] On the other hand, the *anatomai* remind us how, in the very same peripatetic school, simple diagrams upon (presumably) papyrus were used instead of the large-scale *leukōma*.

Closer in nature to the astronomical tables in Delos, Eratosthenes, in the third century BC, set up a mathematical column: an instrument on top, below which was a résumé of a proof, then a diagram and finally an epigram.[19] This diagram was apparently inscribed in stone or marble. But this display may have been the only one of its kind in antiquity.[20]

The development envisaged earlier, from the individual mathematician thinking to himself to the parchment codex, thus collapses into small-scale acts of communication, limited by a small set of media, from the dusted surface, through wax tablets, ostraka and papyri, to the whiteboard. None of these is essentially different from a diagram as it appears in a book. Diagrams, as a rule, were not drawn on site. The limitations of the media available suggest, rather, the preparation of the diagram prior to the communicative act – a consequence of the inability to erase.

[16] See Gardthausen (1911) 32–9.

[17] *ID* 3. 1426 face B. col. II.50ff.; 3. 1442 face B. col. II.40ff.; 3. 1443 face B. col. II.108ff.

[18] It is also useful to see that, in general, wood was an important material in elementary mathematical education, as the archaeological evidence shows; Fowler (1987) 271–9 has 69 items, of which the following are wooden tablets: 14, 16, 18, 24, 25, 39, 42, 44, 45, 59.

[19] Eutocius, *In SC* II.94.8–14.

[20] Allow me a speculation. Archimedes' *Arenarius*, in the manuscript tradition, contains no diagrams. Of course the diagrams were present in some form in the original (which uses the lettered convention of reference to objects). So how were the diagrams lost? The work was addressed to a king, hence, no doubt, it was a luxury product. Perhaps, then, the diagrams were originally on separate *pinakes*, drawn as works of art in their own right?

1.2 Drawing and looking

In terms of optical complexity, there are four types of objects required in ancient mathematics.

1. Simple 2-dimensional configurations, made up entirely by straight lines and arcs;
2. 2-dimensional configurations, requiring more complex lines, the most important being conic sections (ellipse, parabola and hyperbola);
3. 3-dimensional objects, excluding:
4. Situations arising in the theory of spheres ('sphaerics').

Drawings of the first type were obviously mastered easily by the Greeks. There is relatively good papyrological evidence for the use of rulers for drawing diagrams.[21] The extrapolation, that compasses (used for vase-paintings, from early times)[22] were used as well, suggests itself.

On the other hand, the much later manuscripts do not show any technique for drawing non-circular curved lines, which are drawn as if they consist of circular arcs.[23] This use of arcs may well have been a feature of ancient diagrams as well.

Three-dimensional objects do not require perspective in the strict sense, but rather the practice of foreshortening individual objects.[24] This was mastered by some Greek painters in the fifth century BC;[25] an achievement not unnoticed by Greek mathematicians.[26]

Foreshortening, however, does little towards the elucidation of spherical situations. The symmetry of spheres allows the eye no hold on which to base a foreshortened 'reading'. In fact, some of the diagrams for spherical situations are radically different from other, 'normal' diagrams. Rather than providing a direct visual representation, they employ

[21] See Fowler (1987), plates between pp. 202 and 203 – an imperative one should repeat again and again. For this particular point, see especially Turner's personal communication on *PFay.* 9, p. 213.

[22] See, e.g. Noble (1988) 104–5 (with a fascinating reproduction on p. 105).

[23] Toomer (1990) lxxxv.

[24] In fact – as pointed out to me by M. Burnyeat – strictly perspectival diagrams would be less useful. A useful diagram is somewhat schematic, suggesting objective geometric relations rather than subjective optical impressions.

[25] White (1956), first part.

[26] Euclid's *Optics* 36 proved that wheels of chariots appear sometimes as circles, sometimes as elongated. As pointed out by White (1956: 20), Greek painters were especially interested in the foreshortened representation of chariots, sails and shields. Is it a fair assumption that the author of Euclid's theorem has in mind not so much wheels as representations of wheels? Knorr (1992) agrees, while insisting on how difficult the problem really is.

a quasi-symbolical system in which, for instance, instead of a circle whirling around a sphere, its 'hidden' part is shown *outside* the sphere.[27] I suspect that much of the visualisation work was done, in this special context, by watching planetaria, a subject to which I shall return below, in subsection 3.2.2. But the stress should be on the peculiarity of sphaerics. Most three-dimensional objects could have been drawn and 'read' from the drawing in a more direct, pictorial way.[28]

It should not be assumed, however, that, outside sphaerics, diagrams were 'pictures'. Kurt Weitzman offers a theory – of a scope much wider than mathematics – arguing for the opposite. Weitzman (1971, chapter 2) shows how original Greek schematic, rough diagrams (e.g. with little indication of depth and with little ornamentation) are transformed, in some Arabic traditions, into painterly representations. Weitzman's hypothesis is that technical Greek treatises used, in general, schematic, unpainterly diagrams.

The manuscript tradition for Greek mathematical diagrams, I repeat, has not been studied systematically. But superficial observations corroborate Weitzman's theory. Even if depth is sometimes indicated by some foreshortening effects, there is certainly no attempt at painterly effects such as shadowing.[29] The most significant question from a mathematical point of view is whether the diagram was meant to be *metrical*: whether quantitative relations inside the diagram were meant to correspond to such relations between the objects depicted. The alternative is a much more schematic diagram, representing only the qualitative relations of the geometrical configuration. Again, from my acquaintance with the manuscripts, they very often seem to be schematic in this respect as well.[30]

[27] Mogenet (1950). Thanks to Mogenet's work, we may – uniquely – form a hypothesis concerning the genesis of these diagrams. It is difficult to imagine such a system being invented by non-mathematical scribes. Even if it was not Autolycus' own scheme, it must reflect some ancient mathematical system.

[28] While foreshortening is irrelevant in the case of spheres, shading is relevant. In fact, in Roman paintings, shading is systematically used for the creation of the illusion of depth when columns, i.e. cylinders, are painted. The presence of 'strange' representations for spheres shows, therefore, a deliberate avoidance of the practice of shading. This, I think, is related to what I will argue later in the chapter, that Greek diagrams are – from a certain point of view – 'graphs' in the mathematical sense. They are not drawings.

[29] Effects which *do* occur in early editions – and indeed in some modern editions as well.

[30] Compare Jones (1986) 1.76 on the diagrams of Pappus: 'The most apparent . . . convention is a pronounced preference for symmetry and regularization . . . introducing [e.g.] equalities where quantities are not required to be equal.' Such practices (which I have often seen in manuscripts other than Pappus') point to the expectation that the diagram should not be read quantitatively.

To sum up, then: when mathematical results were presented in anything other than the most informal, private contexts, lettered diagrams were used. These would typically have been prepared prior to the mathematical reasoning.[31] Rulers and compasses may have been used. Generally speaking, a Greek viewer would have read into them, directly, the objects depicted, though this would have required some imagination (and, probably, what was seen then was just the schematic configuration); but then, any viewing demands imagination.

2 PRACTICES OF THE LETTERED DIAGRAM

2.1 The mutual dependence of text and diagram

There are several ways in which diagram and text are interdependent. The most important is what I call 'fixation of reference' or 'specification'.[32]

A Greek mathematical proposition is, at face value, a discussion of letters: *alpha*, *bēta*, etc. It says such things as 'AB is bisected at Γ'. There must be some process of fixation of reference, whereby these letters are related to objects. I argue that in this process the diagram is indispensable. This has the surprising result that the diagram is not directly recoverable from the text.

Other ways in which text and diagram are interdependent derive from this central property. First, there are assertions which are directly deduced from the diagram. This is a strong claim, as it seems to threaten the logical validity of the mathematical work. As I shall try to show, the threat is illusory. Then, there is a large and vague field of assertions which are, as it were, 'mediated' via the diagram. I shall try to clarify this concept, and then show how such 'mediations' occur.

[31] *P. Berol.* 17469, presented in Brashear (1994), is a proof of this claim. This papyrus – a second-century AD fragment of unknown provenance – covers *Elements* I.9, with tiny remnants of I.8 and I.10. For each proposition, it has the enuncation together with an unlettered diagram, and nothing else. It is fair to assume that the original papyrus had more propositions, treated in the same way. My guess is that this was a memorandum, or an abridgement, covering the first book of Euclid's *Elements*. Had someone been interested in carrying out the proof, the lettering would have occurred on a copy on, e.g. a wax-tablet. (The same, following Fowler's suggestion (1987) 211–12, can be said of *POxy*. i.29.)

To anticipate: in chapter 2 I shall describe the practices related to the assigning of letters to points, and will argue for a semi-oral dress-rehearsal, during which letters were assigned to points. This is in agreement with the evidence from the papyri.

[32] The word 'specification' is useful, as long as it is clear that the sense is *not* that used by Morrow in his translation of Proclus (a translation of the Greek *diorismos*). I explain my sense below.

2.1.1 Fixation of reference

Suppose you say (fig. 1.1):

Let there be drawn a circle, whose centre is *A*.

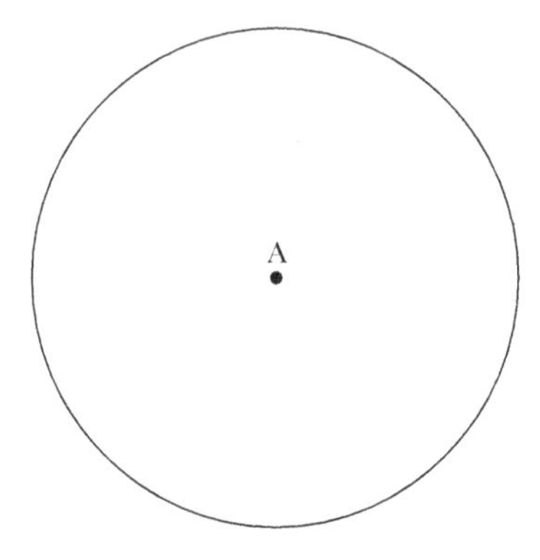

Figure 1.1.

A is thereby completely specified, since a circle can have only one centre.

Another possible case is (figs. 1.2a, 1.2b):

Let there be drawn a circle, whose radius is *BC*.

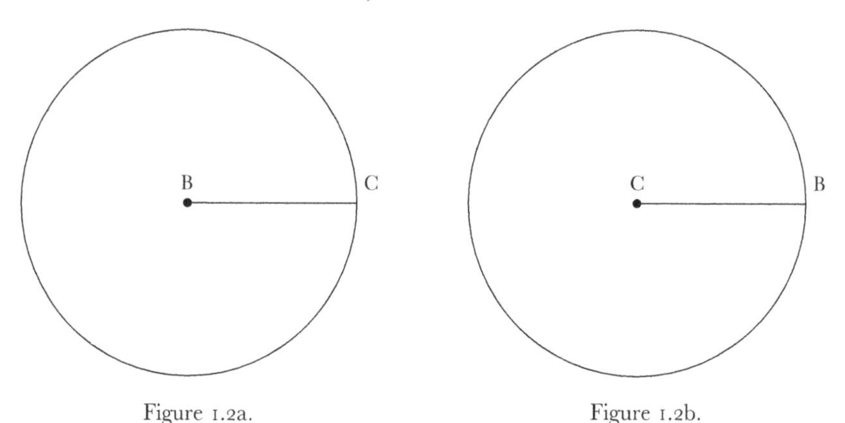

Figure 1.2a. Figure 1.2b.

This is a more complicated case. I do not mean the fact that a circle may have many radii. It may well be that for the purposes of the proof it is immaterial which radius you take, so from this point of view saying 'a radius' may offer all the specification you need. What I mean by 'specification' is shorthand for 'specification for the purposes of the proof'.

But even granted this, a real indeterminacy remains here, for we cannot tell here *which of BC is which*: which is the centre and which touches the circumference. The text of the example is valid with both figures 1.2a and 1.2b. *B* and *C* are therefore underspecified by the text.

Finally, imagine that the example above continues in the following way (fig. 1.3):

> Let there be drawn a circle, whose radius is *BC*. I say that *DB* is twice *BC*.

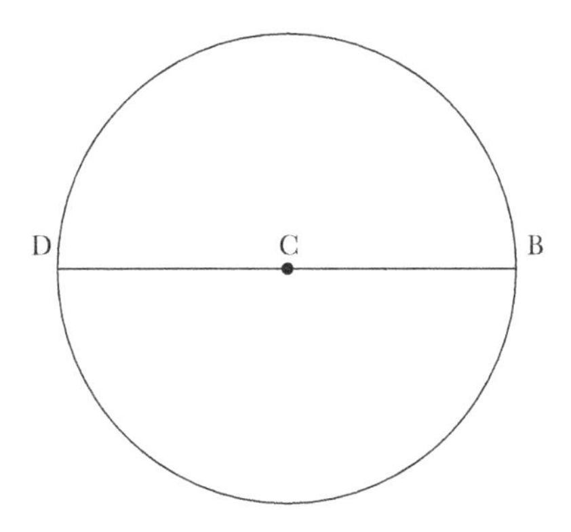

Figure 1.3.

D in this example is neither specified nor underspecified. Here is a letter which gets no specification at all in the text, which simply appears out of the blue. This is a completely unspecified letter.

We have seen three classes: completely specified, underspecified, and completely unspecified. Another and final class is that of letters which change their nature through the proposition. They may first appear as completely unspecified, and then become at least underspecified; or they may first appear as underspecified, and later get complete specification. This is the basic classification into four classes. I have surveyed all the letters in Apollonius' *Conics* I and Euclid's *Elements* XIII, counting how many belong to each class. But before presenting the results, there are a few logical complications.

First, what counts as a possible moment of specification? Consider the following case. Given the figure 1.4, the assertion is made: 'and therefore *AB* is equal to *BC*'. Suppose that nothing in the proposition so far specified *B* as the centre of the circle. Is this assertion then a specification of *B* as the centre? Of course not, because of the 'therefore' in the assertion. The assertion is meant to be a *derivation*, and

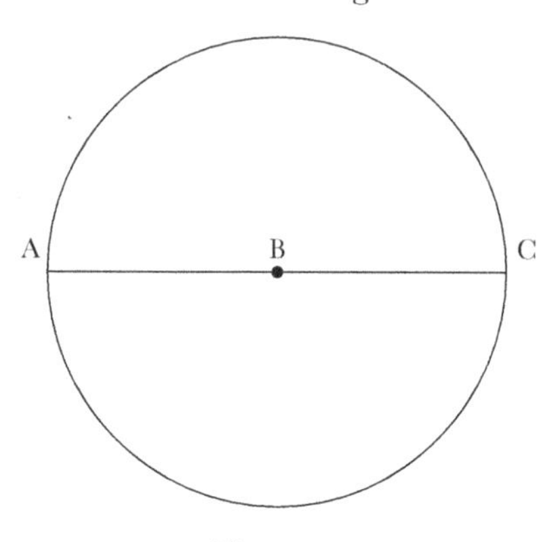

Figure 1.4.

making it into a specification would make it effectively a *definition*, and the derivation would become vacuous. Thus such assertions cannot constitute specifications. Roughly speaking, specifications occur in the imperative, not in the indicative. They are 'let the centre of the circle, *B*, be taken', etc.

Second, letters are specified by other letters. It may happen that those other letters are underspecified themselves. I have ignored this possibility. I have been like a very lenient teacher, who always gives his pupils a chance to reform. At any given moment, I have assumed that all the letters used in any act of specification were fully specified themselves. I have concentrated on *relative* specification, specification of a letter relative to the preceding letters. This has obvious advantages, mainly in that the statistical results are more interesting. Otherwise, practically all letters would turn out to be underspecified in some way.

Third and most important, a point which Grattan-Guinness put before me very forcefully: it must always be remembered, not only what the text specifies, but also what the mathematical sense demands. I have given such an example already, with 'taking a radius'. If the mathematical sense demands that we take *any* radius, then even if the text does not specify which radius we take, still this constitutes no underspecification. This is most clear with cases such as 'Let *some* point be taken on the circle, *A*'. Whenever a point is taken in this way, it is *necessarily* completely specified by the text. The text simply cannot give any better specification than this. So I stress: what I mean by 'underspecified letters' is not at all 'variable letters'. On the contrary: variable points have to be, in fact, completely specified. I mean letters which are left ambiguous by the text – which the text does not specify fully, *given the mathematical purposes*.

Now to the results.[33] In Euclid's *Elements* XIII, about 47% of the letters are completely specified, about 8% are underspecified, about 19% are completely unspecified, and about 25% begin as completely unspecified or underspecified, and get increased specification later. In Apollonius' *Conics* I, about 42% are completely specified, about 37% are underspecified, about 4% are completely unspecified, and about 16% begin as completely unspecified or underspecified, and get increased specification later. The total number of letters in both surveys is 838.

Very often – most often – letters are not completely specified. So how do we know what they stand for? Very simple: we see this in the diagram.

In fact the difficult thing is to 'unsee' the diagram, to teach oneself to disregard it and to imagine that the only information there is is that supplied by the text. Visual information is compelling itself in an unobtrusive way. Here the confessional mode may help to convert my readers. It took me a long time to realise how ubiquitous lack of specification is. The following example came to me as a shock. It is, in fact, a very typical case.

Look at Apollonius' *Conics* I.11 (fig. 1.5). The letter Λ is specified at 38.26, where it is asserted to be on a parallel to ΔE, which passes through K. Λ is thus on a definite line. But as far as the text is concerned, there is no way of knowing that Λ is a very specific point on that line, the one intersecting with the line ZH. But I had never even thought about this insufficiency of the text: I always read the diagram into the text. This moment of shock started me on this survey. Having completed the survey, its implications should be considered.

First, why are there so many cases falling short of full specification? To begin to answer this question, it must be made clear that my results have little quantitative significance. It is clear that the way in which letters in Apollonius fail to get full specification is different from that in Euclid. I expect that there is a strong variability between works by the same author. The *way* in which letters are not fully specified depends upon mathematical situations. Euclid, for instance, in book XIII, may construct a circle, e.g. ΑΒΓΔΕ, and then construct a pentagon within

[33] The complete tables, with a more technical analysis of the semantics of specification, are to appear in Netz (forthcoming).

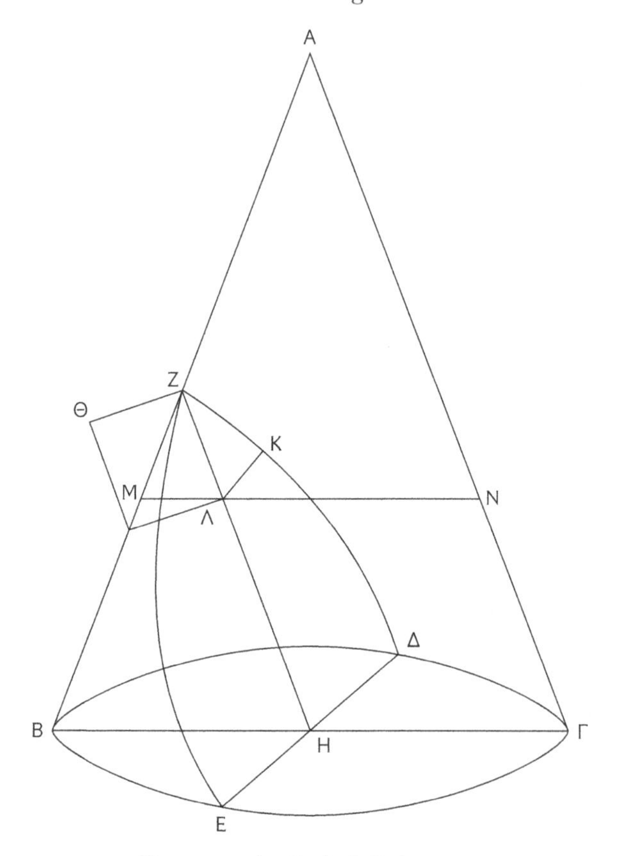

Figure 1.5. Apollonius' *Conics* 1.11.

the same circle, such that its vertices are the very same ΑΒΓΔΕ. This is moving from underspecification to complete specification, and is demanded by the subject matter dealt with in his book. In the *Conics*, parallel lines and ordinates are the common constructions, and letters on them are often underspecified (basically, they are similar to 'BC' in figs. 1.2a, 1.2b above).

What seems to be more stable is the percentage of fully specified letters. Less than half the letters are fully specified – but not much less than half. It is as if the authors were indifferent to the question of whether a letter were specified or not, full specification being left as a random result.

This, I claim, is the case. Nowhere in Greek mathematics do we find a moment of specification *per se*, a moment whose purpose is to make sure that the attribution of letters in the text is fixed. Such moments are very common in modern mathematics, at least since Descartes.[34] But specifications in Greek mathematics are done, literally, *ambulando*. The essence of the 'imperative' element in Greek mathematics – 'let a line be drawn . . .', etc. – is to do some job upon the geometric space, to get things moving there. When a line is drawn from one point to another, the letters corresponding to the start and end positions of movement ought to be mentioned. But they need not be carefully differentiated; one need not know precisely which is the start and which is the end – both would do the same job, produce the same line (hence underspecification); and points traversed through this movement may be left unmentioned (hence complete unspecification).

What we see, in short, is that while the text is being worked through, the diagram is assumed to exist. The text takes the diagram for granted. This reflects the material implementation discussed above. This, in fact, is the simple explanation for the use of *perfect* imperatives in the references to the setting out – 'let the point A have been taken'. It reflects nothing more than the fact that, by the time one comes to discuss the diagram, it has already been drawn.[35]

The next point is that, conversely, the text is not recoverable from the diagram. Of course, the diagram does not tell us what the proposition asserts. It could do so, theoretically, by the aid of some symbolic apparatus; it does not. Further, the diagram does not specify all the objects on its own. For one thing, at least in the case of sphaerics, it does not even look like its object. When the diagram is 'dense', saturated in detail, even the attribution of letters to points may not be obvious from the diagram, and modern readers, at least, reading modern diagrams, use the text, to some extent, in order to elucidate the diagram. The stress of this section is on *inter*-dependence. I have not merely tried to upset the traditional balance between text and

[34] In Descartes, the same thing is both geometric and algebraic: it is a line (called AB), and it is an algebraic variable (called a). When the geometrical configuration is being discussed, 'AB' will be used; when the algebraic relation is being supplied, 'a' is used. The square on the line is 'the square on AB' (if we look at it geometrically) or a^2 (if we look at it algebraically). To make this double-accounting system workable, Descartes must introduce explicit, *per se* specifications, identifying *symbols*. This happens first in Descartes (1637) 300. This may well be the first *per se* moment of specification in the history of mathematics.

[35] The suggestion of Lachterman (1989) 65–7, that past imperatives reflect a certain *horror operandi*, is therefore unmotivated, besides resting on the very unsound methodology of deducing a detailed philosophy, presumably shared by each and every ancient mathematician, from linguistic practices. The methodology adopted in my work is to explain shared linguistic practices by shared situations of communication.

diagram; I have tried to show that they cannot be taken apart, that neither makes sense in the absence of the other.

2.1.2 *The role of text and diagram for derivations*

In general, assertions may be derived from the text alone, from the diagram alone, or from a combination of the two. In chapter 5, I shall discuss grounds for assertions in more detail. What is offered here is an introduction.

First, some assertions do derive from the text alone. For instance, take the following:[36]

> As BE is to EΔ, so are four times the rectangle contained by BE, EA to four times the rectangle contained by AE, EΔ.

One brings to bear here all sorts of facts, for instance the relations between rectangles and sides, and indeed some basic arithmetic. One hardly brings to bear the diagram, for, in fact, 'rectangles' of this type often involve lines which do not stand at right angles to each other; the lines often do not actually have any point in common.[37]

So this is one type of assertion: assertions which may be viewed as verbal and *not* visual.[38] Another class is that of assertions which are based on the visual alone. To say that such assertions exist means that the text hides implicit assumptions that are contained in the diagram.

That such cases occur in Greek mathematics is of course at the heart of the Hilbertian geometric programme. Hilbert, one of the greatest mathematicians of the twentieth century, who repeatedly returned to foundational issues, attempted, in Hilbert (1899), to rewrite geometry without any unarticulated assumptions. Whatever the text assumes in Hilbert (1899), it either proves or explicitly sets as an axiom. This was never done before Hilbert, mainly because much information was taken from the diagram. As is well known, the very first proposition

[36] Apollonius' *Conics* 1.33, 100.7–8. The Greek text is more elliptic than my translation.

[37] Here the lines mentioned do share a point, but they are not at right angles to each other. See, for instance, *Conics* 1.34, 104.3, the rectangle contained by KB, AN – lines which do not share a point.

[38] This class is not exhausted by examples such as the above (so-called 'geometrical algebra'). For instance, any calculation, as e.g. in Aristarchus' *On Sizes and Distances*, owes nothing to the diagram. It should be noted that even 'geometrical algebra' is still 'geometrical': the text does not speak about multiplications, but about rectangles. This of course testifies to the primacy of the visual over the verbal. In general, see Unguru (1975, 1979), Unguru and Rowe (1981–2), Unguru and Fried (forthcoming), Hoyrup (1990a), for a detailed criticism of any interpretation of 'geometrical algebra' which misses its visual motivation. The term itself is misleading, but helps to identify a well-recognised group of propositions, and I therefore use it, quotation marks and all.

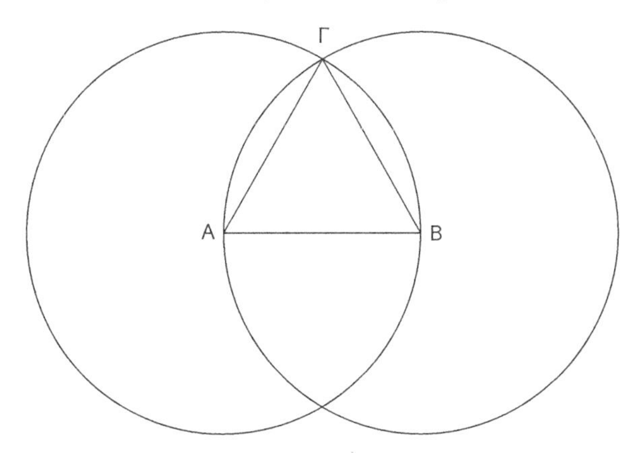

Figure 1.6. Euclid's *Elements* I.I.

of Euclid's *Elements* contains an implicit assumption based on the diagram – that the circles drawn in the proposition meet (fig. 1.6).[39]

There is a whole set of assumptions of this kind, sometimes called 'Pasch axioms'.[40] 'A line touching a triangle and passing inside it touches that triangle at two points' – such assumptions were generally, prior to the nineteenth century, taken to be diagrammatically obvious.

Many assertions are dependent on the diagram alone, and yet involve nothing as high-powered as 'Pasch axioms'. For instance, Apollonius' *Conics* III.I (fig. 1.7):[41] the argument is that AΔBZ is equal to AΓZ and, therefore, subtracting the common AEBZ, the remaining AΔE is equal to ΓBE. Adopting a very grand view, one may say that this involves assumptions of additivity, or the like. This is part of the story, but the essential ground for the assertion is identifying the objects in the diagram.

My argument, that text and diagram are interdependent, means that many assertions derive from the combination of text and diagram. Naturally, such cases, while ubiquitous, are difficult to pin down precisely. For example, take Apollonius' *Conics* I.45 (fig. 1.8). It is asserted – no special grounds are given – that MK:KΓ::ΓΔ:ΔΛ.[42] The implicit ground for this is the similarity of the triangles MKΓ, ΓΔΛ. Now diagrams cannot, in themselves, show satisfactorily the similarity of triangles. But the diagram may be helpful in other ways, for, in fact, the similarity

[39] Most recommended is Russell (1903) 404ff., viciously and in a sense justly criticising Euclid for such logical omissions.

[40] For a discussion of the absence of Pasch axioms from Greek mathematics, see Klein (1939) 201–2.

[41] 318.15–18. [42] 138.10–11.

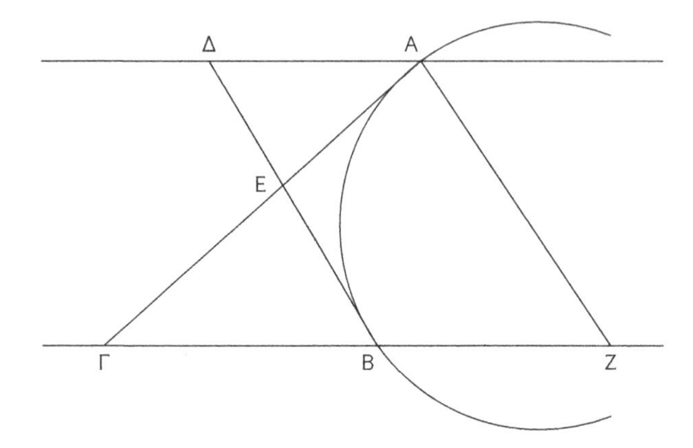

Figure 1.7. Apollonius' *Conics* III.1 (Parabola Case).

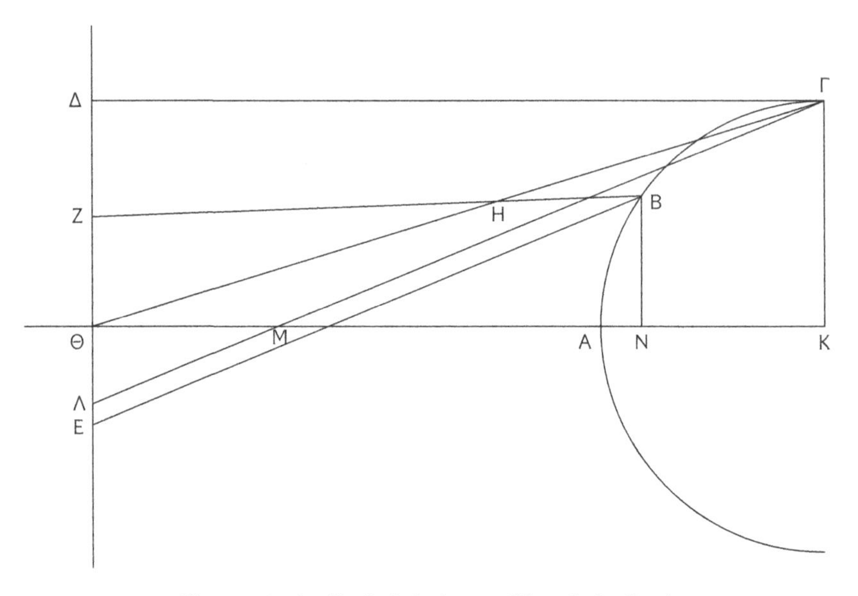

Figure 1.8. Apollonius' *Conics* I.45 (Hyperbola Case).

of the relevant triangles is not asserted in this proposition. To see this similarity, one must piece together a few hints: ΓΔ is parallel to KΘ (136.27); M lies on KΘ (underspecified by the text); ΓK is parallel to ΔΘ (138.6); Λ lies on ΔΘ (underspecified by the text); M lies on ΓΛ (136.26). Putting all of these together, it is possible to prove that the two triangles are similar. In a sense we do piece together those hints. But we are supposed to be able to do so at a glance (a significant phrase!). How do we do it then? We coordinate the various facts involved, and we coordinate them at great ease, because they are all simultaneously available on the diagram. The diagram is synoptic.

Note carefully: it is not the case that the diagram asserts information such as 'ΓΚ is parallel to ΔΘ'. Such assertions cannot be shown to be true in a diagram. But once the *text* secures that the lines are parallel, this piece of knowledge may be encoded into the reader's representation of the diagram. When necessary, such pieces of knowledge may be mobilised to yield, as an ensemble, further results.

2.1.3 *The diagram organises the text*

Even at the strictly linguistic level, it is possible to identify the presence of the diagram. A striking example is the following (fig. 1.9):[43]

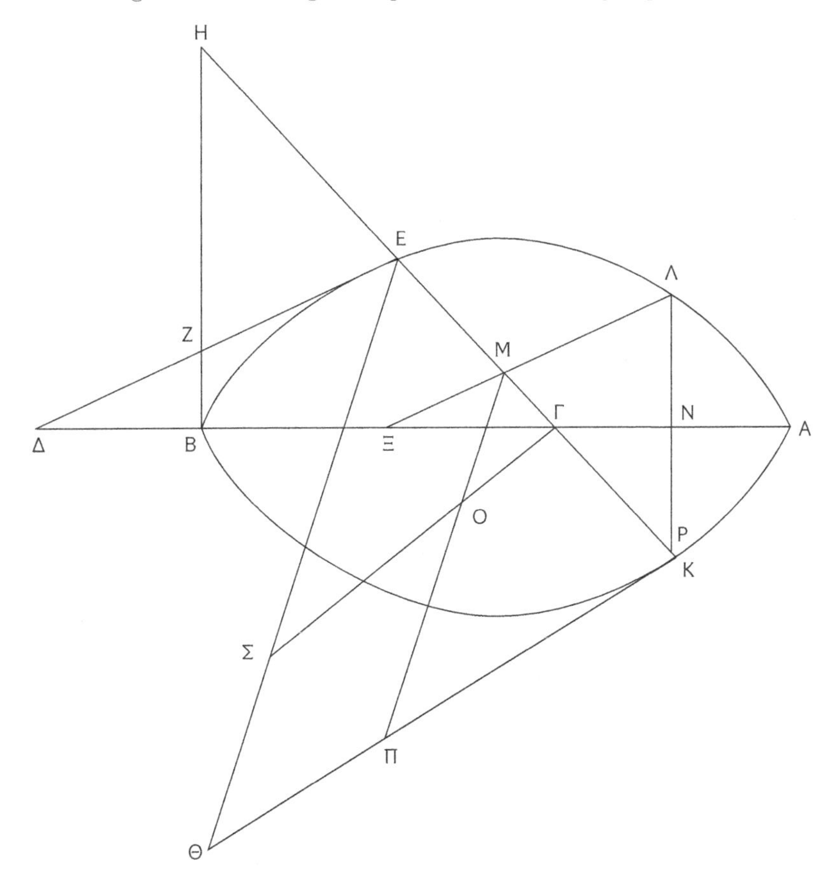

Figure 1.9. Apollonius' *Conics* 1.50 (Ellipse Case).

[43] Apollonius, *Conics* 1.50, 150.23–5: καὶ εἰλήφθω τι ἐπὶ τῆς τομῆς σημεῖον τὸ Λ, καὶ δὶ αὐτοῦ τῇ ΕΔ παράλληλος ἤχθω ἡ ΛΜΞ, τῇ δὲ ΒΗ ἡ ΛΡΝ, τῇ δὲ ΕΘ ἡ ΜΠ.

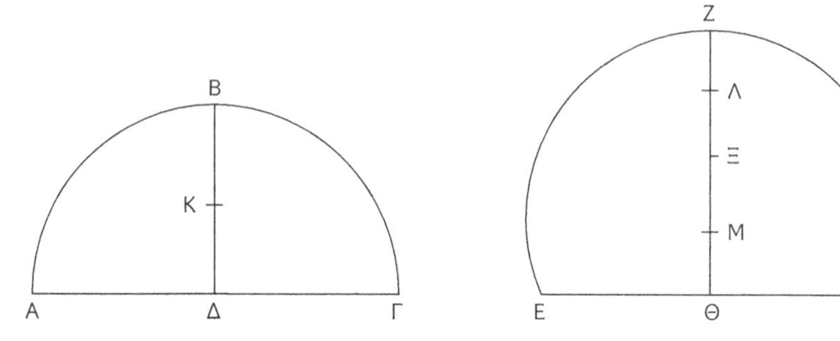

Figure 1.10. Archimedes' *PE* II.7.

> And let some point be taken on the section, Λ, and, through it, let ΛΜΞ be drawn parallel to ΕΔ, ΛΡΝ to ΒΗ, ΜΠ to ΕΘ.

Syntactically, the sentence means that ΜΠ passes through Λ – which ΜΠ does not. The diagram forces one to carry Λ over to a part of the sentence, and to stop carrying it over to another part.[44] The pragmatics of the text is provided by the diagram. The diagram is the framework, the set of presuppositions governing the discourse.

A specific, important way in which the diagram organises the text is the setting of cases. This is a result of the diagrammatic fixation of reference. Consider Archimedes' *PE* II.7: ΕΖΗ, ΑΒΓ are two similar sections; ΖΘ, ΒΔ are, respectively, their diameters; Λ, Κ, respectively, their centres of gravity (fig. 1.10). The proposition proves, through a *reductio*, that ΖΛ:ΛΘ::ΒΚ:ΚΔ. How? By assuming that a different point, Μ, satisfies ΖΜ:ΜΘ::ΒΚ:ΚΔ. Μ could be put either above or below Λ. The cases are asymmetrical. Therefore these are two distinct cases. Archimedes, however, does not distinguish the cases in the text. Only the diagram can settle the question of which case he preferred to discuss.

There are many ways in which it can be seen that the guiding principle in the development of the proof is spatial rather than logical. Take, for instance, Apollonius' *Conics* I.15 (fig. 1.11): the proposition deals with a construction based on an ellipse. This construction has two 'wings', as it were. The development of the proof is the following: first, some work is done on the lower wing; next, the results are reworked on the ellipse itself; finally, the results are transferred to the

[44] Compare also the same work, proposition 31, 94.2–3: the syntax seems to imply that ΔΘ passes through Ε; it does not. In the same proposition, 92.23–4: is Γ on the hyperbola or on the diameter? The syntax, if anything, favours the hyperbola; the diagram makes it stand on the diameter: two chance examples from a chance proposition.

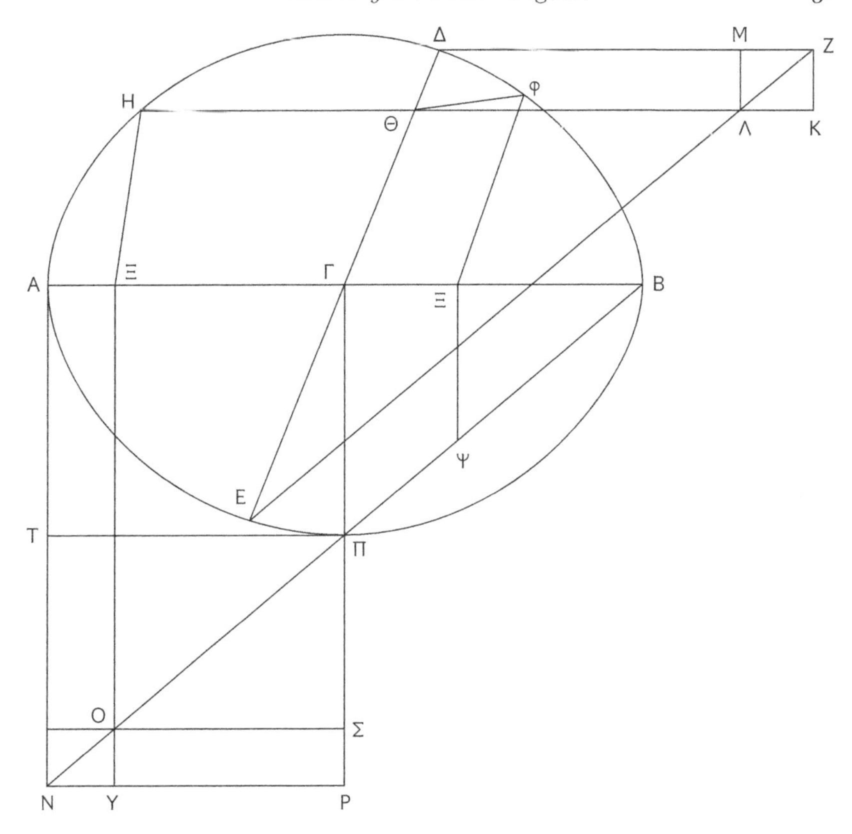

Figure 1.11. Apollonius' *Conics* 1.15.

upper wing. One could, theoretically, proceed otherwise, collecting results from all over the figure simultaneously. Apollonius chose to proceed spatially.[45] There are a number of contexts where the role of spatial visualisation can be shown, on the basis of the practices connected with the assignment of letters to objects, and I shall return to this issue in detail in chapter 2 below. The important general observation is that the diagram sets up a world of reference, which delimits the text. Again, this is a result of the role of the diagram for the process of fixation of reference. Consider a very typical case: Λ in Apollonius' *Conics* 1.6. It is specified in the following way (fig. 1.12):[46] 'From K, let a

[45] The first part is 60.5–19, the second is 60.19–29, the third is 60.29–62.13. That the second part casts a brief glance – seven words – back at the lower wing serves to show the contingency of this spatial organisation.

[46] 22.3.

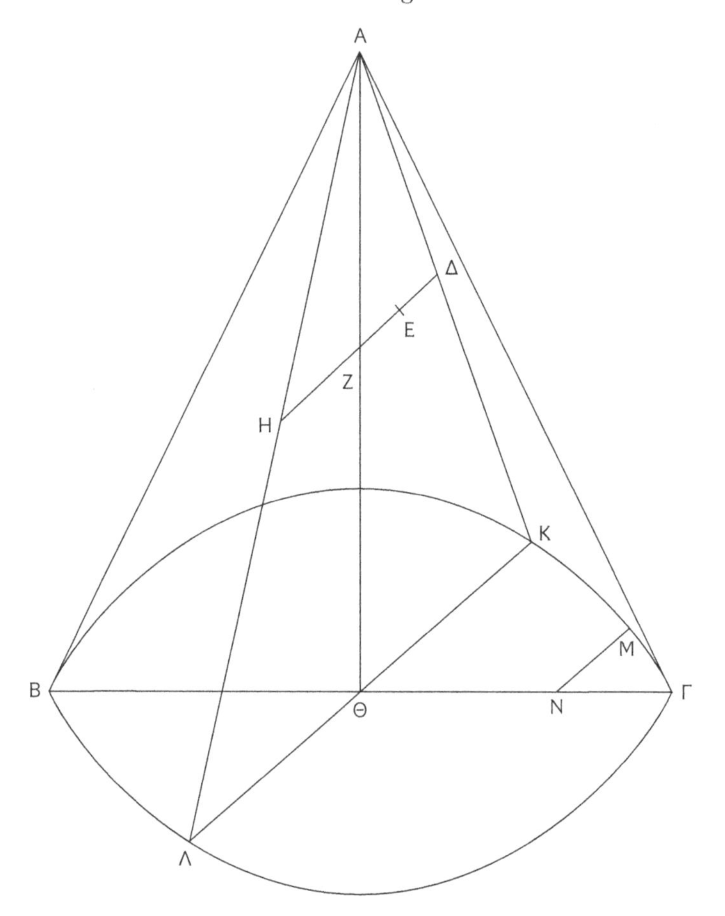

Figure 1.12. Apollonius' *Conics* 1.6 (One of the Cases).

perpendicular, to ΒΓ be drawn (namely) ΚΘΛ.' The locus set up for Λ is a line. How do we know that it is at the limit of that line, on the circle ΓΚΒ? Because Λ is the end point of the action of drawing the line ΚΘΛ – and because this action must terminate on this circle *for this circle is the limit of the universe of this proposition.* There are simply no points outside this circle.

Greek geometrical propositions are not about universal, infinite space. As is well known, lines and planes in Greek mathematics are always finite sections of the infinite line and plane which *we* project. They are, it is true, indefinitely extendable, yet they are finite. Each geometrical proposition sets up its own universe – which is its diagram.

2.1.4 The mutual dependence of text and diagram: a summary

Subsections 2.1.1–2.1.3, taken together, show the use of the diagram as a vehicle for logic. This might be considered a miracle. Are diagrams not essentially misleading aids, to be used with caution?

Mueller, after remarking on Greek implicit assumptions, went on to add that these did not invalidate Greek mathematics, for they were *true*.[47] This is a startling claim to be made by someone who, like Mueller, is versed in modern philosophy of mathematics, where truth is often seen as relative to a body of assumptions. Yet Mueller's claim is correct.

To begin with, a diagram may always be 'true', in the sense that it is there. The most ultra-abstract modern algebra often uses diagrams as representations of logical relations.[48] Diagrams, just like words, may be a way of encoding information. If, then, diagrams are seen in this way, to ask 'how can diagrams be true?' is like asking 'how can language be true?' – not a meaningless question, but clearly a different question from that we started from.

But there is more to this. The problem, of course, is that the diagram, *qua* physical object, does *not* model the assertions made concerning it. The physical diagram and the written text often clash: in one, the text, the lines are parallel; in the other, the diagram, they are not. It is only the diagram perceived in a certain way which may function alongside the text. But this caveat is in fact much less significant than it sounds, since whatever is perceived is perceived in a certain way, not in the totality of its physical presence. Thus the logical usefulness of the diagram *as a psychological object* is unproblematic – the important requirement is that the diagram would be perceived in an intersubjectively consistent way.

Poincaré – having his own axe to grind, no doubt – offered the following interpretation of the diagram:[49] 'It has often been said that geometry is the art of reasoning correctly about figures which are poorly constructed. This is not a quip but a truth which deserves reflection. But what is a poorly constructed figure? It is the type which can be drawn by the clumsy craftsman.'

Immediately following this, Poincaré goes on to characterise the useful diagram: 'He [the clumsy craftsman] distorts proportions more or less flagrantly . . . But [he] must not represent a closed curve by an

[47] Mueller (1981) 5.

[48] See e.g. Maclane and Birkhoff (1968), *passim* (explanation on the diagrammatic technique is found in 5ff.).

[49] I quote from the English translation, Poincaré (1963) 26.

open curve . . . Intuition would not have been impeded by defects in drawing which are of interest only in metric or projective geometry. But intuition will become impossible as soon as these defects involve *analysis situs*.'

The *analysis situs*[50] is Poincaré's hobby-horse, and should be approached with caution. The diagram is not just a graph, in the sense of graph theory. It contains at least one other type of information, namely the straightness of straight lines; that points stand 'on a line' is constantly assumed on the basis of the diagram.[51] This fact is worth a detour.

How can the diagram be relied upon for the distinction between straight and non-straight? The technology of drawing, described in section 1 above, showed that diagrams were drawn, probably, with no other tools than the ruler and compasses. Technology represented no more than the distinction between straight and non-straight. The man-made diagram, unlike nature's shapes, was governed by the distinction between straight and non-straight alone. The infinite range of angles was reduced by technology into a binary distinction.[52] This is hypothetical, of course, but it may serve as an introduction to the following suggestion.

There is an important element of truth in Poincaré's vision of the diagram. The diagram is relied upon as a *finite* system of relations. I have described above the proposition as referring to the finite universe of the diagram. This universe is finite in two ways. It is limited in space, by the boundaries of the figure; and it is discrete. Each geometrical proposition refers to an infinite, continuous set of points. Yet only a limited number of points are referred to, and these are almost always (some of) the points standing at the intersections of lines.[53] The great multitude of proletarian points, which in their combined efforts construct together the mathematical objects, is forgotten. All attention is fixed upon the few intersecting points, which alone are named. This,

[50] Corresponding – as far as it is legitimate to make such correspondences – to our notion of 'topology'.

[51] That the full phrase of the form ἡ εὐθεῖα γραμμὴ AB is almost always contracted to the minimum ἡ AB, even though this may equally well stand for ἡ γραμμὴ AB *simpliciter* – i.e. for a curved rather than a straight line – reflects the fact that this basic distinction, between curved and straight, could generally be *seen* in the diagram.

[52] So far, the technology is not confined to Greece; and Babylonian 'structural diagrams', described by Hoyrup (1990a: 287–8), are useful in this context.

[53] In Archimedes' *SL*, which includes 22 geometrical propositions (i.e. a few hundred letters), there are 24 which do not stand in extremes, or intersections, of lines, namely proposition 14: Β, Γ, Κ; 15: Β, Κ; 16: Β, Γ, Κ, Ν; 17: Β, Κ, Ν; 18: Β, Γ, Κ, Λ; 19: Β, Ε, Κ, Λ; 20: Β, Λ; 21: Δ; 28: Β. I choose this example as a case where there are relatively many such points, the reason being Archimedes' way of naming spirals by many letters, more letters than he can affix to extremes and intersections alone – essentially a reflection of the peculiarity of the spiral.

finally, is the crucial point. The diagram is named – more precisely, it is lettered. It is the lettering of the diagram which turns it into a system of intersections, into a finite, manageable system.

To sum up, there are two elements to the technology of diagrams: the use of ruler and compasses, and the use of letters. Each element redefines the infinite, continuous mass of geometrical figures into a man-made, finite, discrete perception. Of course, this does not mean that the object of Greek mathematics is finite and discrete. The perceived diagram does not exhaust the geometrical object. This object is partly defined by the text, e.g. metric properties are textually defined. But the properties of the perceived diagram form a true subset of the real properties of the mathematical object. This is why diagrams are good to think with.[54]

2.2 *Diagrams as metonyms of propositions*

A natural question to ask here is whether the practices described so far are reflected in the Greek conceptualisation of the role of diagrams. The claim of the title is that this is the case, in a strong sense. Diagrams are considered by the Greeks not as appendages to propositions, but as the core of a proposition.[55]

2.2.1 *Speaking about diagrams*[56]

Our 'diagram' derives from Greek *diagramma* whose principal meaning LSJ define as a 'figure marked out by lines', which is certainly etymologically correct. The word *diagramma* is sandwiched, as it were, between its anterior and posterior etymologies, both referring simply to drawn figures. Actual Greek usage is more complex.

Diagramma is a term often used by Plato – one of the first, among extant authors, to have used it – either as standing for mathematical

[54] A disclaimer: I am not making the philosophical or cognitive claim that the only way in which diagrams can be deductively useful is by being reconceptualised via letters. As always, I am a historian, and I make the historical claim that diagrams came to be useful as deductive tools in Greek mathematics through this reconceptualisation.

[55] That they *put* diagrams as 'appendages' – i.e. at the end of propositions rather than at their beginning or middle – shows something about the relative role of beginning and end, not about the role of the diagram. It should be remembered that the titles of Greek books are also often put at the *end* of treatises. My guess is that, reading a Greek proposition, the user would unroll some of the papyrus to have the entire text of the proposition (presumably a few columns long) ending conveniently with the diagram. It was the advent of the codex which led to today's nightmare of constant backwards-and-forwards glancing, from text to diagram, whenever the text spills from one page to the next.

[56] Part of the argument of this subsection derives from Knorr (1975) 69–75.

proofs or as the *de rigueur* accompaniment of mathematics.[57] With Aristotle, *diagrammata* (the plural of *diagramma*) can practically mean 'mathematics', while *diagramma* itself certainly means 'a mathematical proposition'.[58] Xenophon tells us that Socrates used to advise young friends to study geometry, but not as far as the unintelligible *diagrammata*,[59] and we begin to think that this may mean more than just very intricate diagrams in the modern sense. Further, Knorr has shown that the cognates of *graphein*, 'to draw', must often be taken to carry a logical import.[60] He translates this verb by 'prove by means of diagrams'. Certainly this phrase is the correct translation; however, we should remember that the phrase stands for what, for the Greeks, was a single concept.

Complementary to this, the terminology for 'diagram' in the modern sense is complex. The word *diagramma* is never used by Greek mathematicians in the sense of 'diagram'. When they want to emphasise that a proposition relies upon a diagram, they characterise it as done *dia grammōn* – 'through lines', in various contexts opposed to the only other option, *di' arithmōn* – 'through numbers'.[61]

A word mathematicians may use when referring to diagrams present within a proof is *katagraphē* – best translated as 'drawing'.[62] The verb *katagraphein* is regularly used in the sense of 'completing a figure', when the figure itself is not specified in the text. The verb is always used within this formula, and with a specific figure: a parallelogram (often rectangle) with a diagonal and parallel lines inside it.[63]

[57] As in *Euthd.* 290c; *Phaedr.* 73b; *Theaetet.* 169a; the [pseudo?]-Platonic *Epin.* 991e; and, of course, *Rep.* 510c.

[58] E.g. *APr.* 41b14; *Meteor.* 375b18; *Cat.* 14b1; *Metaph.* 998a25, 1051a22; *SE* 175a27.

[59] *Mem.* IV.7.3.

[60] Knorr (1975) 69–75.

[61] See, e.g. Heron: *Metrica* II.10.3; Ptolemy: *Almagest* I.10, 32.1, VIII.5, 193.19, *Harmonics* I.5, 12.8; Pappus VI.600.9–13. Proclus, *In Rem Publicam* II.23. The treatment of book II by Hero, as preserved in the Codex Leidenensis (Besthorn and Heiberg (1900: 8ff.), is especially curious: it appears that Hero set out to prove various results with as few lines as possible, preferably with none at all, but with a single line if the complete avoidance of lines was impossible (one is reminded of children's puzzles – 'by moving one match only, the train changes into a balloon'). Hero's practice is comparable to the way a modern mathematician would be interested in proving the result X on the basis of fewer axioms than his predecessors. Modern mathematicians prove with axioms; Greek mathematicians proved with lines.

[62] See e.g. Euclid's *Elements* III.33, IV.5, XII.4; Apollonius, *Conics* IV.27. Archimedes usually refers simply to σχήματα (*CF* II.394.6, 406.2, 410.24; *SC* II.224.3). This is 'figure' in the full sense of the word, best understood as a *continuous* system of lines; a single diagram – especially an Archimedean one! – may include more than a single σχῆμα. Finally, Archimedes uses once the verb ὑπογράφειν (*PE* I.5 Cor. 2, 132.12), a relative of καταγράφειν.

[63] The first five propositions of Euclid's *Elements* XIII, and also: II.7, 8; VI.27–9; X.91–6. The formula is a feature of the Euclidean style – though the fact that Apollonius and Archimedes do not use it should be attributed, I think, to the fact that they do not discuss this rectangle.

Aristotle's references to diagrams are even more varied. On several occasions he refers to his own diagrams as *hupographai*, yet another relative of the same etymological family.[64] *Diagraphai* – a large family – are mentioned as well.[65] None of these diagrams are *mathematical* diagrams; when referring to a proof where a mathematical diagram occurs, Aristotle uses the word *diagramma*, and we are left in the dark as to whether this refers to the diagram or to the proof as a whole. What does emerge in Aristotle's case is a certain discrepancy between the standard talk *about* mathematics and the talk *of* mathematics. We will become better acquainted with this discrepancy in chapter 3.

Mathematical commentators may combine the two discourses, of mathematics and about mathematics. What is their usage? Pappus uses *diagramma* as a simple equivalent of our 'proposition'.[66] In several cases, when referring to a diagram inside a proposition, he uses *hupographē*.[67] Proclus never uses *diagramma* when referring to an actual present diagram, to which he refers by using the term *katagraphē* or, once, *hupogegrammenē*.[68] Eutocius uses *katagraphē* quite often.[69] *Schēma*, in the sense of *one* of the diagrams referred to in a proposition, is used as well. It is interesting that one of these uses derives directly from Archimedes,[70] while all the rest occur in – what I believe is a genuine – Eratosthenes fragment.[71]

The evidence is spread over a very long period indeed, but it is coherent. Alongside more technical words signifying a 'diagram' in the modern sense – words which never crystallised into a systematic terminology – the word *diagramma* is the one reserved for signifying that which a mathematical proposition is. Should we simply scrap, then, the notion that *diagramma* had anything at all to do with a 'diagram'? Certainly not. The etymology is too strong, and the semantic situation can be easily understood. *Diagramma* is the metonym of the proposition.

[64] *de Int.* 22a22; *Meteor.* 346a32, 363a26; *HA* 510a30; *EE* 1220b37.

[65] *EE* 1228a28; *EN* 1107a33; *HA* 497a32, 525a9. The γεγραμμέναι of *de Part.* 642a12 is probably relevant as well; I guess that the last mentioned are ἀνατομαί-type diagrams, included in a book, and that diagrams set out in front of an audience (e.g. on wooden tablets) are called ὑπογραφαί; but this is strictly a guess.

[66] E.g. VII.638.17, 670.1–2. When counting propositions in books, Pappus often counts θεωρήματα ἤτοι διαγράμματα, 'theorems, or diagrams' – a nice proof that 'diagrams' may function as metonyms of propositions.

[67] Several cognate expressions occur in IV.200.26, 272.14, 298.6; VI.542.11, 544.19 and, perhaps, III.134.22.

[68] *In Eucl.*: καταγραφή: 340.11, 358.11, 370.14, 400.9–15; ὑπογεγραμμένη: 286.22.

[69] Seventeen times in the commentary to Archimedes, for which see index II to Archimedes vol. III.

[70] 216.24. [71] 88.15, 92.7, 94.13, 19.

It is so strongly entrenched in this role that when one wants to make quite clear that one refers to the diagram and *not* to the proposition – which happens very rarely – one has to use other, more specialised terms.[72]

2.2.2 *Diagrams and the individuation of propositions*

That diagrams may be the metonyms of propositions is surprising for the following reason. The natural candidate from our point of view would be the 'proposition', the enunciation of the content of the proposition – because this enunciation *individuates* the proposition. The hallmark of Euclid's *Elements* I.47 is that it proves 'Pythagoras' theorem' – which no other proposition does. On the other hand, nothing, logically, impedes one from using the same diagram for different propositions.

Even if this were true, it would show not that diagrams cannot be metonyms, but just that they are awkward metonyms. But interestingly this is wrong. The overwhelming rule in Greek mathematics is that propositions *are* individuated by their diagrams. Thus, diagrams are convenient metonyms.[73]

The test for this is the following. It often happens that two separate lines of reasoning employ the same basic geometrical configuration. This may happen either within propositions or between propositions.[74] Identity of configuration need not, however, imply identity of diagram, since the lettering may change while the configuration remains. My claim is that identity of configuration implies identity of diagram *within* propositions, and does not imply such identity *between* propositions.

What is an 'identity' between diagrams? This is a matter of degree – one can give grades, as it were:

1. 'Identity *simpliciter*' – the diagrams may be literally identical.
2.1. 'Inclusion' – the diagrams may not be identical, because the second has some geometrical elements which did not occur in the

[72] Note that I am speaking here not of diachronic evolution, but of a synchronic situation. It is thus useful to note that in contexts which are not strictly mathematical διάγραμμα has clearly the sense 'diagram' – e.g. Bacchius, in *Musici Graeci* ed. Janus, 305.16–17: Διάγραμμα . . . τί εστι; – Συστήματος ὑπόδειγμα. ἤτοι οὕτως, διάγραμμά ἐστι σχῆμα ἐπίπεδον . . .

[73] Here it should be clarified that the 'diagram' of a single proposition may be composed of a number of 'figures', i.e. continuous configurations of lines. When these different figures are not simply different objects discussed by a single proof, but are the same object with different cases (e.g. Euclid's *Elements* I.35), the problem of transmission becomes acute. Given our current level of knowledge on the transmission of diagrams, nothing can be said on such diagrams.

[74] Such continuities may be singled out in the text by the formulae τῶν αὐτῶν ὑποκειμένων/κατασκευασθέντων, καὶ τὰ ἄλλα τὰ αὐτὰ προκείσθω/κατασκευάσθω – see e.g. Euclid's *Elements* III.3, 14; VI.2, 3; Archimedes, *SC* II.6; Apollonius, *Conics* III.6. I will argue below that such continuities do not imply identities. Whether the continuity is explicitly noted or not does not change this.

first (or vice versa). However, the basic configuration remains. Furthermore, all the letters which appear in both diagrams stand next to identical objects (some letters would occur in this diagram but not in the other; but they would stand next to *objects* which occur in this diagram but not in the other). Hence, wherever the two diagrams describe a similar situation they may be used interchangeably.

2.2. 'Defective inclusion' – diagrams may have a shared configuration, but some letters change their objects between the two diagrams. Thus, it is no longer possible to interchange the diagrams, even for a limited domain.

3. 'Similarity' – the configuration is not identical, and letters switch objects, but there is a certain continuity between the two diagrams.

'F'. No identity at all – although the two propositions refer to a mathematical situation which is basically similar, the diagrams are flagrantly different.

Conics III offers many cases of interpropositional continuity of subject matter. I have graded them all.[75] The results are: a single first, seven 2.1, four 2.2, six thirds and four fails. Disappointing; in fact, the results are very heterogeneous and should not be used as a quantitative guide. The important point is the great rarity of the first – which makes it look like a fluke.

To put this evidence in a wider context, it should be noted that *Conics* III is remarkable in having so many cases of continuities. More often, subject matters change between propositions, ruling out identical diagrams. An interesting case in the Archimedean corpus is *CF* 4/5: a 2.2 by my marking system, but the manuscripts are problematic. Euclid's *schēma*, used in the formula 'and let the figure be drawn' to which I have referred in n. 63 above, is usually in the range 3–F.[76] There are no relevant cases in Autolycus; I shall now mention a case from Aristarchus (and, in n. 79, Ptolemy).

The best way to understand the Greek practice in this respect is to compare it with Heath's editions of Archimedes and Apollonius. One of the ways in which Heath mutilated their spirit is by making diagrams as identical as possible. This makes the individuated unit larger

[75] 1: 46 (identical to 45); 2.1: 2 (compared with 1), 14 (13), 29 (28), 47–50 (15); 2.2: 7–8 (6), 10 (8), 21 (20); 3: 3 (1), 6 (5), 9 (6), 25 (24), 35 (34), 38 (37); F: 11 (8), 12 (11), 36 (34), 40 (39).

[76] In this I ignore *Elements* x.91–6, which is a specimen from a strange context. In general, book x works in hexads, units of six propositions proving more or less the same thing. It is difficult to pronounce exactly on the principle of individuation in this book: are propositions individuated, or are hexads?

than a given proposition: it is something like a 'mathematical idea'. But such identities ranging over propositions are Heath's, not Archimedes' nor Apollonius'.

The complementary part of my hypothesis has to do with internal relations. It is not at all rare for a proposition to use the same configuration twice. For instance, this is very common in some versions of the method of exhaustion, where the figure is approached from 'above' and from 'below'. The significance of the diagram changes; yet, there is no evidence that it has been redrawn.[77]

The following case appears very strange at first glance: the construction of Aristarchus 14 begins with ἔστω τὸ αὐτὸ σχῆμα τῷ πρότερον – 'let there be the same figure as before'.[78] Having said that, Aristarchus proceeds to draw a diagram which I would mark 2.2 – not at all the identity suggested by his own words (figs. 1.13a and 1.13b)! How can we account for this? I suggest the following: Aristarchus' motivation is to save space; that is, he does not want to give the entire construction from scratch – that would be tedious. But then, saying 'let *A* and *B* be the same, *C* and *D* be different, and so on' is just as tedious. So he simply says 'let it be the same', knowing that his readers would not be misled, for no reader would expect two diagrams to be literally identical. When you are told somebody's face is 'the same as Woody Allen's', you do not accept this as literally true – the pragmatics of the situation rule this out. Faces are just too individual. Greek diagrams are, as it were, the faces of propositions, their metonyms.[79]

2.2.3 *Diagrams as metonyms of propositions: summary*

I have claimed that diagrams are the metonyms of propositions; in effect, the metonyms of mathematics (as mentioned in n. 58 above).

[77] See, e.g. Archimedes, *CS* 21 352.9, 25 380.16, 26 388.10, 27 402.7–8, 29 420.15; *SC* II.6 204.14; *QP* 16 296.26. For examples from outside the method of exhaustion, see Apollonius' *Conics* I.26 82.20–1; 32 96.23–6; Euclid's *Elements* III.3 172.17, 14 204.11.

[78] Aristarchus 14 398.23. Incidentally, this is another mathematical use of σχῆμα for 'diagram'.

[79] I have not discussed Ptolemy's diagrams in this subsection. Ptolemy often uses expressions like 'using the same diagram'. Often the diagrams involved are very dissimilar (e.g. the first diagram of *Syntaxis* V.6, in 380.18–19, referring to the last diagram of V.5). Sometimes Ptolemy registers the difference between the diagrams by using expressions such as 'using a *similar* diagram' (e.g. the first diagram of XI.5, in 393.1–2, referring to the first diagram of XI.1). Rarely, diagrams are said to be 'the same' and are indeed practically identical (e.g. the fourth diagram of III.5, in 245.6–7, referring to the third diagram of III.5). But this is related to another fact: Ptolemy uses in the *Syntaxis* a limited *type* of diagram. Almost always, whether he does trigonometry or astronomy, Ptolemy works with a diagram based on a single circle with some lines passing through it. A typical Greek mathematical work has a wide range of diagrams; each page looks different. Ptolemy is more repetitive, more schematic. L. Taub suggested to me that this should be related to Ptolemy's wider programme – that of preparing a 'syntaxis', *organised* knowledge.

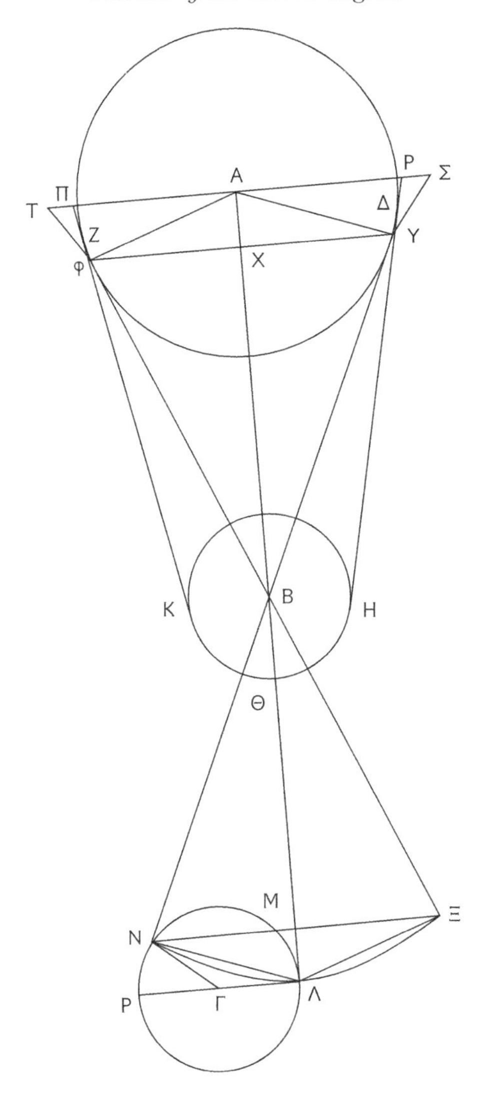

Figure 1.13a. Aristarchus 13.

That diagrams were considered essential for mathematics is proved by books v, vii–ix of Euclid's *Elements*. There, all the propositions are accompanied by diagrams, as individual and – as far as the situations allow – as elaborate as any geometrical diagram. Yet, in a sense, they are redundant, for they no longer represent the situations discussed. As Mueller points out, these diagrams may be helpful in various ways.[80]

[80] Mueller (1981) 67.

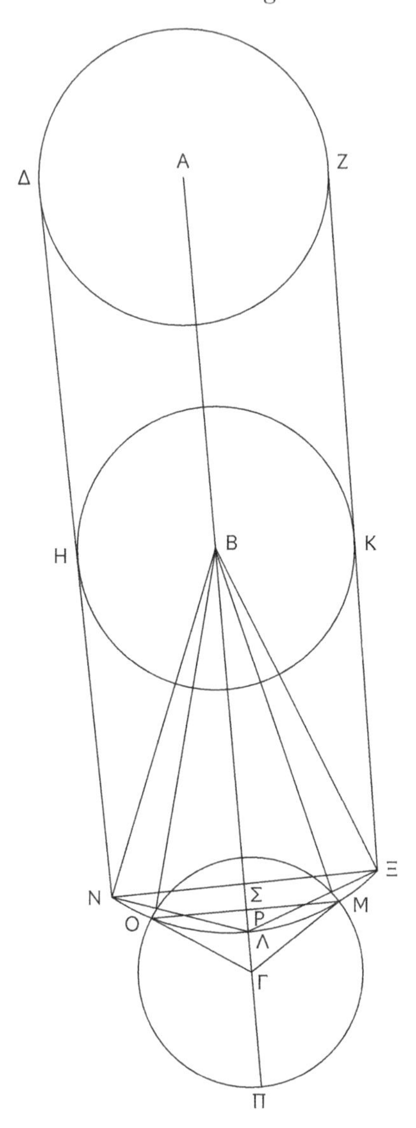

Figure 1.13b. Aristarchus 14.

Yet, as he asserts, they no longer have the same function. They reflect a cultural assumption, that mathematics *ought* to be accompanied by diagrams. Probably line diagrams are not the best way to organise proportion theory and arithmetic. Certainly symbolic conventions such as '=', for instance, may be more useful. The lettered diagram functions here as an obstacle: by demanding one kind of representation, it obstructs the development of other, perhaps more efficient representations.[81] An obstacle or an aid: the diagram was essential.

2.3 *The semiotic situation*

So far I have used neutral expressions such as 'the point *represented* by the letter'. Clearly, however, the cognitive contribution of the diagram cannot be understood without some account of what is involved in those 'representations' being given. This may lead to problems. The semiotic question is a tangent to a central philosophical controversy: what is the object of mathematics? In the following I shall try not to address such general questions. I am interested in the semiotic relation which Greek mathematicians have used, not in the semiotic relations which mathematicians in general ought to use. I shall first discuss the semiotic relations concerning letters, and then the semiotic relations concerning diagrams.

2.3.1 *The semiotics of letters*

Our task is to interpret expressions such as ἔστω τὸ μὲν δοθὲν σημεῖον τὸ A[82] – 'let the given point be the A'. To begin with, expressions such as τὸ A, 'the A', are not shorthand for 'the *letter* A'; A is not a letter here, but a point.[83] The letter in the text refers not to the letter in the diagram, but to a certain point.

Related to this is the following. Consider this example, one of many:[84]

ἔστω εὐθεῖα ἡ AB

(I will give a translation shortly).

This is translated by Heath as 'Let *AB* be a straight line.'[85] This creates the impression that the statement asserts a correlation between a symbol and an object – what I would call 'a moment of specification *per se*'.

[81] By a process which eludes our knowledge, manuscripts for Diophantus developed a limited system of shorthand, very roughly comparable to an abstract symbolic apparatus. Whether this happened in ancient times we can't tell; at any rate, Diophantus requires a separate study.

[82] Euclid's *Elements* I.2, 12.21.

[83] This can be shown through the wider practice of such abbreviations, which I discuss in chapter 4.

[84] Euclid's *Elements* XIII.4, 256.26; Heath's version is vol. III.447.

[85] Heath probably preferred, in this case, a slight unfaithfulness in the translation to a certain stylistic awkwardness. It so happens that this slight unfaithfulness is of great semiotic significance. It should be added that I know of no translation of Euclid which does not commit – what I think is – Heath's mistake. Federspiel (1992), in a context very different from the present one, was the first to suggest the correct translation.

In fact, this translation is untenable, since the article before AB can only be interpreted as standing for the elided phrase 'straight line',[86] so Heath's version reads as 'let the straight line *AB* be a straight line', which is preposterous. In fact the word order facilitates the following translation:

'Let there be a straight line, [viz.] *AB*.'

First, what such clauses do not assert: they do not assert a relation between a symbol and an object. Rather, they assert an action – in the case above, the taking for granted of a certain line – and they proceed to localise that action in the diagram, on the basis of an independently established reference of the letters. The identity of 'the *AB*' as a certain line in the diagram is assumed by Euclid, rather than asserted by him.

So far, expressions use the bare article and a combination of one or more letters. This is the typical group of expressions. There is another, rarer, group of expressions, which may shed some light on the more common one. Take the Hippocratic fragment, our evidence for earliest Greek mathematics[87] (fig. 1.14):

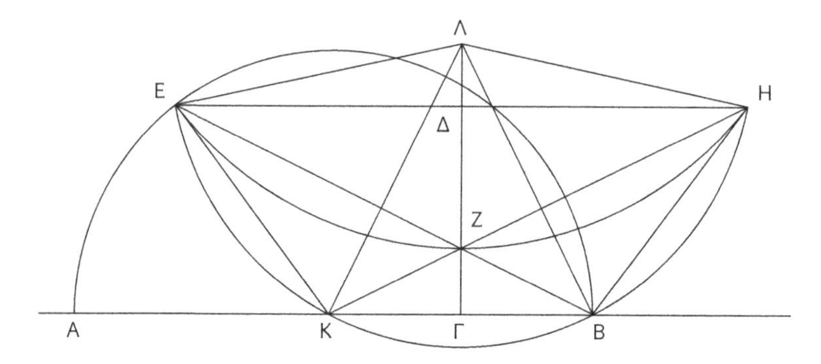

Figure 1.14. Hippocrates' Third Quadrature.

ἔστω κύκλος οὗ διάμετρος ἐφ' ᾗ AB κέντρον δε αὐτοῦ ἐφ' ᾧ K

'Let there be a circle whose diameter [is that] on which AB, its centre [that] on which K'.

[86] While the feminine gender, in itself, does not imply a straight line, the overall practice demands that one reads the bare feminine article, *ceteris paribus*, as referring to a *straight* line.

[87] Becker (1936b) 418.32.

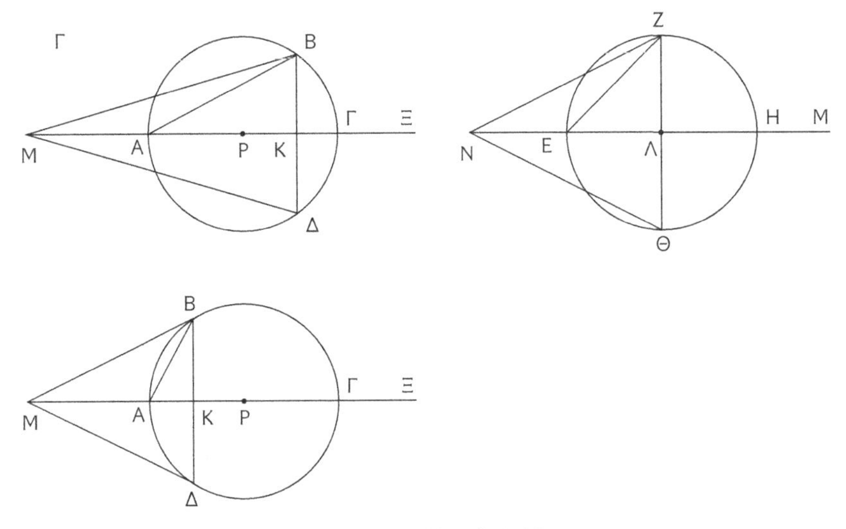

Figure 1.15. Archimedes' *SC* II.9.

I translate by 'on which' a phrase which in the Greek uses the preposition *epi* with the dative (which is interchangeable with the genitive).[88] Our task is to interpret this usage.

Expressions such as that of the Hippocratic fragment are characteristic of the earliest Greek texts which use the lettered diagram, that is, besides the Hippocratic fragment itself, the mathematical texts of Aristotle.[89] However, Aristotle – as ever – has his own, non-mathematical project, which makes him a difficult guide. I shall first try to elucidate this practice out of later, well-understood *mathematical* practice, and then I shall return to Aristotle.

The Archimedean corpus contains several expressions similar to the *epi* + dative. First, at *SC* II.9 Archimedes[90] draws several *schēmata*, and in order to distinguish between them, a Γ (or a special sign, according to another manuscript)[91] is written next to that *schēma* (fig. 1.15). Later

88 For the genitive in the Hippocratic fragment, see Simplicius, *In Phys.* 65.9, 16; 67.21–2; 68.14. It is interesting to see that in a number of cases the manuscripts have either genitive or dative, and Diels, the editor, *always* chooses the dative: 64.13, 15; 67.29, 37 – which gives the text a dative-oriented aspect stronger than it would have otherwise (though Diels, of course, may be right).

89 E.g. *Meteor.* 375b22, 376a5, 15, b5, 13, etc.; as well as many examples in contexts which are not strictly mathematical, e.g. *Meteor.* 363a34; *HA* 510a31, 550a25; *Metaph.* 1092b34. The presence of a diagram cannot always be proved, and probably is not the universal case.

90 Or some ancient mathematical reader; for our immediate purposes, the identification is not so important.

91 The same sign (astronomical sun) is used to indicate a scholion, in *PE* II.7, 188.18.

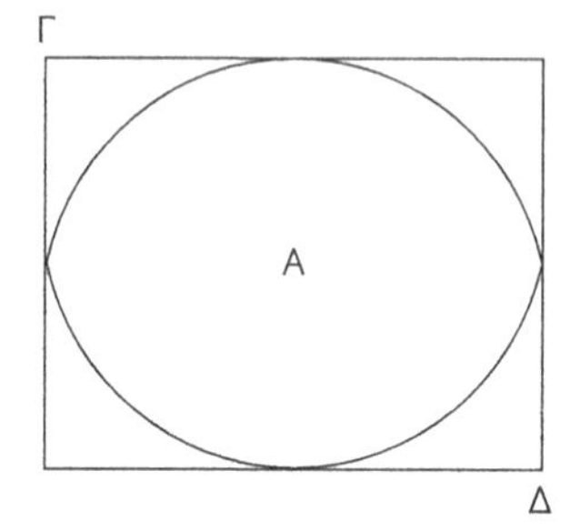

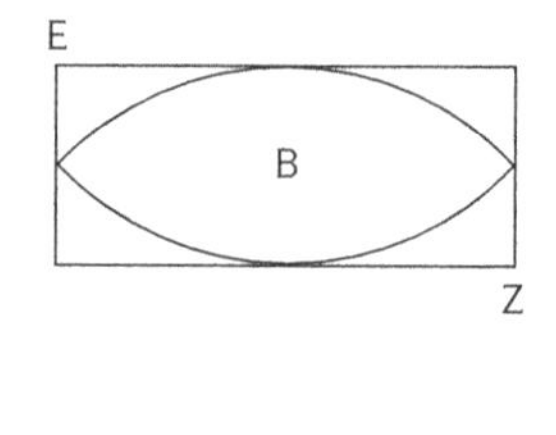

Figure 1.16. Archimedes' *CS* 6.

Λ	Λ	Λ	Λ	Λ	Λ
K	K	K	K	K	K
I	I	I	I	I	I
Θ	Θ	Θ	Θ	Θ	Θ

Figure 1.17. Archimedes' *CS* 2.

in the same proposition, at 250.6–7, when referring to that *schēma*, the expression used is πρὸς ᾧ το Γ σημεῖον – 'that, next to which is the sign[92] Γ'. This uses the preposition *pros* with the dative. I shall take *CS* 6 282.17–18 next. In order to refer to areas bounded by ellipses, in turn surrounded by rectangles, Archimedes writes the letters A, B inside the ellipses (fig. 1.16), then describes them in the following way: ἔστω περιεχόμενα χωρία ὑπὸ ὀξυγωνίου κώνου τομᾶς, ἐν οἷς τὰ A, B – 'let there be areas bounded by ellipses, in which are A, B'. This uses the preposition *en* with the dative. Proposition 2 in the same work refers, first, to signs which stand near lines and, consequently, within rectangles (fig. 1.17). It comes as no surprise now that the rectangles are mentioned at 268.1 as ἐν οἷς τὰ Θ, I, K, Λ – 'in which the Θ, I, K, Λ'. More interestingly, the lines in question are referred to at, e.g. 266.22–3 as ἐφ' ὧν . . . Θ, I, K, Λ – 'on which Θ, I, K, Λ' – where we finally get as far as the *epi* + genitive.[93] A certain order begins to emerge.

[92] Undoubtedly this is the sense of σημεῖον here. That the word becomes homonymous is not surprising: we shall see in chapter 3 that, in the border between first-order and second-order language, many such homonyms occur.

[93] For further examples of prepositions with letters, see Archimedes, *SL* 46.27, 48.20, 52.22, 72.18, 76.21, 80.24, 84.5, 86.11, 92.24, 102.13, 24 116.22; *CS* 266.17, 268.7, 19, 270.2, 276.7–9, 276.14, 280.23–5, 282.18, 24, 372.18, 276.3, 12–13, 378.13, 19, 390.10, 394.9, 396.12, 400.12, 410.27, 414.5, 418.10, 426.20; Apollonius, *Conics* III.13, 338.12; Pappus, book II, *passim* (in the context 'ἀριθμοὶ ἐφ' ὧν τὰ A . . .').

When Archimedes deviates from the normal letter-per-point convention, he often has to clarify what he refers to. A fuller expression is needed, and this is made up of prepositions, relatives and letters. Now the important fact is that the prepositions are used in a spatial sense – as is shown by their structured diversity. Different prepositions and cases are used in different spatial configurations. They describe various *spatial* relationships between the letters in the diagram and the objects referred to.

There is a well-known distinction, offered by Peirce, between three types of signs. Some signs are *indices*, signifying by virtue of some deictic relation with their object: an index finger is a good example. Other signs are *icons*, signifying by virtue of a similarity with their object: a portrait is a good example. Finally, some signs are *symbols*, signifying by virtue of arbitrary conventions: most words are symbols. We have gradually acquired evidence that in some contexts the letters in Greek diagrams may be seen as indices rather than symbols. My theory is that this is the case generally, i.e. the letter *alpha* signifies the point next to which it stands, not by virtue of its being a symbol for it, but simply because it stands next to it. The letters in the diagram are useful signposts. They do not stand *for* objects, they stand *on* them.

There are two different questions here. First, is this the correct interpretation of *epi* + dative/genitive in the earliest sources? Second, should this interpretation be universally extended?

The answer to the first question should, I think, be relatively straightforward. The most natural reading of *epi* is spatial, so, given the presence of a diagram which makes a spatial reading possible, I think such a reading cannot be avoided. It is true that many spatial terms are used metaphorically (if this is the right word), probably in all languages. In English, one can debate whether 'Britain should be inside the European Union', and it is clear that no spatial reading is intended: 'European Union' is (in a sense) an abstract, non-spatial object. The debate can be understood only in terms of inclusion in a wide, non-spatial sense. But if you ask whether 'the plate should be inside the cupboard', it is very difficult to interpret this in non-spatial terms. When a spatial reading suggests itself at all, it is irresistible. I have argued that the mathematical text is focused on the strictly spatial object of the diagram. It is as spatial as the world of plates and cupboards; and a spatial reading of the expressions relating to it is therefore the natural reading.

The case of Aristotle is difficult.[94] Setting aside cases where a reference to a diagram is clear, the main body of evidence is from the *Analytics*. There, letters are used very often.[95] When the use of those letters is of the form '*A* applies to all *B*', etc., the bare article + letter is used, i.e. the *epi* + dative/genitive is never used in such contexts. From time to time, Aristotle establishes a relation between such letters and 'real' objects – *A* becomes man, *B* becomes animal, etc. Usually, when this happens, the *epi* + dative/genitive is used at least with one of the correlations, and should probably be assumed to govern all the rest.[96] A typical example is 30a30:

> ἐφ' ᾧ δὲ τὸ Γ ἄνθρωπος
>
> 'And [if that] on which Γ [is] man' / 'and [if that] which Γ stands for [is] man'.

I have offered two alternative translations, but the second should probably be preferred, for after all Γ does not, spatially speaking, stand on the class of all human beings. It's true that the antecedent of the relative clause need not be taken here to be 'man'. Indeed, often it cannot, when the genders of the relative pronoun and the signified object clash.[97] But there are other cases, where the gender, or more often the number of the relative pronoun do change according to the signified object.[98] The most consistent feature of this Aristotelian usage is its inconsistency – not a paradox, but a helpful hint on the nature of the usage. Aristotle, I suggest, uses language in a strange, forced way. That his usage of letters is borrowed from mathematics is extremely likely. That in such contexts the sense of the *epi* + dative/genitive would have been spatial is as probable. In a very definite context – that of establishing external references to letters of the syllogism – Aristotle uses this expression in a non-spatial sense. Remember that Aristotle had to start logic from scratch, the notions of referentiality included. I suggest that the use of the *epi* + dative/genitive in the *Analytics* is a bold *metaphor*, departing from the spatial mathematical

[94] Readers unfamiliar with Greek or Aristotle may prefer to skip the following discussion, which is relatively technical.

[95] The letter A is used more than 1,200 times; generally, the density of letters is almost comparable to a mathematical treatise.

[96] There are about – very roughly – a hundred such examples in the *Analytics*, which I will not list here. In pages 30–49 of *APr.* the examples are: 30a30, 31b5, 28, 34a7–8, b34, 39, 37b1, 38a31, b20, 44a13–17, 26, b3, 46b3–5, 13, 34, 47b21–2, 30–1, 48a3, 33–4, 48b6, 49a15–16, 32, 34, 39, b1.

[97] E.g. *APr.* 64a24: ἰατρικὴ δ' ἐφ' οὗ Δ.

[98] E.g. *APr.* 44a13: ἑπόμενα τῷ Α ἐφ' ὧν Β; *APo.* 94a29: ἡμισεῖα δυοῖν ὀρθαῖν ἐφ' ἧς Β.

usage. Aristotle says, 'let A stand on "man"', implying 'as mathematical letters stand on their objects and thus signify them', meaning 'let A signify "man"'. The index is the metaphor through which the general concept of the sign is broached.[99] This, I admit, is a hypothesis. At any rate, the *contents* referred to by Aristotle are like 'Britain' and 'European Union', not like 'plates' and 'cupboards'; hence a non-spatial reading becomes more natural.

Moving now to the next question: should the mathematical letters be seen as indices even in the absence of the *epi* + dative/genitive and its relatives?

The first and most important general argument in favour of this theory is the correction offered above to Heath's translation of expressions such as ἔστω εὐθεῖα ἡ AB, 'let there be a line, <namely> AB'. If the signification of the 'AB' is settled independently, and antecedently to the text, then it could be settled only via the letters as indices. The setting of symbols requires speech; indices are visual. The whole line of argument, according to which specification of objects in Greek mathematics is visual rather than verbal, supports, therefore, the indices theory.

Next, consider the following. In the first proposition of the *Conics* – any other example with a similar combination of genders will do – a point is specified in the following way:[100]

ἔστω κωνικὴ ἐπιφάνεια, ἧς κορυφὴ τὸ A σημεῖον

'Let there be a conic surface, whose vertex is the point A'.

The point A has been defined as a vertex, and it will function in the proposition *qua* vertex, not *qua* point. Yet it will always be called, as in the specification itself, τὸ A, in the neuter ('point' in Greek is neuter, while 'vertex' is feminine). This is the general rule: points, even when acquiring a special significance, are always called simply 'points', never, e.g. 'vertices'. The reason is simple: the expression τὸ A is a periphrastic reference to an object, using the letter in the diagram, A, as a signpost useful for its spatial relations. This letter in the diagram, the actual shape of ink, stands in a spatial relation to a *point*, not to a vertex – the point is spatial, while the vertex is conceptual.

[99] Another argument for the 'metaphor' hypothesis is the fact that the *epi* + dative/genitive is not used freely by Aristotle, but only within a definite formula: he never uses more direct expressions such as καὶ Γ ἐπ' ἀνθρώπῳ – 'and [if] Γ stands for man' – instead he sticks to the cumbersome relative phrase. Could this reflect the fact that the expression is a metaphor, and thus cannot be used outside the context which makes the metaphor work?

[100] 8.25.

Third, an index (but not a symbol) can represent simultaneously several objects; all it needs to do so is to point to all of them. Some mathematical letters are polyvalent in exactly this way: e.g. in Archimedes' *SC* 32, the letters O, Ξ, stand for both the circles and for the cones whose bases those circles are.[101]

Fourth, my interpretation would predict that the letters in the text would be considered as radically different from other items, whereas otherwise they should be considered as names, as good as any. There is some palaeographic evidence for this.[102]

Fifth, a central thesis concerning Greek mathematics is that offered by Klein (1934–6), according to which Greek mathematics does not employ variables. I quote:[103] 'The Euclidean presentation is not symbolic. It always intends determinate numbers of units of measurements, and it does this without any detour through a "general notion" or a concept of a "general magnitude".'

This is by no means unanimously accepted. Klein's argument is philosophical, having to do with fine conceptual issues.[104] He takes it for granted that *A* is, in the Peircean sense, a symbol, and insists that it is a symbol of something determinate. Quite rightly, the opposition cannot see why (the symbolhood of *A* taken for granted) it cannot refer to whatever it applies to. My semiotic hypothesis shows why *A* must be determinate: because it was never a symbol to begin with. It is a signpost, and signposts are tied to their immediate objects.

Finally, my interpretation is the 'natural' interpretation – as soon as one rids oneself of twentieth-century philosophy of mathematics. My proof is simple, namely that Peirce actually took letters in diagrams as *examples* of what he meant by 'indices':[105] '[W]e find that indices are

[101] Or a somewhat different case: Archimedes' *PE* I.3, where A, B are simultaneously planes, and the planes' centres of gravity.

[102] It should be remembered that, as a rule, Greek papyri do not space words. *P. Berol.* 12609, from *c.*350–325 BC (Mau and Mueller 1962, table II): the continuous text is, as usual, unspaced. Letters referring to the diagram are spaced from the rest of the text. *P. Herc.* 1061, from the last century BC, contains no marking off of letters, but the context is non-mathematical. *PFay.* 9, later still, marks letters by superscribed lines, as does the *In Theaetet.* (early AD? *CPF* III 341, n. ad XXIX.42–XXXI.28). This practice can often be seen in manuscripts. Generally, letters are comparable to *nomina sacra*. Perhaps it all boils down to the fact that letters, just as *nomina sacra*, are not read phonetically (i.e. 'AB' was read *'alpha-bēta'*, not 'ab')?

[103] The quotation is from the English translation (Klein 1968: 123). Klein has predominantly arithmetic in mind, but if this is true of arithmetic, it must *a fortiori* be true of geometry.

[104] Unguru and Rowe (1981–2: the synthetic nature of so-called 'geometric algebra'), Unguru (1991: the absence of mathematical induction; I shall comment on this in chapter 6, subsection 2.6) and Unguru and Fried (forthcoming: the synthetic nature of Apollonius' *Conics*), taken together, afford a picture of Greek mathematics where the absence of variables can be shown to affect mathematical contents.

[105] Peirce (1932) 171.

absolutely indispensable in mathematics . . . So are the letters *A*, *B*, *C* etc., attached to a geometrical figure.'

The context from which the quotation is taken is richer, and one need not subscribe to all aspects of Peirce's philosophy of mathematics there. But I ask a descriptive, not a prescriptive question. What sense did people make of letters in diagrams? Peirce, at least, understood them as indices. I consider this a helpful piece of evidence. After all, why not take Peirce himself as our guide in semiotics?

2.3.2 *The semiotics of diagrams*

So far, I have argued that letters are primarily indices, so that representations employing them cannot but refer to the concrete diagram. A further question is the semiotics of the diagram itself: does it refer to anything else, or is it the ultimate subject matter?

First, the option that the diagram points towards an ideal mathematical object can be disposed of. Greek mathematics cannot be about squares-as-such, that is, objects which have no other property except squareness, simply because many of the properties of squares are not properties of squares-as-such; e.g. the square on the diagonal of the square-as-such is the square-as-such, not its double.[106] It is not that speaking about objects-as-such is fundamentally wrong. It is simply not the same as speaking about objects. The case is clearer in algebra. One can speak about the even-as-such and the odd-as-such: this is a version of Boolean algebra.[107] Modern mathematics (that is, roughly, that of the last century or so) is characterised by an interest in the theories of objects-as-such; Greek mathematics was not.[108]

So what *is* the object of the proof? As usual, I look to the practices for a guide. We take off from the following. The proposition contains imperatives describing various geometrically defined operations, e.g.: κύκλος γεγράφθω – 'let a circle have been drawn'.[109] This is a certain action, the drawing of a circle. A different verb is 'to be', as in the

[106] The impossibility of Greek mathematics being about Platonic objects has been argued by Lear (1982), Burnyeat (1987).

[107] As the above may seem cryptic to a non-mathematician, I explain briefly. What is 'the essence' of the odd and the even? One good answer may be, for instance, to provide their table of addition: Odd + Odd = Even, O + E = O, E + O = O, E + E = E. One may then assume the existence of objects which are characterised by this feature only. One would thus 'abstract' odd-as-such and even-as-such from numbers. Such abstractions are typical of modern mathematics.

[108] Of course, the import of Greek proofs is general. This, however, need not mean that the proof itself is about a universal object. This issue will form the subject of chapter 6.

[109] Euclid's *Elements* 1.1, 10.19–12.1.

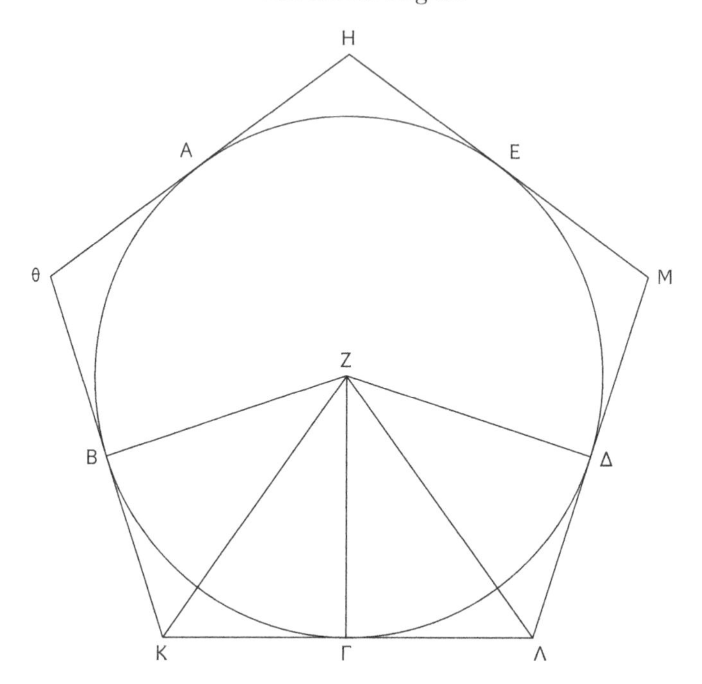

Figure 1.18. Euclid's *Elements* IV.12.

following:[110] ἔστω ἡ δοθεῖσα εὐθεῖα πεπερασμένη ἡ AB – 'let the given bounded straight line be AB'. The sense is that you identify the bounded given straight line (demanded earlier in the proposition) as AB. So this is another action, though here the activity is that of visually identifying an object instead of constructing it.

A verb which does not fit into this system of actions is *noein*, which may be translated here as 'to imagine', as in the following:[111]

> νενοήσθω τοῦ ἐγγεγραμμένου πενταγώνου τῶν γωνίων σημεῖα τὰ A, κτλ
>
> 'Let the points A, etc. be imagined as the points of the angles of the inscribed pentagon'.

What is the point of imagination here? The one noticeable thing is that the inscribed pentagon does not occur in the diagram, which for once should, with all the difficulties involved, be taken to reflect Euclid's diagram (fig. 1.18). On the logical plane, this means that

[110] Ibid. 10.16. [111] Euclid's *Elements* IV.12, 302.10–11.

the pentagon was taken for granted rather than constructed (its constructability, however, has been proved, so no falsity results).

Though not as common as some other verbs, *noein* is used quite often in Greek mathematics.[112] It is used when objects are either not drawn at all, as in the example above, or when the diagram, for some reason, fails to evoke them properly. The verb is relatively rare because such cases, of under-representation by the diagram, are relatively rare. It is most common with three-dimensional objects (especially the sphere, whose Greek representation is indeed indistinguishable from a circle). Another set of cases is in 'applied' mathematics, e.g. when a line is meant to be identified as a balance. Obviously the line is not a balance, it is a line, and therefore the verb *noein* is used.[113]

However, if the diagram is meant as a representation of some ideal mathematical object, then one should have said that any object whatever was 'imagined'. By delegating some, but not all, action to 'imagination', the mathematicians imply that, in the ordinary run of things, they literally mean what they say: the circle of the proof is drawn, not imagined to be drawn. It will not do to say that the circle was drawn in some ideal geometrical space; for in that geometrical space one might as easily draw a sphere. Thus, the action of the proof is literal, and the object of the proof must be the diagram itself, for it is only in the diagram that the acts of the construction literally can be said to have taken place.

This was one line of argument, showing that the diagram is the object of the proposition. In true Greek fashion, I shall now show that it is not the object of the proposition.

An obvious point, perhaps, is that the diagram must be false to some extent. This is indeed obvious for many moderns,[114] but at bottom this

[112] There are at least ten occurrences in Euclid's *Elements*, namely IV.12 302.10, XI.12 34.22, XII.4 lemma 162.21, 13 216.20, 14 220.7, 15 222.22 (that's a nice page and line reference!), 17 228.12, 228.20, 18 242.16, 244.6. There are three occurrences in Apollonius' *Conics* I, namely 52 160.18, 54 168.14, 56 178.12. Archimedes' works contain 38 occurrences of the verb in geometrical contexts, which may be hunted down through Heiberg's index. The verb is regularly used in Ptolemy's *Harmonics*. Lachterman (1989) claims on p. 89 that the verb is used by Euclid in book XII alone (the existence of Greek mathematicians other than Euclid is not registered), to mitigate, by its noetic function, the operationality involved in the generation of the sphere and the cylinder. We all make mistakes, and mine are probably worse than Lachterman's; but, as I disagree with Lachterman's picture of Greek mathematics as non-operational, I find it useful to note that this argument of his is false.

[113] E.g. Archimedes, *Meth.* 434.23 – one of many examples. The use of the verb in Ptolemy's *Harmonics* belongs to this class.

[114] E.g. Mill (1973), vol. VII 225: 'Their [sc. geometrical lines'] existence, as far as we can form any judgement, would seem to be inconsistent with the physical constitution of our planet at least, if not of the universe.' For this claim, Mill offers no argument.

is an empirical question. I imagine our own conviction may reflect some deeply held atomistic vision of the world; there is some reason to believe that atomism was already seen as inimical to mathematics in antiquity.[115] An ancient continuum theorist could well believe in the physical constructability of geometrical objects, and Lear (1982) thinks Aristotle did. This, however, does not alter the fact that the actual diagrams in front of the mathematician are not instantiations of the mathematical situation.

That diagrams were not considered as exact instantiations of the object constructed in the proposition can, I think, be proved. The argument is that 'construction' corresponds, in Greek mathematics, to a precise practice. The first proposition of Euclid's *Elements*, for instance, shows how to construct an equilateral triangle. This is mediated by the construction of two auxiliary circles. Now there simply is no way, if one is given only proposition I.1 of the *Elements*, to construct this triangle without the auxiliary circles. So, in the second proposition, when an equilateral triangle is constructed in the course of the proposition,[116] one is faced with a dilemma. Either one assumes that the two auxiliary circles have been constructed as well – but how many steps further can this be carried, as one goes on to ever more complex constructions? Or, alternatively, one must conclude that the so-called equilateral triangle of the diagram is a fake. Thus the equilateral triangle of proposition I.2 is a token gesture, a make-believe. It acknowledges the shadow of a possible construction without actually performing it.

We seem to have reached a certain impasse. On the one hand, the Greeks speak as if the object of the proposition is the diagram. Verbs signifying spatial action must be taken literally. On the other hand, Greeks act in a way which precludes this possibility (quite regardless of what their ontology may have been!), and the verbs signifying spatial action must, therefore, be counted as metaphors.

To resolve this impasse, the 'make-believe' element should be stressed. Take Euclid's *Elements* III.10. This proves that a circle does not cut a circle at more than two points. This is proved – as is the regular

[115] Plato's peculiar atomism involved, apparently, some anti-geometrical attitudes (surprisingly enough), for which see Aristotle, *Metaph.* 992a20ff. Somewhat more clear is the Epicurean case, discussed in Mueller (1982) 92–5. The evidence is thin, but Mueller's educated guess is that Epicureans, as a rule, did assume that mathematics is false.

[116] Euclid's *Elements* I.2, 12.25–6. Needless to say, the text simply says 'let an equilateral triangle have been set up on [the line]', no hint being made of the problem I raise.

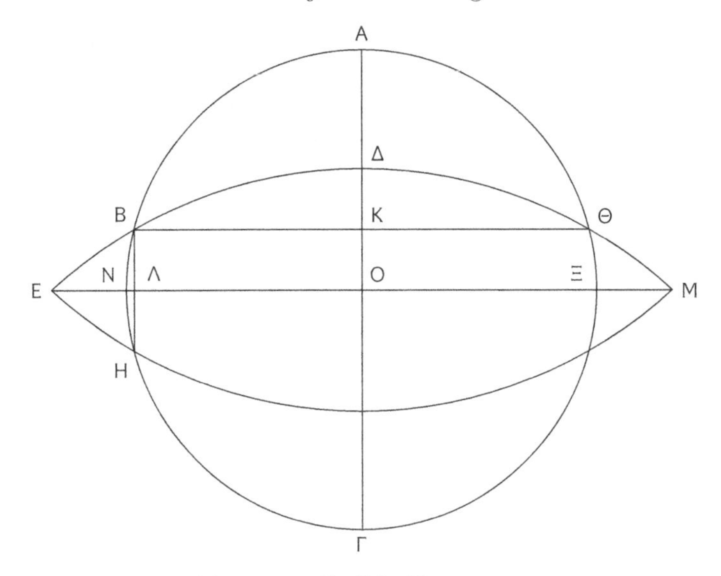

Figure 1.19. Euclid's *Elements* III.10.

practice in propositions of this nature – through a *reductio ad absurdum*: Euclid assumes that two circles cut each other at more than two points (more precisely, at four points), and then derives an absurdity. The proof, of course, proceeds with the aid of a diagram. But this is a strange diagram (fig. 1.19): for good geometrical reasons, proved *in this very proposition*, such a diagram is impossible. Euclid draws what is impossible; worse, what is patently impossible. For, let us remember, there is reason to believe a circle is one of the few geometrical objects a Greek diagram could represent in a satisfying manner. The diagram cannot be; it can only survive thanks to the make-believe which calls a 'circle' something which is similar to the oval figure in fig. 1.19. By the force of the make-believe, this oval shape is invested with circlehood for the course of the *reductio* argument. The make-believe is discarded at the end of the argument, the bells of midnight toll and the circle reverts to a pumpkin. With the *reductio* diagrams, the illusion is dropped already at the end of the *reductio* move. Elsewhere, the illusion is maintained for the duration of the proof.

Take Pünktchen for instance.[117] Her dog is lying in her bed, and she stands next to it, addressing it: 'But grandmother, why have you got such large teeth?' What is the semiotic role of 'grandmother'? It is not

[117] Kästner (1959), beginning of chapter 2 (and elsewhere for similar phenomena, very ably described. See also the general discussion following chapter 3).

metaphorical – Pünktchen is not trying to insinuate anything about the grandmother-like (or wolf-like) characteristics of her dog. But neither is it literal, and Pünktchen knows this. Make-believe is a *tertium* between literality and metaphor: it is literality, but an as-if kind of literality. My theory is that the Greek diagram is an instantiation of its object in the sense in which Pünktchen's dog is the wolf – that the diagram is a make-believe object. It shares with Pünktchen's dog the following characteristics: it is similar to the intended object; it is functionally identical to it; what is perhaps most important, it is never questioned.

2.4 *The practices of the lettered diagram: a summary*

What we have seen so far is a series of procedures through which the text maintains a certain implicitness. It does not identify its objects, and leaves the identification to the visual imagination (the argument of 2.2). It does not *name* its objects – it simply points to them, via indices (the argument of 2.3.1). Finally, it does not even hint *what*, ultimately, its objects are; it simply works with an ersatz, as if it were the real thing (the argument of 2.3.2). Obviously there is a certain vague assumption that some of the properties of the 'real thing' are somehow captured by the diagram, otherwise the mediation of the proposition via the diagram would collapse. But my argument explaining why the diagram is useful (because it is redefined, especially through its letters, as a discrete object, and therefore a manageable one) did not deal with the ontological question of why it is assumed that the diagram could in principle correspond to the geometrical object. Undoubtedly, many mathematicians would simply assume that geometry is about spatial, physical objects, the *sort* of a thing a diagram is. Others could have assumed the existence of mathematicals. The centrality of the diagram, however, and the roundabout way in which it was referred to, meant that the Greek mathematician would not have to speak up for his ontology.[118]

[118] Let me explain briefly why the indexical nature of letters is significant. This is because indices signify references, not senses. Suppose you watch a production of *Hamlet*, with the cast wearing soccer shirts. John, let's say, is the name of the actor who plays Hamlet, and he is wearing shirt number 5. Then asking 'what's your opinion of John?' would refer, probably, to his acting; asking 'what's your opinion of Hamlet?' would refer, probably, to his indecision; but asking 'what's your opinion of no. 5?' would refer ambiguously to both. Greek letters are like numbers on soccer shirts, points in diagrams are like actors and mathematical objects are like Hamlet.

Plato, in the seventh book of the *Republic*, prized the ontological ambiguity of mathematics, especially of its diagrams. An ontological borderline, it could confuse the philosophically minded, and lead from one side of the border to the other. He was right. However, this very ambiguity meant also that the mathematicians could choose not to engage in the philosophical argument, to stick with their proofs and mutual agreements – a point (as claimed above) conceded by Plato.

To conclude, then: there are two main ways in which the lettered diagram takes part in the shaping of deduction. First, there is the whole set of procedures for argumentation based on the diagram. No other single source of evidence is comparable in importance to the diagram. Essentially, this centrality reverts to the fact that the specification of objects is done visually. I shall return to this subject in detail in chapter 5. Second, and more complex, is this. The lettered diagram supplies a universe of discourse. Speaking of their diagrams, Greek mathematicians need not speak about their ontological principles. This is a characteristic feature of Greek mathematics. Proofs were done at an object-level, other questions being pushed aside. One went directly to diagrams, did the dirty work, and, when asked what the ontology behind it was, one mumbled something about the weather and went back to work. This is not meant as a sociological picture, of course. I am speaking not of the mathematician, but of the mathematical proposition. And this proposition acts in complete isolation, hermetically sealed off from any second-order discourse.[119] There is a certain single-mindedness about Greek mathematics, a deliberate choice to do mathematics and nothing else. That this was at all possible is partly explicable through the role of the diagram, which acted, effectively, as a *substitute* for ontology.[120]

It is the essence of cognitive tools to carve a more specialised niche within general cognitive processes. Within that niche, much is automatised, much is elided. The lettered diagram, specifically, contributed to both elision (of the semiotic problems involved with mathematical discourse) and automatisation (of the obtaining of a model through which problems are processed).

[119] I will discuss this in chapter 3 below.

[120] I am not saying, of course, that the *only* reason why Greek mathematics became sealed off from philosophy is the existence of the lettered diagram. The lettered diagram is not a cause for sealing mathematics off from philosophy; it is an important explanation of how such a sealing off was possible. I shall return to discussing the single-mindedness of Greek mathematics in the final chapter.

3 CONTEXTS FOR THE EMERGENCE OF THE LETTERED DIAGRAM

The lettered diagram is a distinctive mark of Greek mathematics, partly because no other culture developed it independently.[121] Indeed, it would have been impossible in a pre-literate society and, obvious as this may sound, this is an important truth.[122] An explanatory strategy may suggest itself, then: to explain the originality of the lettered diagram by the originality of the Phoenician script. The suggestion might be that alphabetic letters are more suitable, for the purpose of the lettered diagram, than pictograms, since pictograms suggest their symbolic content. The coloured constituents of some Chinese figures may be relevant here.[123]

But of course such technological reductionism – everything the result of a single tool! – is unconvincing. The important question is *how* the tool is used. This is obvious in our case, since the technology involved the combination of two different tools. Minimally, the contexts of diagrams and of letters had to intersect.

The plan of this section is therefore as follows. First, the contexts of diagrams and letters outside mathematics are described. Next, I discuss two other mathematical tools, abaci and planetaria. These, too, are 'contexts' within which the lettered diagram emerged, and understanding their limitations will help to explain the ascendancy of the lettered diagram.

3.1 Non-mathematical contexts for the lettered diagram

3.1.1 Contexts of the diagram

As Beard puts it,[124] 'It is difficult now to recapture the sheer profusion of visual images that surrounded the inhabitants of most Greek cities.' Greeks were prepared for the visual.

[121] Babylonian and Chinese diagrams exist, of course – though Babylonian diagrams are less central for Babylonian mathematics, or at least for Babylonian mathematical texts (Hoyrup 1990a), while Chinese diagrams belong to a different context altogether, of representations endowed with rich symbolic significance (Lackner 1992). Neither refers to the diagram with a system similar to the Greek use of letters. Typically, in the Babylonian case, the figure is referred to through its geometric elements (e.g. breadth and width of rectangles), or it is inscribed with numbers giving measurements of some of its elements (e.g. *YBC* 7289, 8633: Neugebauer 1945).

[122] Also, while this point may sound obvious, it would have been impossible to make without Goody (1977), Goody and Watt (1963) on the role of writing for the historical development of cognition and, generally, Goody's *œuvre*; this debt applies to my work as a whole.

[123] See Chemla (1994), however, for an analysis of this practice: what is important is not the *individual* colours, but their existence as a *system*. In fact, one can say that the Chinese took colours as a convenient metaphor for a system.

[124] Beard (1991) 14.

This is true, however, only in a limited sense. Greek elite education included literacy, numeracy, music and gymnastics, but not drawing or indeed any other specialised art.[125] The educated Greek was experienced in looking, not in drawing. Furthermore, the profusion of the visual was limited to the visual as an aesthetic object, not as an informative medium. There is an important difference between the two. The visual as an aesthetic object sets a barrier between craftsman and client: the passive and active processes may be different in kind. But in the visual as a medium of information, the coding and decoding principles are reciprocal and related. To the extent that I can do anything at all with maps I must understand some of the principles underlying them. On the other hand, while the 'readers' of art who know nothing about its production may be deemed philistines, they are possible. The visual as information demands some exchange between craftsmen and clients, which art does not.

Two areas where the use of the visual *qua* information is expected are maps and architectural designs. Herodotus gives evidence for world maps, designed for intellectual (IV.36ff.) and practical (V.49–51) purposes. Such maps could go as far back as Anaximander.[126] Herodotus' maps were exotic items, but we are told by Plutarch that average Athenians had a sufficiently clear grasp of maps to be able to draw them during the euphoric stage of the expedition to Sicily, in 415 BC.[127] Earlier, in 427, a passage in Aristophanes' comedy *The Clouds* shows an understanding of what a map is: schematic rather than pictorial,[128] preserving shapes, but not distances.[129] The main point of Aristophanes' passage is clear: though diagrammatic representations were understood by at least some members of the audience, they were a technical, specialised form. It may be significant that the passage follows immediately upon astronomy and geometry.

Our later evidence remains thin. There is a map in Aristotle's *Meteorology*,[130] and *periodoi gēs* – apparently world maps – are included, as

[125] Excluding mathematics itself – to the extent that it actually gained a foothold in education (see chapter 7).

[126] Agathemerus I.1; D.L. II.1–2; Herodotus II.109. Anaxagoras may have added some visual element to his book (D.L. II.11 – the first to do so? See also DK 59A18 (Plutarch), A35 (Clement)). I guess – and I can do no more – that this was a cosmological map (both Plutarch's and Clement's reference come from a cosmological context).

[127] *Vit. Alc.* XVII.3. The context is historically worthless, but the next piece of evidence could give it a shade of plausibility.

[128] 208–9: a viewer of the map is surprised to see Athens without juries!

[129] Shapes: 212, the 'stretched' island Euboea leads to a pun. Distances: 215–17, the naive viewer is worried about Sparta, which is too near.

[130] 363a26ff.

mentioned already, in Theophrastus' will.[131] There is also some – very little – epigraphic and numismatic evidence, discussed by Dilke.[132] Most interestingly, it seems that certain coins, struck in a military campaign, showed a relief-map of its terrain.[133] All these maps come from either intellectual or propaganda contexts. As early as Herodotus, the drawing of a map in pragmatic contexts was meant to impress rather than to inform. Otherwise, much of the evidence comes from sources influenced by mathematics.

Surprisingly, the same may be true of architectural designs.[134] The main tools of such design in classical times were either verbal descriptions (*sungraphai*), or actual three-dimensional and sometimes full-scale models of repeated elements in the design (*paradeigmata*). Rules of trade, especially a modifiable system of accepted proportions, allowed the transition from the verbal to the physical. There is a strong *e silentio* argument against any common use of plans in early times. From Hellenistic times onwards, these began to be more common, especially – once again – in the contexts of persuasion rather than of information. This happened when competition between architects forced them to evolve some method of conveying their intentions beforehand, in an impressive manner. Interestingly, the use of visual representations in architecture is earliest attested in mechanics, which may show a mathematical influence.

What is made clear by this brief survey is that Greek geometry did not evolve as a reflection upon, say, architecture. The mathematical diagram did not evolve as a modification of other practical diagrams, becoming more and more theoretical until finally the abstract geometrical diagram was drawn. Mathematical diagrams may well have been the first diagrams. The diagram is not a representation of something else; it is the thing itself. It is not like a representation of a building, it is like a building, acted upon and constructed. Greek geometry is the study of spatial action, not of visual representation.

However speculative the following point may be, it must be made. The first Greeks who used diagrams had, according to the argument above, to do something similar to building rather than to reflect upon building. As mentioned above, the actual drawing involved a practical skill, not an obvious part of a Greek education. Later, of course, the lettered diagram would be just the symbol of mathematics, firmly

[131] D.L. v.51–2. [132] Dilke (1985) chapter 2.
[133] Johnston (1967). [134] The following is based on Coulton (1977) chapter 3.

situated there; but at first, some contamination with the craftsman-like, the 'banausic', must be hypothesised. I am not saying that the first Greek mathematicians were, e.g. carpenters. I am quite certain they were not. But they may have felt uneasily close to the banausic, a point to which I shall return in the final chapter.

3.1.2 *Contexts of letters as used in the lettered diagram*

Our earliest direct evidence for the lettered diagram comes from outside mathematics proper, namely, from Aristotle. There are no obvious antecedents to Aristotle's practice. Furthermore, he remained an isolated phenomenon, even within the peripatetic school which he founded. Of course, logical treatises in the Aristotelian tradition employed letters, as did a few quasi-mathematical works, such as the pseudo-Aristotelian *Mechanics*. But otherwise (excluding the mathematically inclined Eudemus) the use of letters disappeared. The great musician Aristoxenus, just like the great mechanician Strato – both in some sense followers of Aristotle – do not seem to have used letters. The same is true more generally: the Aristotelian phenomenon does not recur. And, of course, nothing similar to our common language use of '*X*' and '*Y*' ever emerged in the Greek language.[135]

Otherwise, few cases of special sign systems occur. At some date between the fifth and the third centuries BC someone inserted an acrophonic shorthand into the Hippocratic *Epidemics* III.[136] Galen tells us about another shorthand designed for pharmaceutical purposes, this time based, in part, upon iconic principles (e.g. *omicron* for 'rounded').[137] A refined symbolic system was developed for the purposes of textual criticism. Referring as it did to letters, the system employed *ad hoc* symbols.[138] This system evolved in third-century Alexandria. Another case of a special symbolic system is musical notation, attested from the third century BC but probably invented earlier.[139] Letters, grouped and repeated in various ways, are among other symbols considered to have magical significance.[140] Finally, many systems

[135] Which should not surprise us: the Greek letters as used in diagrams, being indices, were inseparable from specific situations, unlike the modern symbolic '*X*'.
[136] This is not a feature of the manuscripts alone – which might have suggested a Byzantine origin – since Galen reports the system, XVII 600ff.
[137] Galen XIII 995–6. The system is due to Menecrates, of an early AD provenance.
[138] See Turner (1968) esp. 112–18. [139] West (1992) chapter 9.
[140] See Betz (1992) for many examples, e.g. 3, 17 (letters), 191 (other symbols). For a discussion, see Dornseiff (1925).

of abbreviation are attested in our manuscripts, and while the vast majority are Byzantine, 'shorthand' was known already in antiquity.[141] The common characteristic of all the above is their reflective, written context. These are all second-order signs: signs used to refer to other signs. Being indices to diagrams, the letters of Greek mathematics form part of the same pattern.

What we learn is that the introduction of a special sign-system is a highly literate act – this indeed should have been obvious to start with. The introduction of letters as tools is a reflective use of literacy. Certainly the social context within which such an introduction could take place was the literate elite.

3.2 Mathematical non-verbal contexts

Generally speaking, mathematical tools are among the most widespread cultural phenomena of all, beginning with the numerical system itself and going through finger-reckoning, abaci, etc., up to the computer.[142] Many of these tools have to do with calculation rather than proof and are thus less important for my purposes here. Two tools used by Greek mathematics, besides the lettered diagram, may have been of some relevance to proof, and are therefore discussed in the following subsections: these are abaci and planetaria.

It is natural to assume that not all tools can lead equally well to the elaboration of scientific theories. To make a simple point, science demands a certain intersubjectivity, which is probably best assisted through language. A completely non-verbalised tool is thus unlikely to lead to science.[143] On the other hand, intersubjectivity may be aided by the presence of a material object around which communication is organised. Both grounds for intersubjectivity operate with the lettered diagram; I shall now try to consider the case for other tools.

[141] See, e.g. Milne (1934). The compendia used in mathematical manuscripts are usually restricted to the scholia. It doesn't seem that abbreviations were important in Greek mathematics, as, indeed, is shown by the survival of Archimedes in Doric.

[142] See, e.g. Dantzig (1967). Schmandt-Besserat (1992, vol. I: 184ff.) is very useful.

[143] I am thinking of the Inca *quipu* (where strings represent arithmetical operations) as a tool where verbalisation is not represented at all (as shown by the problematic deciphering) (Ascher 1981).

3.2.1 The abacus in Greek mathematics

The evidence is:

(a) Greeks used pebbles for calculations on abaci.[144]
(b) Some very few hints suggest that something more theoretical in nature was done with the aid of pebbles.[145]
(c) It has been argued that a certain strand in early Greek arithmetic becomes natural if viewed as employing pebbles. According to this theory, some Greeks represented numbers by configurations of pebbles or (when written) configurations of dots on the page: three dots represent the number three, etc.[146] However:
(d) Not a single arithmetical BC text refers to pebbles or assumes a dot representation of an arithmetical situation.

Philip[147] argued that we should not pass too quickly from (b) to (c). Certainly, Eurytus' pebbles need not be associated with anything the Greeks themselves would deem arithmetical. I shall argue below[148] that what is sometimes brought as evidence, Epicharmus' fragment 2, belongs to (a) and not to (b), let alone (c). Similarly, Plato's analogy of mathematical arts and *petteutikē* – pebble games[149] – need not involve any high-powered notion of mathematics.

This leaves us with two Aristotelian passages:

> 'Like those who arrange numbers in shapes [such as] triangle and square';[150]
> 'For putting gnomons around the unit, and without it, in this [case] the figure will always become different, in the other it [will be] unity'.[151]

[144] Lang (1957).
[145] The only substantial early hints are the two passages from Aristotle quoted below (which can be somewhat amplified for Eurytus by *DK* 45A2: he somehow related animals(?) to numbers, via pebble-representations).
[146] Becker (1936a). Knorr (1975) goes much further, and Lefevre (1988) adds the vital operational dimension.
[147] Philip (1951), appendix II, esp. 202–3. [148] Chapter 7, subsection 1.1 272–5.
[149] *Grg.* 450cd; *Lgs.* 819d–820d; also relevant is *Euthyph.* 12d.
[150] *Metaph.* 1092b11–12: ὥσπερ οἱ τοὺς ἀριθμοὺς ἄγοντες εἰς τὰ σχήματα τρίγωνον καὶ τετράγωνον.
[151] *Phys.* 203a13–15: περιτιθεμένων γὰρ τῶν γνωμόνων περὶ τὸ ἓν καὶ χωρὶς ὁτὲ μὲν ἄλλο ἀεὶ γίγνεσθαι τὸ εἶδος, ὁτὲ δὲ ἕν. Both passages are mere clauses within larger contexts, and are very difficult to translate.

Philip maintained that, however arithmetical these passages may sound, they are relatively late fourth-century and therefore might be due to the great mathematical progress of that century, and so need have nothing to do with the late fifth century. Knorr[152] quite rightly objected that this makes no evolutionary sense: could that progress lead to mathematics at the pebbles level? Knorr must be right, but he does not come to terms with the fact that our evidence is indeed late fourth-century. Moreover, the texts refer to Pythagoreans, in connection with Plato, and the natural reading would be that Aristotle refers to someone roughly contemporary with Plato. Thus, our only evidence for an arithmetical use of pebbles comes from a time when we know that mathematically stronger types of arithmetic were available.

I certainly would not deny the role of the abacus for Greek arithmetical concept-formation.[153] The question is different: whether any arithmetical proof, oral or written, was ever conducted with the aid of pebbles. The evidence suggests, perhaps, oral proofs. Aristotle talks about people *doing* things, not about anything he has *read*. Why this should be the case is immediately obvious. Pebble manipulations admit a transference to a written medium, as is amply attested in modern discussions. However, the special advantage of pebbles over other types of arithmetical representations is a result of their direct, physical manipulations, which are essentially tied up with actual operations. It is not the mere passive looking at pebbles which our sources mention: they mention pebbles being moved and added. This must be lost in the written medium, which is divorced from specific actions. Thus, it is only natural that pebbles would lose their significance as the written mode gained in centrality. They would stay, but in a marginal role, emerging in a few asides by Plato and Aristotle, never as the centre of mathematical activity.[154]

[152] Knorr (1975) 135–7.

[153] Lefevre (1988) offers a theory of such concept formation, with a stress on the general role of operations for concept-formation.

[154] An important comparison is the following, which, however, being no Assyriologist, I will express tentatively and in this footnote alone. The geometrical reconstructions offered by Hoyrup (1990a) for Babylonian 'algebra' take the shape of *operations* upon spatial objects, moved, torn and appended – following the verbs of the Akkadian text. I would say:

1. The loss of (most) diagrams from Babylonian mathematics is related to this manner in which Babylonian mathematics was visualised. The texts refer to objects which were actually moved, not to inscribed diagrams.
2. The visualisation was operational because the role of the text was different from what it is in the Greek case. Babylonian mathematical texts are not context-independent; they are

3.2.2 *Planetaria in Greek mathematics*

The earliest and most extensive piece of evidence on planetaria in Greek astronomy is Epicurus' – biased – description of astronomical practices, in *On Nature* XI.[155] The description is of a school in Cyzicus, where astronomers are portrayed as using *organa*, 'instruments', while *sullogizesthai*, *dialegesthai* (i.e. reasoning in various ways), having *dianoia* (translated by Sedley in context by 'a mental model') and *epinoēsis* ('thought-process') and referring to a *legomenon* (something 'said' or 'asserted'). What is the exact relation between these two aspects of their practice, the instrument and the thought? One clue is the fact that Epicurus claims that the aspects are irreconcilable because, according to him, the assumption of a heavens/model analogy is indefensible. This assumes that some dependence of the verbal upon the mechanical is necessary. This dependence might be merely the thesis that 'the heavens are a mechanism identical to the one in front of us', or it might be more like 'setting the model going, we see [e.g.] that some stars are never visible, QED'. Where in the spectrum between these options should we place the mathematicians of Cyzicus?[156]

My following guess starts from Autolycus, a mathematician contemporary with this Epicurean text. Two of his astronomical works survive – *The Moving Sphere* and *The Risings and Settings*. He never mentions any apparatus, or even hints at such, even though *The Risings and Settings* are practical astronomy rather than pure spherical mathematics. Neither, however, does he give many definitions or, generally, conceptual hints.[157] Furthermore, as mentioned above, his diagrams – belonging

the internal working documents of scribes, who know the operational context in which these texts are meant to be used.

3. The different contexts and technologies of writing meant that in one case (Mesopotamia) we have lost the visualisations alone, while in the other (Greek pebble arithmetic) we have lost both visualisations and text.
4. Babylonian mathematics is limited, compared to Greek mathematics, by being tied to the particular operation upon the particular case; which reflects the difference mentioned above.

[155] Sedley (1976) 32–4. The text survives only on papyrus.

[156] And not only them: the evidence for the use of planetaria (and related star-modelling mechanisms) in antiquity goes beyond any other archaeological evidence for mathematics. A truly remarkable piece of evidence is the Antikytheran 'planetarium', described in Price (1974). See there the evidence for sundials (51), and for other planetaria (55–60).

[157] That the definitions of *The Moving Sphere* are spurious is probable, though not certain. See Aujac (1979) 40 (in the edition of Autolycus used in this study: see Appendix, p. 314), who rejects them. If they are spurious, then they are the result of a perplexity similar to that which the modern reader must feel. The definitions of *The Risings and Settings* explain the terminology of observation, not the spatial objects discussed.

as they do to the theory of spheres – are sometimes only very roughly iconic. The reader – who may be assumed to be a beginner – is immediately plunged into a text where there is a very serious difficulty in visualising, in conceptualising. No doubt much of the difficulty would have been solved by the Greek acquaintance with the sky. But a model would certainly be helpful as well, at such a stage. After all, you cannot turn the sky in your hands and trace lines on its surface. An object which can be manipulated would contribute to concept-formation.[158] This acquaintance is more than the mere analogy claim – the model is used to understand the heavens – yet this is weaker than actually using the model for the sake of proof.[159]

Timaeus excuses himself from astronomy by claiming that τὸ λέγειν ἄνευ δι' ὄψεως τούτων αὖ τῶν μιμημάτων μάταιος ἂν εἴη πόνος – 'again, explaining this without watching models would be a pointless task'.[160] This, written by the staunch defender of mathematical astronomy! It seems that models were almost indispensable for the pedagogic level of astronomy. The actual setting out in writing of mathematical astronomy, however, does not register planetaria. Again, just as in the case of the abacus, the tool may have played a part in concept-formation. And a further parallelism with the abacus is clear. Why is it difficult for Timaeus to explain his astronomy? Why indeed could he not have brought his planetaria? The answer is clear: the written text filtered out the physical model.

In Plato's case, of course, not only physical models were out of the question: so were diagrams, since the text was not merely written, but also the (supposed) reflection of conversation, so that diagrams used by the speakers must be reconstructed from their speeches (as is well known, e.g. for the *Meno*). Plato's text is double-filtered. More generally, however, we see that the centrality of the written form functions as a filter. The lettered diagram is the tool which, instead of being filtered out by the written mode, was made more central and, with the marginalisation of other tools, became the metonym of mathematics.

[158] For whatever its worth, it should be pointed out that Epicurus' criticisms fasten upon the concept-formation stage.

[159] This is certainly not the only purpose of building planetaria. Planetaria could do what maps did: impress. Epicurus is setting out to persuade students away from Cyzicus. The planetarium seems to have been set up in order to persuade them to come.

[160] Plato, *Tim.* 40d2–3.

4 SUMMARY

Much of the argument of this chapter can be set out as a list of ways in which the lettered diagram is a combination of different elements, in different planes.

(a) On the logical plane, it is a combination of the continuous (diagram) and the discrete (letters), which implies,
(b) On the cognitive plane, a combination of visual resources (diagram) and finite, manageable models (letters).
(c) On the semiotic plane, the lettered diagram is a combination of an icon (diagram) and indices (letters), allowing the – constructive – ambiguity characteristic of Greek mathematical ontology.
(d) On the historical plane, it is a combination of an art, almost perhaps a banausic art (diagram) and a hyper-literate reflexivity (letters).

The line of thought suggested here, that it is the fertile intersection of different, almost antagonistic elements which is responsible for the shaping of deduction, will be pursued in the rest of the book.

CHAPTER 2

The pragmatics of letters

PLAN OF THE CHAPTER

Much has already been said on letters representing objects in the diagram. The interest has so far focused on the light which such letters throw upon the use of diagrams. Here we concentrate on the letters themselves as they occur in the text. This is a convenient object: a definite, small set, and the combinatorial possibilities are limited. The results can therefore obtain an almost quantitative precision. The practices described here are interesting, then, mainly as a case-study for Greek mathematical pragmatics in general. It is clear that Greek mathematics follows many conventions of presentation, some of which we have noticed already. What can the origin of these conventions be? I will take the conventions regarding letters as a first case-study.

Following an explanation of the nature of the practices involved, I offer a hypothesis concerning the origins of those practices (section 1). Then, in section 2, a few cognitive implications of the practices will be spelled out. These implications are generally a development of points raised in the preceding chapter, but they already suggest the issues concerning the use of language, to be developed in chapters 3–4.

1 THE ORIGINS OF THE PRACTICES

1.1 A preliminary description

The practices with regard to letters fall into two kinds. The first is what I call 'baptism' – the process of attaching individual letters to individual objects. Second, given the distribution of individual letters to individual objects, many-lettered names can be made, e.g. if A and B represent points, AB may be a many-lettered name representing the straight line between those two points.

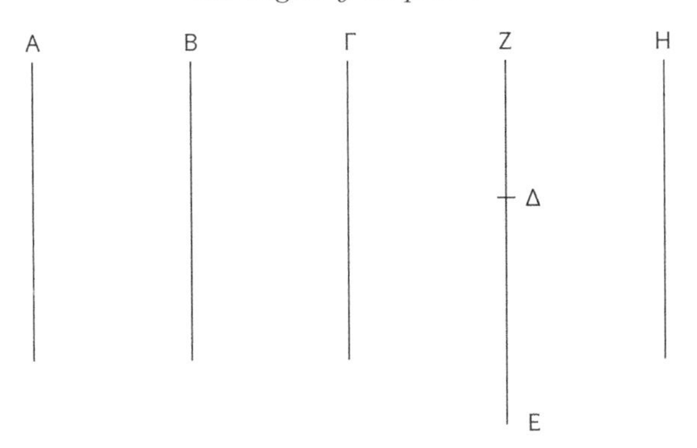

Figure 2.1. Euclid's *Elements* IX.20.

What do letters stand for? The baptised object in Greek mathematics is most often a point. In non-geometrical contexts, the baptised object is often the term directly under discussion, e.g. a number. However, through the assimilation of arithmetic to geometry, numbers, which are represented by lines, may sometimes be baptised indirectly. Their 'extreme' points will be baptised instead. Thus, in Euclid's *Elements* IX.20, the letters A, B, Γ, H represent numbers, while the letters Δ, E, Z represent extremes of line-segments (in turn representing numbers) (fig. 2.1).[1] The decision on which approach to take is not arbitrary. In Euclid, letters represent 'extremes' if and only if the represented objects are meant to combine. This can be seen in fig. 2.1: the object represented by ΔE + ΔZ will be used in the proposition, hence the special treatment of these three letters. This is comparable with the other type of situation where the principle of letter-per-point is neglected. This is when merely 'quantitative' objects, which do not take part in the geometrical configuration, are introduced, e.g. in Archimedes' *SC* I.9, Θ stands for an area which does not itself take part in any geometrical configuration. It is introduced as the difference between other areas, in themselves meaningful in terms of the geometrical configuration (fig. 2.2).[2] The general principle, then, is that whenever there is a scheme of interacting objects, they will be designated through their points of contact. Other, independent objects will be designated directly. What this system avoids is the use of more than one letter for an independent object (fig. 2.3) and the use of letters as

[1] The treatment of time in Archimedes' *SL* I is similar: points represent 'extremes' of time.
[2] *SC* I.30.16–18.

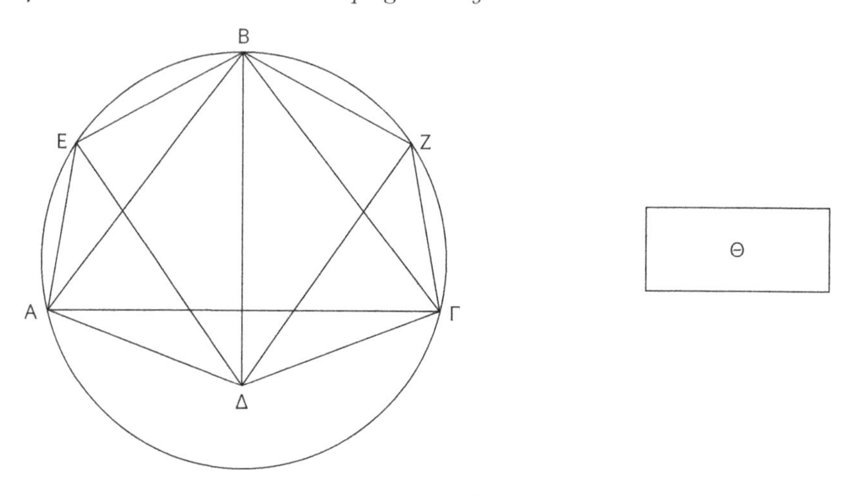

Figure 2.2. Archimedes' *SC* I.9.

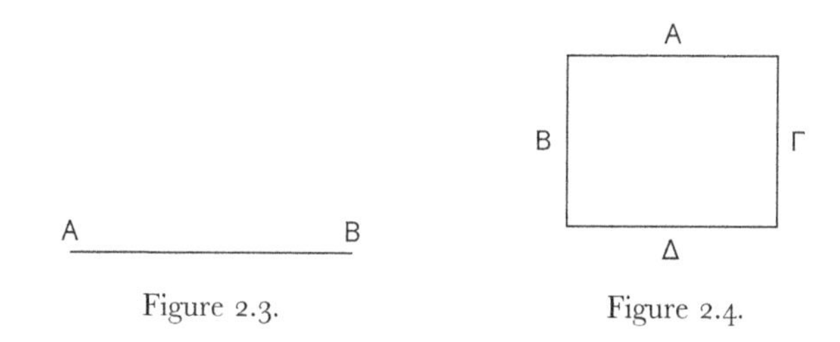

Figure 2.3.

Figure 2.4.

non-punctual in interacting schemes (fig. 2.4).[3] This is another proof of a central argument of the preceding chapter: that the diagram is conceptualised as a discrete set of interacting objects (designated by letters).[4]

So what we look for is, generally speaking, individual letters, standing for points, and many-lettered names, representing objects through their points of intersection.

[3] I ignore the extremely rare cases where the same point is used *twice* in a single diagram: see the diagram to Archytas' duplication of the cube (Eutocius, *In SC* III.87; Δ is repeated to represent movement), and Euclid's *Elements* II.5 (M is repeated: probably – as Heiberg hints – a well-entrenched textual corruption).

[4] See chapter 1, subsection 2.1.4 above. There are some complications in this system, e.g. the use of single letters to stand for either of the opposite sections, in Apollonius (e.g. A, B in *Conics* III.4). In this case the letter designates primarily a point, and, by extension, it is also taken to designate the section on which the point stands. Other complications can often be explained in similar ways.

I follow two practices:

(a) The sequence of baptisms in a given proposition can be plotted. It turns out to be often, though not always, alphabetic, i.e. the first object tends to be called A, the second B, etc.
(b) Objects may change their many-lettered names. An AB may turn into a BA, and then return to an AB. When this occurs, I say that a 'switch' took place (there are thus two switches in the example just given: a switch and a counter-switch, as it were). When, on the other hand, the same object is referred to on two consecutive occasions, and its name is the same, I call this a 'repetition'.[5] The class consisting of both switches and repetitions is that of 'reappearances'. Most reappearances tend to be repetitions.

These rough estimates are based on precise surveys. I shall now describe them.

1.2 Quantitative results

I have surveyed 100 sequences of baptisms: most of Apollonius' *Conics* 1, as well as Archimedes' *SC* 1.[6] This means that I derive a string for each proposition in these books. Consider a proposition in which five letters are introduced, the first five letters of the Latin alphabet. Say it starts 'Let there be a triangle *ABD*, and let *CE* be some other line intersecting with it'. Then the string I derive for this proposition is *ABDCE*.

Twenty-nine of the 56 Apollonian propositions I have surveyed are strictly alphabetical, as are 4 of the 44 Archimedean propositions. Many sequences, while not strictly alphabetical, are a permutation of the alphabetical: i.e. as in the example above, they employ n letters, which are the first n letters of the alphabet.[7] Fifty-one of the 56 Apollonian propositions are at least a permutation of the alphabetical, as are 30 out of the 44 Archimedean propositions. Elsewhere, usually no more than a few letters are 'missing'.

The 'distance' of a sequence from the strictly alphabetical can be measured by the minimal number of permutations of a given elementary

[5] Thus, the following sequence, *AB*, *BA*, *BA* consists of one switch (*AB* → *BA*) and one repetition (*BA* → *BA*).
[6] I will describe the results and not give (for reasons of space) my table, which can be sent to interested readers.
[7] Ignoring I, which tends to be avoided.

form.[8] It is very difficult to calculate this quantity in a general way. However, an acceptable rough first approximation is the number of cases where $\text{letter}_n > \text{letter}_{n+1}$[9] (subscript n represents the ordinal number of the letter in the string: letter_2, for instance, is the second letter in the string). I will call such a case '>', and say that a sequence where this occurs twice 'has two >'.[10]

Over my entire survey, the number of possible > is 921, while the number of actual > is 141. The average percentage is therefore approximately 15%. I found only one proposition where the percentage is above 50%: Archimedes' *SC* I.1. This is in fact a very rare proposition. The diagram is not reconstructed verbally at all, and one plunges directly into the proof – so that all the letters are, in the terms of the preceding chapter, completely unspecified. As a result, the principle of baptism is not alphabetical but, uniquely, spatial, namely counter-clockwise (fig. 2.5).

The situation regarding many-lettered names is similar. I have gathered some data, all of which cohere around a single picture.[11] It is very rare to find propositions without any switches at all, i.e. in which, whenever a many-lettered name is repeated, it is never changed. The usual level of switches is 20% or less.[12] I have not come across any

[8] The discussion will become now slightly mathematical. Readers may skip to my results. What I mean by 'elementary' permutations are those which involve transplanting a whole sequence *in toto* from one point to another, e.g. *abCDef* → *abefCD* (the transplanted sequence marked by upper case).

[9] I.e. cases where a letter from 'higher up' in the alphabet is immediately followed by a letter from 'below' in the alphabet. In the example above, *ABDCE*, the only such case is $D > C$.

[10] The maximum possible number of > in any sequence of n letters is, of course, $(n-1)$, and the distribution of different numbers of > over all possible sequences is symmetric (by mirror-inverting any sequence of $n+1$ length with k >, one gets a sequence with $n-k$ >. Therefore the number of sequences with k > is equal to that of sequences of $n-k$ >). Therefore the percentage of actual > out of possible > is a directly measurable, acceptable estimate of the distance of a sequence from the alphabetical.

[11] Again, for reasons of space, I will merely summarise here the results:

Apollonius' *Conics* I has about 2300 reappearances and 358 switches: about 15%.

Autolycus' works comprise 43 propositions. I have counted only 11 switches within that corpus.

In Aristarchus' *On the Sizes and Distances of the Sun and the Moon* there are 12 relevant propositions. There, 87 switches occur. I have not counted the number of reappearances, but the general profile seems comparable to Apollonius' *Conics* I (in *Conics* I there are about 6 switches per proposition, as against more than 7 in Aristarchus; but the propositions of Aristarchus are more complicated).

I have also made a detailed study of 30 propositions from many other sources, which confirm the same picture.

[12] It should be noted that the chance result here is not 50%, but higher: this is because some many-lettered objects can be represented by many names (e.g. six in the case of triangles, so chance level approximates 83%).

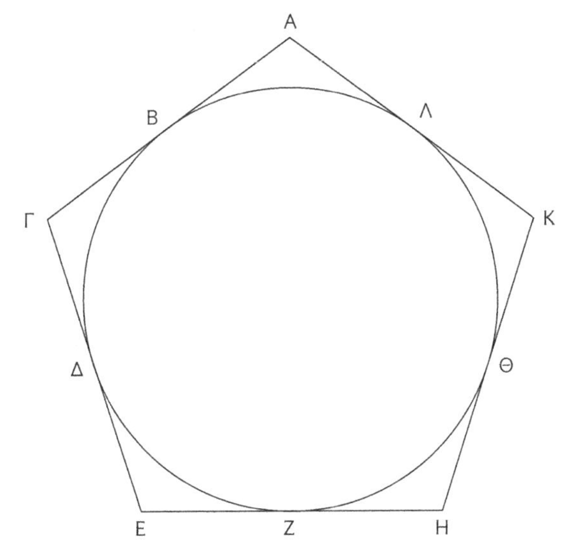

Figure 2.5. Archimedes' *SC* I.I.

proposition where the percentage is above 50%. About 30% is the highest level that is still common.[13]

This is interesting, then: both sets of practice show similar behaviour. Two options are ruled out. First, the results are not random.

[13] The issue of switches in many-lettered names involves a difficult textual question. How trustworthy are the many-lettered names as printed in the standard editions? I checked the first 70 propositions in Euclid's *Data* (an edition rare in giving variations of many-lettered names). We can distinguish two sorts of variation between manuscripts. One is the mathematically innocent one, where both manuscripts represent the same object with different letters (an *AB*/*BA* difference). The other is the mathematically significant one, where one manuscript seems to refer to a different object altogether (an *AB*/*AC* difference). The two types of mistake point at very different cognitive situations. In the *AB*/*BA* variation, one of the copiers may (not necessarily) have internalised the reference of the letters, and freely introduced his own designation of the same object. In the *AB*/*AC* variation, it is clear that one of the copiers did not have a sense of what the text meant, and misread, or miscopied through some other error, his original.

I have counted no more than 79 variations of the *AB*/*BA* type in the first 70 propositions of the *Data*. This is around one variation per proposition. The number of many-lettered names in each proposition is usually around 80, with five manuscripts: 79 variations of the *AB*/*BA* sort in about 28,000 many-lettered names, i.e. less than 3 per mille. (I have not counted variations of the *AB*/*AC* sort, but they are clearly more common than *AB*/*BA* variations.)

A significant result, then, on the way in which mathematical texts were transmitted. The copiers did not have a sense of the meaning of the text. They definitely did not internalise the text; they did not creatively rewrite it by changing such trivial matters as the order of letters in many-lettered names. They copied, slavishly. We may safely cross the ocean of the middle ages and disembark, with almost precisely the same text, on the shores of antiquity. There, of course, the terrain becomes more difficult. But what we have is at least *some* ancient text, saying something about antiquity.

Chance sequences of letters, or many-lettered names, do not look like this. Second, explicit codification is ruled out. Explicit codification, something like a style manual, would have yielded simpler results: perhaps not 100% (mistakes occur, of course), but it would be a much more compact scatter. These practices were neither random nor regulated. Is there a third option between the two? I argue that there is, indeed that it is characteristic of the conventionality of Greek mathematics in general.

1.3 Self-regulating conventionality: the suggestion

The image towards which I am striving is that of some sort of order, of pattern which, however, does not result from a deliberate attempt to create that pattern.

Such patterns are legion. Language, for instance, is such: extremely well regulated, yet not introduced by anyone. Patterns of this kind result from the regularity of our cognitive system, not from any cultural innovation. This could, theoretically, be extended directly to our case, like this: when an object reappears, there is a natural tendency – no more – to repeat its name. However, memory being inexact, many repetitions will be inexact. The interaction of the natural tendency towards conservatism, on the one hand, and of memory, on the other hand, result in the pattern described above. A similar story can be told for baptism. The alphabetical sequence is memorised, and there is thus a tendency to employ it. But this can be no more than a tendency, hence our results.

I think it is wrong to go that far. Self-regulating conventions are the track we should follow. But this should not lead us into believing that the practices are the result of blind, totally impersonal cognitive forces.

It is difficult, first, to see what cognitive forces could lead to just the pattern we observe. For instance, I once thought that short-term memory might be the main factor in the reappearances of many-lettered names. The assumption was that switches are strongly correlated with reappearances over longer stretches. Two lines later, the exact name is still remembered. Five lines later, and it is already processed away: the reference is registered, the precise name is forgotten. I have therefore surveyed the lengths of switches and of repetitions in some propositions. The survey was not very large in terms of propositions – no more than 11 – but the number of reappearances was quite large, 785, of which 161 were switches. The average length of switches turned

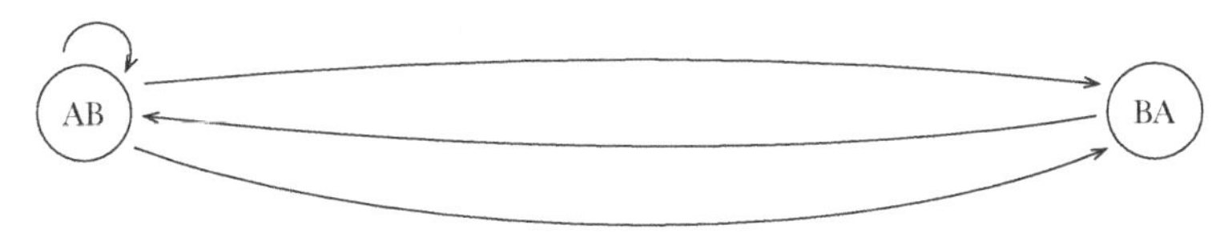

Figure 2.6.

out to be larger than that of repetitions – by about a single line. The average length of a repetition is about 5 to 6 lines, while the average length of switches is about 6 to 7 lines.[14] The breakdown of this into cases revealed that a great many switches are over a very small range indeed, often within the same line, while most reappearances over a long range were repetitions, not switches.

This suggests two conclusions: first, the slight tendency of switches to be larger may be statistically meaningful, and it does show that memory is a factor. It is probably true that one reason why the pattern is not 100% perfect is that the pattern is not followed with great care and attention. In other words, it is clear that the text was not proofread to expunge deviations. Such 'blind' cognitive factors can therefore go a long way towards explaining a non-regular pattern. However, they cannot be the key factor. Whoever switched a name within the very same line did so on purpose. While most often names are not switched, they are sometimes switched on purpose: i.e. the behaviour is *differential*, and this differentiality is apparently intentional.

The way in which the behaviour is differential is significant. Commonly, what we see is that while several objects are switched often, others are switched rarely, or not switched at all. This is differentiation between different objects of the proposition. I sometimes plot the 'histories' of individual many-lettered names in a proposition. Say, in a proposition a name appears first as *AB*, then, again, as *AB*, then *BA*, then *AB* and then finally *BA*. Its 'history' is depicted in figure 2.6.

Look at the history of ΕΣ in Apollonius' *Conics* 1.50 (fig. 2.7). This is the history of a chaotic journey between two poles – a metal more or less repelled by two magnets. Such names occur sometimes. They are so changeable that they lose all identity. Some other objects, however, just cannot be random. ΛΜ, ΛΜΡ and ΜΟ, in the same proposition, are repeated six times in a row, each without change. The probability for this is about 1/250,000. Apollonius took care not to switch them.

[14] I calculated in Heiberg lines, which are not a constant unit, but are good enough for such statistical purposes.

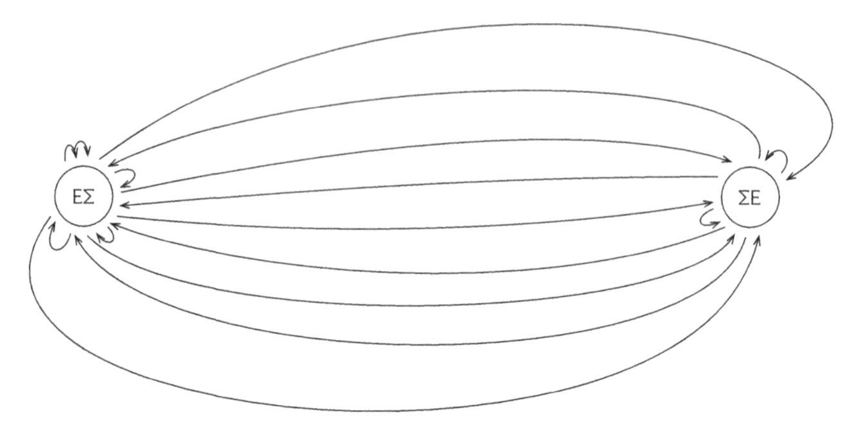

Figure 2.7. History of ΕΣ, Apollonius' *Conics* 1.50.

Apollonius deliberately chose not to switch a few objects, which means that he deliberately chose to switch some others. More precisely, since the more common practice is that of repetition, it is the switched objects that become marked. Against the background of a general tendency not to switch, switched objects are marked. And similarly, against the background of alphabetical baptisms, non-alphabetical objects are marked.

In some cases, it is possible to see the role of such marking. For instance, a major reason for the non-alphabetical sequences in Archimedes' *SC* is his frequent use of auxiliary objects which are not part of the main geometrical configuration. Such objects are often introduced non-alphabetically. By being out of order, their being out of the main scheme is enhanced.[15]

'Marking' does not have its own signification. It is interpreted in context, signifying what, in the given discourse, is worth being marked. In Archimedes, what is being marked is often discontinuities in the same proposition. In Apollonius, what is being marked is often continuities between different propositions, e.g. in the first 14 propositions of the *Conics*, ΒΓ always refers to the base of the axial triangle. In propositions 1–13 this is dictated by the alphabetical sequence. In proposition 14 (where the sequence of the construction is different) this is already a deviation from the alphabetical. Thus, the letters are marked in 14, and this serves to signal a continuity.[16]

[15] E.g. 1.4: ΕΖ; 1.7: ΕΖΗΘ are preceded by ΚΛΜ and are thus made to be out of their normal sequence; 1.32: ΞΟ.

[16] A similar case is the letters ΑΓ in *Conics* 1.46, repeating the closely related 1.42.

Many switches are explicable as markings of this sort. For instance, *Conics* I.45: three triangles are mentioned, BEZ, HΘZ and ΛΓΘ. The last two reappear immediately, but switched, as ZHΘ and ΓΛΘ. Why is that? We must remember that Apollonius most often uses three-lettered names for rectangles, in expressions such as 'the rectangle by *ABC* [= the rectangle contained by the lines *AB*, *BC*]'. Because of its deep structure, the three-lettered name 'the rectangle by *ABC*' can be permuted in only two ways, *ABC* or *CBA* ('the rectangle by *ACB*' would already be that by the lines *AC*, *CB*). On the other hand, a triangle is a three-lettered name which can be freely permuted, in all six ways. By permuting the triangles in *exactly* the way impossible for rectangles, the objects are marked – to stress further that they are triangles and not rectangles.

That switches are semantically meaningful is clear from the following: if two objects share the same letters, they are often distinguished by the sequence of their letters (besides the articles and prepositions differentiating between them).[17] And again, switches, just like non-alphabetical baptisms, may mark continuities between propositions. In *Conics* I.48, EΔ from l. 20 is switched into ΔE in l. 22. This is in the context of the reference to a result of the preceding proposition, which was expressed in terms of ΔE.

We can begin, then, to identify the various factors at work. There is a general conservative tendency to follow a pattern. This, however, is not followed with great care, and thus negative performance factors such as short-term memory, or positive factors such as visualisation, make this pattern far short of 100% perfection. However, this still leaves a recognisable pattern, against which deviations may be informative. So there are three main forces going against strict adherence to the regularity:

* negative performance factors (such as limitations of short-term memory);
* positive, interacting, cognitive factors (such as visualisation);
* intentional, informative deviations.

Each factor is in itself limited, responsible for no more than a few per cent of deviation. They combine, however, in complex ways (thus, for instance, once objects begin to get switched, for some positive

[17] For instance, this is very common with circles and arcs on circles in Autolycus, e.g. in *The Moving Sphere* 9, the circle EHK and the arc HKE.

reason, it becomes more difficult to memorise them, and hence there is an increased tendency to switch because of negative performance reasons). This combination finally leads to the general result of approximately 85% regularity.

The conventionality, therefore, is not the result of any explicit attempt to conventionalise a practice. Further, the conventionality is not just the operation of blind, general cognitive forces. It does have a component of intentional agency. This agency, however, does not shape the overall structure of the regularity. Rather, it takes this overall structure and uses it for its own purposes. The overall structure is 'usually – but not always', and it is possible to use such a structure to convey information by deviations from the pattern. The deviations are allowed by the 'but not always' clause, and are informative because of the 'usually' clause. I now proceed to generalise.

1.4 Self-regulating conventionality

We imagine a group of speakers, and a corpus of texts unified by subject matter, produced by those speakers.[18] Because of the unity of subject matter, the cognitive forces shaping the texts have a certain unity. Hence, some characteristic patterns will emerge in the corpus. Now we add a further assumption: that the speakers are well acquainted with the texts. Thus, whether they articulate this to themselves explicitly or not, they are aware of the patterns. This means that they are able to operate with those patterns, either by deviating from them informatively or, alternatively, by trying to remain close to them. What starts as the raw material of conventions, what is merely the work of natural forces, becomes, by being tinkered with, a meaningful tool, a convention. Can such a tool be transmitted? Certainly – simply by a deep acquaintance with the texts. More than this: the very use of the tool assumes a deep acquaintance with the texts on the part of the readers.

I shall now use the word 'professional' to mean a speaker with a deep acquaintance with the texts. What I have argued can be translated into the following claim: that in any professionalised group, conventions of the sort described above will necessarily emerge. They will

[18] The written medium itself is immaterial here; I shall use 'texts' and related terms in a general sense.

be spontaneously transmitted, as a natural result of the process of professionalisation, of getting acquainted with the texts. And the professionalised group would come to be ever more marked, as its texts could reveal their full meaning only to professionals, to members of that group. This is the process of self-regulating conventionality. It will necessarily emerge wherever professionalisation (in my very limited sense) occurs. And it is self-regulating, in the sense that it is self-perpetuating and in need of no explicit, external introduction. This is the general principle accounting for the tendency towards perpetuation and growing differentiation of professional groups and genres.

Of course, conventions may be introduced by other routes as well. They may be explicitly introduced. I shall argue in detail, as we proceed, that explicit introductions played a very minor role in Greek mathematics, and that its conventionality was self-regulating.

The practices described in this chapter are interesting as examples of a self-regulating conventionality. They have other, specific, cognitive meanings, and I proceed now to discuss these.

2 THE IMPLICATIONS OF THE PRACTICES

Some implications of the practices are direct consequences of features discussed sufficiently already. First, we saw how second-order information is carried implicitly by first-order language. For instance, the use of letters may imply a continuity between propositions or a discontinuity between parts in the same proposition. This is second-order information: it is not about the geometrical objects, but about the discourse about those geometric objects. However, this is conveyed without recourse to any second-order language. Generally, within a professional group, the manner will say something about the approach taken towards the matter. Specifically, the tendency of Greek mathematicians to avoid second-order language, and its isolation from first-order language, will be a main theme of the next chapter.

Further, we saw that there is no regular relation between signifier and signified in Greek mathematics: specific letters do not represent stereotypical objects (no '*r* for *r*adius'). Rather, the principle of baptism is mainly alphabetical. It may happen that, for a stretch of a few propositions, a local regular relation between signifier and signified is established. But this is strictly local. Consider, for instance, Ψ in Archimedes' *CS*. It is very late in the alphabet and therefore its use in any normal proposition (one without a huge cast) would constitute a dramatic break from the alphabetical convention. It can thus be used meaningfully. However, in *CS* 1, where there are 24 objects, Ψ appears normally and does not play any particular role. Later, in propositions 4–6, it represents a specific circle – so here we hit upon a sign–denotation relation. But as we move on, the stage is meanwhile silently cleared, the stage-workers rearrange the requisites, and when the lights return in proposition 21 Ψ has become a specific cone, in which function it will continue to serve, on and off, until the end of the treatise. The same object may serve, even in the same treatise, in different roles. In itself, Ψ means nothing. There is nothing particularly Ψ-ish about circles or cones (which are rather K-ish, if anything). So it is not Ψ which means a particular circle or cone; it is the entire structure of conventional practice which implies that there is something being referred to. Ψ is a mere cipher in that system.

There is one type of point which does have a meaning of its own, and this did evolve a rudimentary relation between signifier and signified. I refer to the pair (letter K)/(centre of circle). K is the centre in Eudemus' (?) version of the fourth Hippocratic quadrature;[19] also in the second mathematical proposition of the *Meteorology*;[20] and it is consistently the same in a whole treatise, Theon's (or Euclid's?) edition of the *Optics*.[21] On the other hand, such a consistent application of the relation K/centre is very rare. If we move from Aristotle's second to his first theorem, we see that K is not a centre there. Furthermore, even when representing the centre by some non-alphabetical letter, this letter often is not K. The K/centre relation comes naturally, given that letters represent points, and that the centre is such an obviously significant point. But this relation is only very rudimentary in Greek mathematics – and even so, it is unique. The overwhelming principle is that there is no regular sign/denotatum relation. This of course is natural, given that letters are indices, not symbols, as argued in the preceding chapter, and we have seen now an important indication of this. The very principle of alphabetic sequence points in this direction. Letters are assigned to objects in the way in which serial numbers are attached to car plates, or identity cards. And, like car plates indeed, they are essentially arbitrary signs fixed in space upon the objects they are meant to represent.

[19] The textual problems of this treatise are enormous; the most trustworthy reconstruction remains Becker (1936b).
[20] *Meteor.* 375b20. [21] See Knorr (1994) for the problems of ascription.

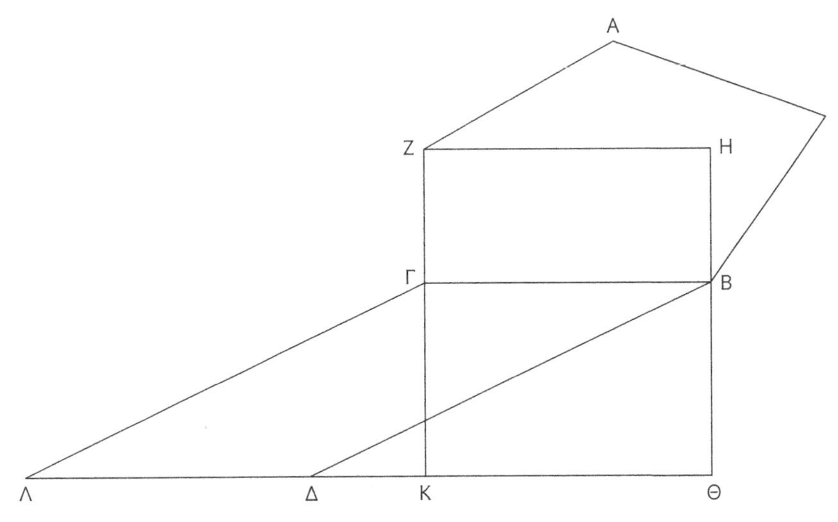

Figure 2.8. Euclid's *Data* 61.

2.1 *The role of visualisation*

There are many cases where it is clear that many-lettered names are switched because of a visualisation, but most important is the simple observation that switches occur. The identity of a Greek object, a line, a triangle, a circle, is not that of a fixed symbol (AB remaining always AB). It is that of the visual object itself, represented by (possibly) alternating combinations of letters. Greeks take it for granted that AB and BA are exactly the same object, and this is because that object is stabilised not through a stable nomenclature but through a stable physical presence in the diagram.

An especially fascinating case is the following derivation:[22]

> '. . . the ratio of ZB to ΓΔ is given . . . but ΓΔ is equal to KB; therefore the ratio of KB to ΓH is also given'.

(I have omitted a few details, with no bearing on the structure of this particular argument.)

The structure seems to be: 'a:b is given. $b = c$. Therefore c:d is given'. This does not make any sense, unless we have the diagram (fig. 2.8). Then it becomes clear that ΓH is equivalent to ZB. They are two ways of representing the same quadrilateral by its opposite vertices. Such

[22] *Data* 61, 112.3–6.

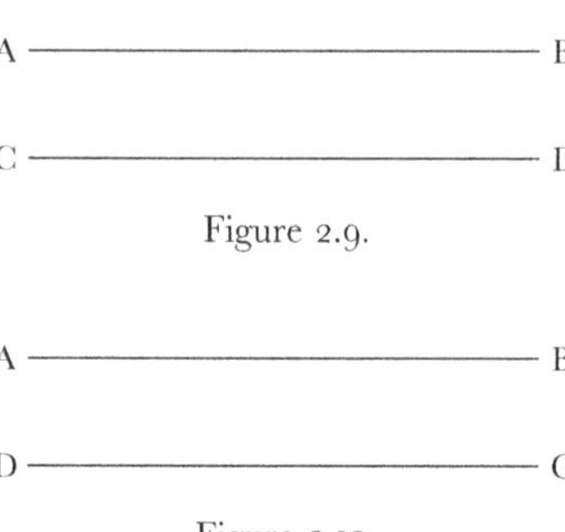

Figure 2.9.

Figure 2.10.

cases are not very common – after all, even simple switches tend to be avoided – but this case is not unique.[23] And the reason for this case is instructive. By switching from ZB to ΓH, the directionality of the parallelogram has switched from south-eastern to north-eastern, the same as KB, the parallelogram mentioned in the conclusion of the argument. The switch is motivated by the desire to represent a pair isodirectionally. This is of course a visual motivation. As usual, the textual difficulties are considerable. But for the sake of the argument, it is worth trusting our manuscripts (as by and large I think we should) to check such questions in general. One place to look for such visual motivations is the isodirectionality of parallels. When two lines are presented as parallel, the text says something about this, say, '*AB* is parallel to *CD*'. This expression may be either isodirectional (as with fig. 2.9: the flow *AB* is the same as *CD*) or as counter-isodirectional (as with fig. 2.10: the flow *AB* is opposite *CD*). I have checked the first 100 cases in Apollonius' *Conics*, as well as 28 Archimedean cases, chosen at random through the index. Of the 100 Apollonian cases, 74 were isodirectional; of the 28 Archimedean cases, 22 were isodirectional. This is clearly meaningful and shows a visual bias.

Even more regular is the phenomenon of linearity. This practice is the following. The sequence of letters in a many-lettered name may be either linear (i.e. corresponding with some continuous survey of the line, e.g. *ABC* for the line *A*—*B*—*C*) or non-linear (e.g. *ACB* for the same line). Out of the 142 relevant cases[24] in Apollonius' *Conics* I, 136 are linear. As usual in Greek mathematics, no rule is sacred, and the

[23] E.g. the equivalence AB ⇔ ΔΓ in *Data* 69.

[24] Most many-lettered names are irrelevant: a line called *AB*, or a circle called *ABC*, for instance, are automatically linear. I do not count such irrelevant cases.

six counter-examples are meaningful.[25] But the tendency is overwhelming. This is the strongest convention we have seen so far, and it is visual in nature. This is a useful indication of the role of visualisation.[26]

We have seen a number of ways in which the practices of many-lettered names betray a strong visualisation. Baptism, pursued point by point, cannot be used in exactly the same way – a point is a point is a point, and there is little to say about its visualisation. But the sequence of baptisms must tell us a lot about the integration of the diagram into the text of the proposition, which is our next topic.

2.2 *The drawing of the diagram*

What we are looking at now is the relation between a number of processes, at least in principle separate:

(a) the formulation of the text prior to its being written down;
(b) the writing down of the text;
(c) the drawing of an unlettered diagram;
(d) the lettering of the diagram.

The processes may interact in many different ways. For instance, it is possible that the drawing of the figure, and its lettering, are simultaneous (the author letters immediately what he has just drawn) or drawing may completely precede lettering. The total number of possible combinations is very large, and it is probable that diagrams were drawn in more than one way. However, it is clearly possible to rule out a number of options.

First, the alphabetical nature of baptism proves that, in general, lettering the diagram did not precede the formulation of the argument. When lettering the diagram, the author must have known at least the order in which objects were to be introduced in the proposition. Result 1: *lettering does not precede the formulation of the argument.*

Further: when the proposition is not strictly alphabetical, it usually tends to be a permutation of the alphabetical: all of the *n* first letters of the alphabet are used (and when this is not the case, it can often be explained as a meaningful deviation), i.e. the author has to know the

[25] Naming a closed shape in a non-linear way connotes the plane of the shape (represented by two non-parallel straight lines joining vertices on the shape). This seems to be the point of counter-examples such *Conics* 1.32.4, 9; 34.16.

[26] Of course, the results are not unique to Apollonius. I have checked Euclid's *Elements* II: all 13 relevant cases are linear.

number of letters to be used in the text. This shows that, when the diagram is being lettered, at least the number of letters which will be required in the proposition must be known in advance. Now this rules out another option, namely: it is impossible that the lettering is done simultaneously with the final writing down of the proposition. For if the lettering were to be done simultaneously with the final writing down of the proof, then it would have to be 'blind', just as the text is, to what comes next. At least a census of the number of the letters required in the proposition must precede its being finally written down; and if a census, then probably also the actual lettering. Of course, it is strictly impossible that lettering should come *after* the final writing down of the proposition – the final written text assumes those letters. It is thus almost certain that as a rule lettering preceded the final writing down of the proposition. Result 2: *lettering precedes writing down the text.*

Furthermore, it is clear that the formulation of the argument had to refer to some diagram (certainly following what we saw concerning the role of the diagram in Greek mathematics). Thus the drawing of the figure itself probably preceded the formulation of the argument. Result 3: *drawing precedes the formulation of the argument.*

I now bring in another sort of argument: completely unspecified letters. We have seen in the preceding chapter that letters may be introduced into the proposition without their being specified anywhere. In such cases, the specification of those letters is completely dependent upon the diagram. Thus, it is clear that at least while the final version of the proposition is being written down, the letters must have been already present in the diagram. All this is clear and based on evidence already marshalled. Another piece of evidence is the following.

Take two examples. Suppose you say, with fig. 2.11:

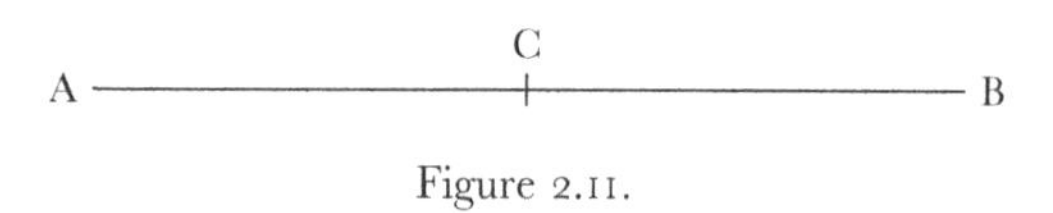

Figure 2.11.

'Draw the line *AB*. Then *AC* equals *CB*.'

Or you say, with fig. 2.12:

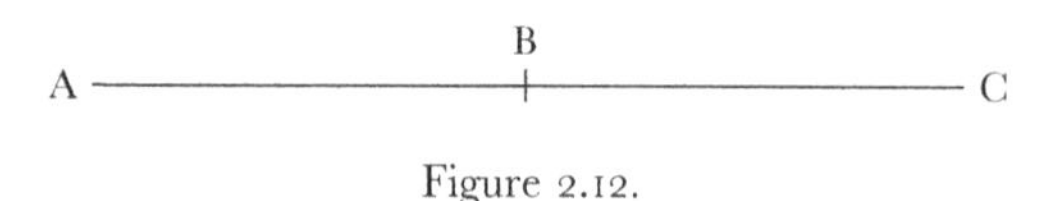

Figure 2.12.

'Draw the line *AC*. Then *AB* equals *AC*.'

In both cases there is a completely unspecified letter. In the first case this is *C*; in the second case it is *B*. In the first case the completely unspecified letter is alphabetical: it is introduced into the proposition in the right order (*C* follows *A* and *B*). In the second case, it is not alphabetical, *but would have become alphabetical had the text been complete at exactly the moment B was produced.* If we assume that, while *AC* was being drawn, *B* was inserted as well (but tacitly), then the sequence becomes alphabetical again. In Greek mathematics, completely unspecified letters behave like *B*, not like *C*. They seem to have been introduced in the text at the right moment, tacitly.[27]

Unspecified letters furnish some evidence, therefore, for a process of lettering which takes place according to the spatial sequence of the proposition, but not according to the final written version of the proposition. That such letters were introduced into the diagram before their introduction to the text has always been clear; what we have learned is that those letters were often introduced into the diagram according to the sequence of spatial events in the proposition. Result 4: *lettering may be structured following the sequence of geometrical actions (instead of the sequence of the text).*

So let us recapitulate our results so far:

Result 1: lettering does not precede the formulation of the argument.
Result 2: lettering precedes writing down the text.
Result 3: drawing precedes the formulation of the argument.
Result 4: lettering may be structured following the sequence of geometrical actions (instead of the sequence of the text).

None of our pieces of evidence is conclusive, but each is convincing, and they all lead to a similar picture, to a similar sequence of events:

1. When a rough idea for the proposition has been formulated, a diagram is drawn.
2. It is lettered, simultaneously with a (possibly oral) dress rehearsal of the text of the proposition.
3. Finally, a text, which assumes this lettered diagram, is written down. Because this text assumes this lettered diagram, it will be different from the dress rehearsal, e.g. in the phenomenon of unspecified letters.

[27] In Apollonius' *Conics* I, of 19 relevant letters, only 4 behave differently. In Euclid's *Elements* XIII, of 42 relevant letters, 14 behave differently (these, however, are mostly from the first two propositions, whose diagrams are interconnected, thus introducing much 'noise' – not a typical case).

A final question remains: what shape can this dress rehearsal take? The options are an informal, oral rehearsal (perhaps aided by a few written notes); or a fully written version. The second option is problematic. This is because it implies a transformation from one fully written version to another, i.e. very much like proofreading of the kind we have ruled out already. If one already has a completely written version, why rewrite it completely? A few modifications, perhaps; but changes such as the omission of reference to unspecified letters involve a major transformation of the text, which the written medium makes less probable.

Such changes, however, are very natural in the transformation from the oral to the written. And in general, there were very few written cognitive aids available to the Greek mathematician (apart, that is, from the lettered diagram itself). He had no machinery of mathematical symbolism. In all the stages prior to the writing down of the proof, there were no written short-cuts the Greek mathematician could use. On the other hand, as we shall see in chapter 4, there was an aural mechanism, of formulae.

My hypothesis can therefore be reformulated in very precise terms. Greek propositions originated in many ways, but the most common was to draw a diagram, to letter it, accompanied by an oral dress rehearsal – an internal monologue, perhaps – corresponding to the main outline of the argument; and then to proceed to write down the proposition as we have it.

This is a hypothesis. What it reflects is the complicated structure of interrelation between text and diagram. The text assumes the diagram, and the diagram assumes the text. The visual and the verbal are closely interrelated – and we begin to see the possible role of orality. The issue of orality will be discussed mainly in the context of the mathematical language. But there is one further piece of evidence concerning letters, which may be relevant here.

2.3 *'Memory' and many-lettered names*

I have suggested above that one factor relevant for the explanation of the distribution of switches is that of memory. It is very difficult to formulate what memory exactly is at work here. Is it the unaided short-term memory, i.e. essentially an oral phenomenon? Or is the use of writing an aid to memory? One obvious point is that the willingness to switch so often shows that the written medium was not used as an aid to memory. Had the writer wanted, he could refrain completely from switching, simply by using the possibility of referring backwards while writing. This clearly he did not do. This may lead us to think, then, that the writing down of the text was done *as if* oral; that the writing was done *on the model* of the oral. The process of writing down was akin to dictation, and indeed may have been that of dictation.

The evidence from the behaviour of many-lettered names is conflicting. The tendency of switches to be somewhat longer than repetitions could be the result of short-term memory effects. Yet in writing, too, physical distance makes a difference: looking five lines up is fractionally more difficult than looking four lines up. On the other hand, unaided short-term memory would have broken down completely over long distances. But we see that there is in fact a tendency not to switch, even in long distances. Clearly in some cases, therefore, the author actually 'looked up his text'.

In a textually minded context, it is natural to think of one form as the 'canonic', presumably the one first struck. Thus, even if switches occur, for some reason, we should expect a tendency for counter-switches to occur immediately afterwards, to return to the original canonic form. You write *AB*; for some reason, to stress some point, you switch to *BA*; but then the 'canonic' *AB* (so one would expect) will be resumed. On the other hand, in the absence of a canonic form, the tendency would be simply to use the last-mentioned form, and any switch would immediately create a locally preferred form. Here, again, the evidence is difficult to interpret in a consistent manner. Of the 109 relevant cases I have identified in Apollonius' *Conics* 1, 59 have counter-switches following switches, while 50 have a locally preferred form established. That the evidence is confusing in this way is, I believe, not an accident. Goody has stressed the ways in which a 'fixed text' would be much more elastic in an oral context, much more static in a written context.[28] He has also stressed that the cognitive implications of literacy will be secondary: they will not be the result of an exposure to the very process of writing, but rather the results of the entire culture derived from writing, the entire set of practices and assumptions concomitant to writing.[29] Our text is written, doubly so, as it includes the reflective use of letters as signs in the diagram. But it does not display the fixity we expect from a written culture. Writing itself is present, but

[28] E.g. Goody (1977).
[29] Goody (1987) 221–3: the term there for my 'secondary' is 'mediated'.

the implications, the possible uses of literacy, are not there in any consistent manner. While the author no doubt looked up his text from time to time (the text as a physical object was there, after all), such procedures did not become part of his ordinary method, which was still partly akin to a more oral method.

The suggestion is, then, that the origins of the conventions governing our texts were in a transitional, ambiguous stage, which, while literate, did not see writing as canonic. We shall pursue this suggestion in more detail in the following chapters.

3 SUMMARY

This chapter had three main themes. First, it looked backwards, to the preceding chapter. It complemented the results of chapter 1, showing that the anchor, the object whose fixity sustains the assertions made in the proposition, was the diagram. We have seen further evidence for the indexicality of the letters, and for the fact of visual specification. Another main theme looked forwards, to the following two chapters, which discuss the use of language. I have offered the hypothesis of the oral dress rehearsal, and I have described the variability of one linguistic object (many-lettered names). In the following two chapters, I shall discuss in detail the variability of other linguistic units, and the oral/written nature of the linguistic practices. I have also pointed out how second-order information may be implicit in first-order language. The relation between first- and second-order languages will also be treated in the following chapter. Finally, an important theme is that of the nature of the conventionality employed in Greek mathematics. This will be picked up again and again in the rest of the book. We will see many more conventions, and most of them, I will argue, are the result of self-regulating conventionality.

CHAPTER 3

The mathematical lexicon

INTRODUCTION AND PLAN OF THE CHAPTER

Greek mathematical deduction was shaped by two tools: the lettered diagram and the mathematical language. Having described the first tool, we move on to the second.

Before starting, a few clarifications. First, my subject matter is not exhausted by that part of the mathematical language which is exclusively mathematical. The bulk of Greek mathematical texts is made up of ordinary Greek words. I am interested in those words no less than in 'technical' words – because ordinary words, used in a technical way, are no less significant as part of a technical terminology.

Second, I am not interested in specific achievements in the development of the lexicon such as, say, Apollonius' definitions of the conic sections. Such are the fruits of deduction, and as such they interest me only marginally. When one is looking for the *prerequisites* of deduction, the language is interesting in a different way. It is clear that (a) a language may be more or less transparent, more or less amenable to manipulation in ways helpful from the point of view of deduction. It is also clear that (b) a language is influenced by the communication-situation. The focus of this chapter is on the ways in which (a) the lexicon served deduction; I shall also try to make some remarks here concerning (b) the probable contexts for the shaping of the lexicon.

Third, there is much more to any lexicon than just one-word-long units. The short phrase is – I argue – no less important in Greek mathematics. I will call such short phrases 'formulae' (following a practice established in Homeric studies) – which of course should not be confused with 'mathematical formulae'. This chapter is devoted mainly to one-word-length items; the next chapter focuses on formulae.

The linguistic tool is unlike the diagram in many ways, and one way in which the difference forces upon me a different approach is that the linguistic tool was, relatively, more visible. Words are manipulated by Greek mathematicians, in definitions. In these moments before the start of the works proper, words stand briefly in the spotlight of attention. Such words, such moments, are then illuminated by another spotlight, that of modern (and ancient) scholarship. The scholarship was always fascinated by definitions – mistakenly, I will argue: even here, in terminology, explicit codification is of minor importance.

Section 1 discusses definitions. I describe what Greek definitions were actually like, and then show their limitations as tools in the shaping of deduction. Section 2 goes on to the actual functioning of the lexicon as a tool for deduction. It consists of descriptions of global features of the lexicon, as well as descriptions of some test-cases (e.g. the lexicon for logical connectors), and what I call 'local lexica' (e.g. the lexicon of Archimedes' *Floating Bodies*). Section 3 offers, briefly, some comparisons with other disciplines, mainly as a further background for the question of the emergence of the lexicon. Section 4 is an interim summary – 'interim', since chapter 4, on formulae, is required as well, before the results on mathematical language can be summarised.

Of course, more Greek is necessary than elsewhere in this book. In a few cases I have left untransliterated Greek in the main text, and I apologise for this. The chapter is readable without any knowledge of Greek. All the non-Greek reader will miss is the detail of some examples, while the argument, I hope, will be clear to non-Greek readers as well.[1]

1 DEFINITIONS: WHAT THEY DO AND WHAT THEY DON'T

The question of the title can get us into deep philosophical waters. This is not my intention. I concentrate on much simpler questions: What are those stretches of text in Greek mathematics which we call 'definitions'? How do they appear in their context? What do they define? Even those simpler questions are difficult and, I find, the answers are surprising.

I have collected some definitions: all the definitions in Euclid's *Elements* and *Data*, and all the definitions in the works of Autolycus,

[1] At this point I would like to bring to the readers' attention the extraordinary study by Herreman (1996), where the lexicon of homology theory, 1895–1935, is rigorously analysed. Herreman's goals are different from mine: not explaining the mathematical achievement but, almost the opposite, showing how mathematics does not differ fundamentally from other discourses. For this reason, Herreman stresses the absolute complexity of the semiotic structures and not (as I do) their simplicity, relative to other, non-mathematical discourses. Allowing for such different goals, Herreman's results can be seen, I believe, as complementary to those offered here.

Archimedes and Apollonius. The list contains somewhat fewer than 300 definitions (counting definitions, as I will explain in subsection 1.2 below, is not as simple as it seems). Most of my arguments are based on checks performed on this list.

1.1 What is a definition?

That which comes first tends to gain canonical status, and 'a point is that which has no part'[2] is perhaps generally taken as an ideal-type definition. And indeed it is a convenient case, one where a thing – a noun – is given a definition. But already definition '4' is more difficult (more on the quotation marks around the numeral, below). 'A straight line is . . .'[3] What is being defined here? 'Line' has been defined in the preceding definition '2'. Or is 'straight' defined here?

Or look at the first definition of *Elements* III: 'equal circles are <those> whose diameters are equal, or <those> whose radii are equal'.[4] This, of course, defines neither 'circles' nor 'equal'. The phrase 'equal circles' is defined – if, indeed, the connotations of the word 'define' are not too misleading. What is the function of such a 'definition'? Clearly neither to abbreviate nor to explicate!

The problem is that of the logical and syntactical form of definitions – and logic and syntax are hard to tell apart here. Perhaps the most useful preliminary classification is into the following four classes of definienda, based on syntactic considerations.[5]

1. The definiendum may be a noun, as in 'a *point* is that which has no part'.
2. The definiendum may be a noun phrase consisting of a noun plus an adjective, as in 'a *straight line* is a line which lies evenly with the points on itself'.
3. The definiendum may be a noun phrase other than a noun plus an adjective, as in 'a *segment of a circle* is the figure contained by a straight line and a circumference of a circle'.[6]

[2] The first definition in Euclid's *Elements*: σημεῖόν ἐστιν, οὗ μέρος οὐθέν.

[3] εὐθεῖα γραμμή ἐστιν . . .

[4] ἴσοι κύκλοι εἰσίν, ὧν αἱ διάμετροι ἴσαι εἰσίν, ἢ ὧν αἱ ἐκ τῶν κέντρων ἴσαι εἰσίν. Note the last formula which is translated by the single English word 'radii'.

[5] I use the following established terms: *definiendum* for the term defined (as 'point' in 'a point is that which has no part'), *definiens* for the defining term (as 'that which has no part' in the same example).

[6] *Elements* III. Def. 6.

4. Finally, the definiendum may not be a noun phrase, as in 'a straight line is said *to touch a circle*: which, meeting the circle and being produced, does not cut the circle'.[7]

First of all, quantitative results. I have counted 33 nouns, 105 nouns+adjectives, 80 other noun phrases and 55 non-noun phrases. It is clear that the first group – which, as noted above, I suspect is usually taken to be the ideal type – is the least important in quantitative terms. It is of course an important group in other ways. These are often the primitive terms of their respective fields, and appear at strategic starting-points: especially the two geometrical starting-points of the *Elements*, books I (point, line, surface, boundary, figure, circle, semicircle, square, oblong, rhombus, rhomboid, trapezia) and XI (solid, pyramid, prism, sphere, cone, cylinder, octahedron, icosahedron, dodecahedron): 21 nouns, most of the nouns in my survey. Defined nouns are mostly things in space. They appear as the subjects in sentences in general enunciations. They are what geometry teaches about.

However, they are not so much what geometry *speaks* about. Geometrical texts speak mostly about specific lettered objects, and references to them often abbreviate away the noun, leaving only articles, prepositions and diagrammatic letters. For instance, as noted already in chapter 1, Greeks often say 'the A' for 'the point A'. More on this later.

The second group is the most numerous. What is defined here? To repeat the example of III.1: '*Equal circles* are those the diameters of which are equal, or the radii of which are equal.' This does not state an equivalence between expressions. This is an *assertion*, as is often the case with definitions (see subsection 1.2 below). And the assertion is not about the composite whole 'equal circles'. It is about the relation between the two components in the definiendum. The assertion amounts to saying that the adjective 'equal' applies to 'circles' under given conditions. Such definitions, then, in general specify the conditions under which an adjective may apply to a noun. More precisely, such definitions specify the conditions under which a property may be assumed to apply to an object. Hence the importance of this group. Greek mathematics is the trading of properties between objects. Arguments often start from the existence of a set of properties, to conclude that another property obtains as well. Theorems in general claim that

[7] *Elements* III. Def. 2.

when a certain property obtains, so does another.[8] Definitions are at their most practical where they supply the building blocks for such structures.

The third group consists most typically of two nouns (together with appropriate articles), one in the nominative, the other in the genitive. Often the focus of the definition is on the noun in the nominative. When defining 'a segment of a circle', the definition regulates the use of the noun 'segment'. The noun in the genitive, in some simple cases as this one, serves simply to delimit the scope in which the noun in the nominative should be employed. In other, more interesting cases, the genitive represents the fact that the definition is an *extension* of a concept, e.g. in Archimedes' *CS*, where 'axis', 'vertex' and 'base' repeatedly appear in the nominative in definitions with different nouns in the genitive: the two types of conoid and the spheroid and their (as well as the cylinder's) segments, so we have defined such objects as the 'axis of a spheroid'.[9] So when definitions are extensions, they serve to codify a new lexicon. Most importantly, by extending words already used in earlier contexts, the definitions help to assimilate the new lexicon to the old, and thus serve to conserve the overall shape of the lexicon. This, the formation of new lexica, is an important phenomenon which we shall look at in more detail later on.

The fourth group – where the definiendum is not a noun phrase – is made of formulae alone. Most often the definiendum in this group is a verb phrase. Most definitions in the *Data*, for instance, belong to this category: e.g. definition 5: 'A circle is said to be given in magnitude whose radius is given in magnitude'[10] (harking back to definition 1, which defined 'being given in magnitude' for areas, lines and angles). In a sense, then, this is like the second group (and verb phrases are, after all, like adjectives: they are complements to nouns). Such definitions specify when a certain property, expressed by a verb (rather than by an adjective), is said to belong to an object (Euclid invariably uses 'is said', *legetai*, in such definitions).

To sum up: there are two preliminary questions concerning definitions. First: does the definition define, or does it specify, instead, conditions where a property (independently understood) is assumed to belong to an object (independently understood)? Second, is the object of the definition a single word, or is it a phrase? I attempt no precision on the

[8] I return to this subject in chapter 6 below.

[9] 246.20, 21; 248.3, 6, 7; 250.1, 2, 11, 14, 15; 252.23, 24; 254.2, 4, 5; 258.27, 28, 29; 260.12, 14.

[10] κύκλος τῷ μεγέθει δεδόσθαι λέγεται, οὗ δέδοται ἡ ἐκ τοῦ κέντρου τῷ μεγέθει.

first question, because the borderline in question is fuzzy; but clearly most definitions seem to belong to the second type. As for the second question, it is simpler: defined words are the 'nouns' in the classification above, about 12% of the definitions in my survey. Most definitions do not define individual words. 'A point is that which has no part' is the exception, not the rule.

2.2 *How do definitions appear?*

David Fowler had to wake me from my dogmatic slumbers. I trusted Heiberg and assumed that Greek mathematical texts often started with neatly numbered sequences of definitions. Seeing a specimen first page of a manuscript of the *Elements*, and glancing at Heiberg's apparatus, should dispel this myth. Numbers were not part of the original text.[11]

That numbers are absent from the original is not just an accident, the absence of a tool we find useful but the Greeks did not require. The absence signifies a different approach to definitions. The text of the definitions appears as a continuous piece of prose, not as a discrete juxtaposition of so many definitions. This is most clear in the Archimedean and Apollonian corpora. There, a special genre was developed, the mathematical introduction, which was not confined to these authors. Hypsicles' *Elements* XIV contains a similar introduction, and the *Sectio canonis* (a mathematico-musical text transmitted in the Euclidean corpus) also has some sort of introduction. So the principle is this. Mathematical texts start, most commonly, with some piece of prose preceding the sequence of proved results. Often, this is developed into a full 'introduction', usually in the form of a letter (prime examples: Archimedes or Apollonius). Elsewhere, the prose is very terse, and supplies no more than some reflections on the mathematical objects (prime example: Euclid).

I suggest that we see the shorter, Euclid-type introduction as an extremely abbreviated, impersonal variation upon the theme offered more richly in Archimedes or Apollonius. Then it becomes possible to understand such baffling 'definitions' as, e.g., *Elements* I.3: 'and the limits of a line are points'.[12] This 'definition' is not a definition of any of the three nouns it contains (lines and points are defined elsewhere,

[11] The editorial notes on the numbers of definitions include 'Numeros definitionum om. PFBb' (*Elements* I), 'Numeros om. PBF' (*Elements* II), 'Numeros om. PBFV' (*Elements* III); or 'numeros om. codd.' (*Data*); or 'numeros add. Torellius' (Archimedes' *SC*). Elsewhere in Archimedes (and in Apollonius and Autolycus) not even the modern editors add numbers.

[12] γραμμῆς δὲ πέρατα σημεῖα. The δὲ relates to the previous definitions, revealing the inappropriateness of division into numbered definitions.

and no definition of limits is required here). It is a brief second-order commentary, following the definitions of 'line' and 'point'.[13] Greek mathematical works do not start with definitions. They start with second-order statements, in which the goals and the means of the work are settled. Often, this includes material we identify as 'definitions'. In counting definitions, snatches of text must be taken out of context, and the decision concerning where they start is somewhat arbitrary. (Bear in mind of course that the text was written – even in late manuscripts – as a continuous, practically unparagraphed whole.)[14]

So far I have played the role of definitions down, noting that they are just an element within a wider fabric of introductory material. The next point, however, stresses their great importance in another system, that of axiomatic starting-points. As I owe the previous point to Fowler, so I owe the next one to Mueller. In Mueller (1991) it is noted that, outside the extraordinary introduction to book I of the *Elements*, 'axioms' in Greek mathematics are definitions.[15] We have seen partly why this may be the case. Most definitions do not prescribe equivalences between expressions (which can then serve to abbreviate, no more). They specify the situations under which properties are considered to belong to objects. Now that we see that most definitions are simply part of the introductory prose, this makes sense. There is no meta-mathematical theory of definition at work here. Before getting down to work, the mathematician describes what he is doing – that's all. Fuzziness between 'definition' and 'axiom' is therefore to be expected.

The reason for giving definitions thus becomes an open question. Definitions cannot be separated from a wider field, that of metamathematical interests. And indeed it is clear, from the attention accorded to definitions by commentators[16] and later mathematicians,[17] that

[13] Compare the similar ending of definition 17: ἥτις <= the diameter> καὶ δίχα τέμνει τὸν κύκλον.

[14] It should be noted also that this second-order material is not confined to the *beginning* of works – though it is much more common there. Rarely, mathematicians explicitly take stock of their results so far: e.g. following Archimedes' *SC* I.12, or following *Conics* I.51. Sometimes, such second-order interventions include definitions: following *Elements* X.47, 84; following Archimedes' *SL* 11; following *Conics* I.16. Worse still, definitions appear sometimes – not often, admittedly – inside propositions: e.g., the sequence *Elements* X.73–8; or 'diameter' in Archimedes' *CS* 272.3–6; most notably, the conic sections themselves, *Conics* I.11–14. The whole issue of the second order has, of course, wider significance, and I will return to it in section 2 below.

[15] Going outside pure mathematics, it is possible to add the postulates at the beginning of Aristarchus' treatise, as well as the *Optics*; in pure mathematics, but later in time, add Archimedes' *SC*. But these exceptions do not invalidate Mueller's point.

[16] 114 pages of Proclus' *In Eucl.* are dedicated to the axiomatic material (mainly to the definitions), and 234 pages are dedicated to the propositions themselves. The proportion in Euclid's own text is less than 5 pages (dense with apparatus) for the axiomatic material, to 54 pages of propositions.

[17] I refer to Hero's (or pseudo-Hero's) work, *Definitions*, given wholly to a compilation of definitions.

definitions may be a focus of interest in themselves. They answer such questions as the 'what is . . . ?' question. Only such an interest can explain such notorious definitions as *Elements* VII.1: 'A unit is that by virtue of which each of the things that exist is called one.' No use can be made of such definitions in the course of the first-order, demonstrative discourse. Such definitions belong to the second-order discourse alone.

In general, we should view the Greek definition enterprise as belonging to the discourse *about* mathematics – the discourse where mathematicians meet with non-mathematicians – precisely the discourse least important for the mathematical demonstration.

1.3 What is not defined?

My survey is far from exhaustive, but it is fair to estimate the number of extant definitions as a few hundred; and definitions are very well represented in the manuscript tradition. I suspect the total number of definitions offered in antiquity was in the hundreds.

These definitions cover a smaller number of word-types. Even ignoring the rare cases of double definition (e.g. 'solid angle', defined in two different ways in *Elements* XI. Def. 11), definitions often return to the same word-type, though in different combinations (for instance, to pick up 'solid angle' again: this is a combination of 'angle', defined in *Elements* I. Def. 8, and 'solid', XI. Def. 1). It is thus to be expected that some, perhaps most, of the words used in Greek mathematics will be undefined. This is the case, for two separate reasons. First, the role of formulae. Return to the first definition: 'A point is that which has no part.' The definiendum is the Greek word *sēmeion*. As I repeatedly explained above, this is not what a Greek mathematician would normally use when discussing points. Much more often, he would use an expression such as τὸ Α, 'the A' (the gender of 'point' supplied by the article). This is a very short formula indeed – the minimum formula – yet a formula. But – and here is the crucial point – τὸ Α is nowhere defined. It was only *sēmeion* which was defined. The concept was defined, conceptually, but the really functional unit was left undefined. The same may be said of the most important words (often, the defined nouns): words such as 'line', 'triangle', 'rectangle', 'circle'. Beneath the process of defining such concepts explicitly, there runs a much more powerful silent current, establishing the real semantic usage through formulae. I shall return to this subject in the next chapter.

A second issue is that defined words belong to a very specific category. Mueller's survey, in the light of modern definition theory, of those words which are used in Euclid's postulates,[18] is useful here. Mueller has found five classes:[19]

(i) Verbs describing mathematical activity;
(ii) Expressions of relative position [mainly prepositions and prepositional phrases];
(iii) Expressions of size comparison [mainly adjectives];
(iv) Expressions of extent;
(v) General terms designating geometric properties and objects.

Mueller observes that all the defined terms belong to (v), and that all the terms in (v), in this particular passage, excluding *diastēma*, are defined.

The main difference between (i)–(iv), on the one hand, and (v), on the other, seems to be that (v) is much more distinctively 'geometrical'. Words such as *agō*, 'to draw' (from class i); *apo*, 'from' (ii); *meizōn*, 'greater' (iii); or *apeiron*, 'unlimited' (iv) occur in non-mathematical texts (though not as often as in mathematical texts). *Sēmeion*, 'point', occurs only minimally in non-mathematical texts. Definitions therefore do define the prominent words in the mathematical discourse, but the 'prominence' is misleading. The defined terms of Greek mathematics cover a small part of the word-tokens in the mathematical texts. These word-tokens are specifically 'mathematical' not because they are responsible for much of the mathematical texts but because they are responsible for no more than a tiny fraction of non-mathematical texts. Cuisines are characterised by their spices, not by their varying use of salt and water. I study the salt, even the water, not so much the pepper (which, I admit, is in itself very interesting!).

Mueller did not list (justifiably, from his point of view) two classes occurring in Euclid's postulates:

(*vi) grammatical words: 'and', 'to be', articles;
(*vii) 'second-order' words, here represented by the verb *aiteō*, 'I postulate'.

[18] There are 70 such word-tokens, 36 word-types. I will use the terms 'word-tokens' and 'word-types' in the following way: by 'word-tokens' I mean words counted separately for each occurrence (in the phrase 'to be or not to be' there are thus 6 word-tokens). By 'word-types' I mean words counted once only for each occurrence (in the phrase 'to be or not to be' there are thus 4 word-types: 'to', 'be', 'or', 'not').

[19] Mueller (1981) 39.

One sees why Mueller preferred not to discuss these classes. They cannot be expected to be defined fully in any context, be it the most axiomatically stringent. One must have a metalanguage, some of whose terms are understood without definition. Naturally, such a metalanguage borrows some of its terms from common language. The points I would like to stress, however, are:

(a) The borrowing of common language grammatical items (class (*vi)) is not limited, in Greek mathematics, to metalanguage passages such as the postulates. The entire discussion is conducted in normal language. Instead of saying that $A = B$, Greeks say that <u>ὁ</u> Α ἴσον <u>ἐστὶ</u> <u>τῷ</u> Β, 'A is equal to B'.

(b) The behaviour of second-order terms is consistently different in Greek mathematics from that of first-order terms, as I have already suggested. Here we see an example. There is a logical difficulty about defining *all* second-order terms, but one could easily define *some* second-order terms (indeed philosophers, Greek philosophers included, hardly do anything else); Greek mathematicians defined none. Again, we see an area where definitions could be used, but were not.

So a dual process renders definitions relatively unimportant as regulators of the actual texts. First, they are confined to a small group of concepts, namely the saliently mathematical objects; and then those saliently mathematical objects themselves are referred to in the text not through those defined words, but through formulae, whose pragmatics develop independently of any explicit codification.

I have made a census of the words used in Apollonius' *Conics* I.15. There are 783 word-tokens, of which 648 are either grammatical words[20] or lettered combinations, such as 'the A' (meaning 'the point A'). Needless to say, no grammatical word is defined.

There are 40 non-grammatical word-types, responsible between them for 135 word-tokens. Defined[21] words are (in brackets, number of word-tokens):

analogon (2), *dielonti* (1), *diametros* (5), *elleipsis* (2), *eutheia* (2), *homoios* (3), *orthos* (1), *parallēlos* (7), *sēmeion* (3)

[20] I refer to what is known in linguistics as 'closed class' words: words such as the article, pronouns, prepositions, etc. These words lack the open-ended productivity of genuine nouns, adjectives, verbs or adverbs (which are therefore known as 'open class').

[21] I.e. defined in the extant literature (though I do not think more surviving works would change the results).

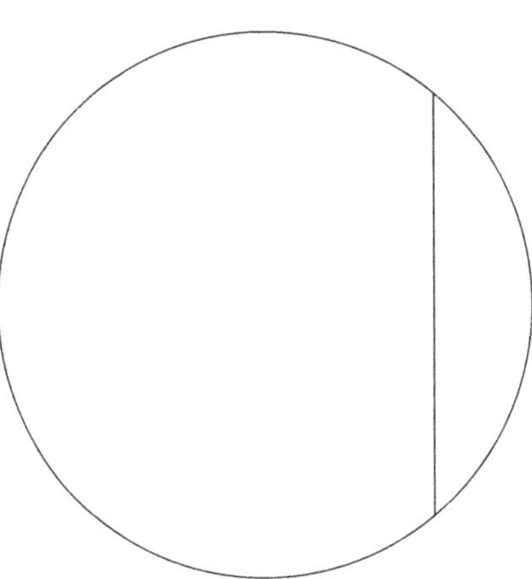

Figure 3.1. Two segments, not one.

Nine word-types, responsible for 26 word-tokens, constitute 3 per cent of the text. This is the order of magnitude for the quantitative role of definitions in the Greek mathematical lexicon.

1.4 What don't definitions do?

One thing a definition can do, in principle, is to disambiguate – to set out clearly what a term means. There are a number of cases where Greeks seem strangely content with definitions falling short of this ideal.

1. *Elements* III. Def. 6: 'A segment of a circle is *the* figure contained by a straight line and a circumference of a circle'.[22] I have stressed the article. It shows that the segment is supposed to be *unique*. The problem is obvious: there are two such segments – which is meant by the definition (see fig. 3.1)? Heiberg argues (convincingly) that the words added in the manuscripts, 'whether greater than a semicircle or smaller than a semicircle',[23] are a late scholion. Someone in (late, probably very late) antiquity felt the ambiguity strongly enough to add a scholion; Euclid and his readers either did not feel the ambiguity (which is very doubtful) or did not care.[24]

2. Is a zig-zag a *line* (rather than a sequence of lines) (fig. 3.2)? Eutocius explains, correctly, that the answer must be affirmative if we

[22] τμῆμα κύκλου ἐστὶ τὸ περιεχόμενον σχῆμα ὑπό τε εὐθείας καὶ κύκλου περιφερείας.

[23] ἢ μείζονος ἡμικυκλίου ἢ ἐλάττονος ἡμικυκλίου.

[24] The same ambiguity is then repeated in III. Defs. 7–8.

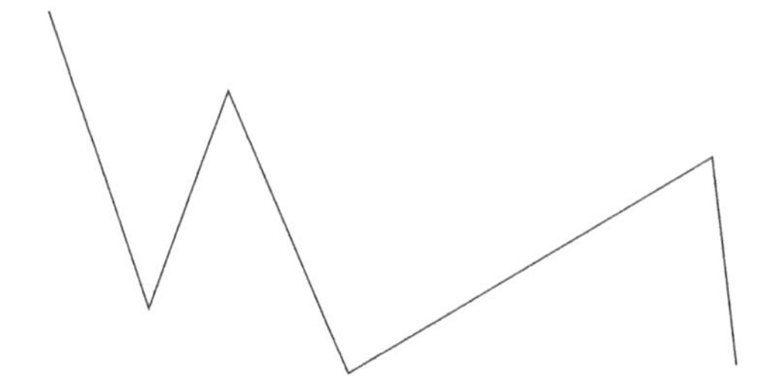

Figure 3.2. A 'curved line'?

are to make sense of a definition in Archimedes *SC*.[25] Archimedes does not explain this, and no one before Eutocius did. To complicate matters further, Archimedes implicitly subsumes zig-zag lines under *curved* lines – clearly a counter-intuitive move. This occurs in one of the greatest meta-geometrical moments in history, where convexity and concavity are first correctly defined, and where 'Archimedes' axiom' is best articulated. The original introduction of new concepts is carefully executed. Details, such as possible remaining ambiguities, do not tax Archimedes' mind.[26]

3. Conic sections are the result of an intersection between planes and surfaces of cones. A parabola, for instance, is the section resulting when the plane is parallel to a line in the surface of the cone. Further, conic sections are characterised by a certain quantitative property (which happens to be equivalent to their definition in modern, analytical geometry). This property is known as their 'symptom', and Apollonius *names* the conic sections after their symptoms: the parabola, for instance, is that section whose symptom involves equality (*parabolē* is Greek for (equal) 'application of areas'). But does he also *define* conic sections after their symptom? In other words, what is the definiens, the construction or the symptom?

Conics I.11–13, where the sections together with their symptoms are introduced, are ambiguous. After introducing both construction and symptom, Apollonius proceeds to say 'let this be called parabola' (or other sections as the case may be). So either, or both, may serve as definiens. However, when he is solving the problem of constructing the sections, in propositions 52–60, the problem is consistently considered solved as soon as the construction-requirement is met. So what *is* the

[25] Eutocius *In SC* I.4.8–13. Archimedes' text is *SC* 6.2–5.

[26] A similar ambiguity in the same text is with zig-zag surfaces, implicitly understood by Archimedes as a kind of curved surface.

definiens of the conic sections? As far as the texts go, this is an idle question.

Again, the context is not that of a careless discussion. On the contrary, this is the only case where a massive *re*definition occurred in Greek mathematics. Apollonius redefined the cone and its sections – and was proud of it. This is a unique case, where scoring points in a mathematical competition involved not only new demonstrations but also a new definition. Where to worry more about ambiguities, if not here? But Apollonius is explicit about his goals: the first book is better than its predecessors in its *constructions* (*geneseis*) and its *prime symptoms* (*archika sumptōmata*).[27] Apollonius does not mention definitions. Again, we see the interests: specifically geometrical results, not meta-geometrical definitions.

4. Euclid's definition of a tangent (III. Def. 2) is a line which 'meeting the circle and being produced, does not cut the circle'.[28]

The definiendum *ephaptesthai*, 'being a tangent', is no more than a compound of *haptesthai*, 'meeting', the verb used in the definition. *Haptesthai*, in turn, is nowhere defined; and as far as the natural senses of the words go, there is not much distinction between the definiens and the definiendum (both mean 'to fasten'). The funny thing is that Euclid himself uses the two verbs interchangeably, as if he never defined the one by means of the other.[29] Clearly the tangle of the *haptesthai* family was inextricable, and post-Euclidean mathematicians evaded the tangle by using (as a rule) a third, unrelated verb, *epipsauein*.[30] This verb originally meant 'to touch lightly'. One wonders why Euclid did not choose it himself. At any rate, a regular expression for tangents in post-Euclidean mathematics was a non-defined term, whose reference was derived from its connotations in ordinary language.

This shows how ambiguous tangents were, even with circles alone. Another problem is that of extension. We do not possess a definition

[27] 4.1–5: περιέχει δὲ τὸ μὲν πρῶτον τὰς γενέσεις τῶν τριῶν τομῶν καὶ τῶν ἀντικειμένων καὶ τὰ ἐν αὐταῖς ἀρχικὰ συμπτώματα ἐπὶ πλέον καὶ καθόλου μᾶλλον ἐξειργασμένα παρὰ τὰ ὑπὸ τῶν ἄλλων γεγραμμένα.

[28] ἁπτομένη τοῦ κύκλου καὶ ἐκβαλλομένη οὐ τέμνει τὸν κύκλον.

[29] The definition in book IV would be wrong if ἅπτεσθαι did not mean a tangent. Archimedes, too, often uses ἅπτεσθαι in this sense (judging by the index, at least 25 times; while ἅπτεσθαι in the wider sense is used no more than 33 times). Autolycus uses ἅπτεσθαι regularly for tangents. It is also interesting to note that Aristotle, in a mathematical context, uses ἐφάπτεσθαι in the general sense of 'being in contact' (*Meteor.* 376a5, 376b8).

[30] See Mugler (1958) s.v. As mentioned in the preceding note, Archimedes uses ἅπτεσθαι for 'being a tangent' at least 25 times. ἐφάπτεσθαι is used for the same purpose 31 times. ἐπιψαύειν is used at least 263 times.

of 'tangent' for the conic sections, but if Apollonius did not offer one, probably no one did. However, the definition cannot be extended directly from circles to parabolas or hyperbolas. This reflects a deeper ambiguity, of the term 'cut' used in the definition above. We know what cutting a circle is. This is like cutting bread: it produces two slices. Our intuitions fail us with parabolas and hyperboles, which encompass either infinite space or no space at all.

Apollonius has two different properties which he obviously associates with being a tangent: 'intersecting and produced both sides falls outside the section',[31] and 'in the space between the conic section and the line [sc. the tangent] no other line may pass'.[32] None is offered as a definition. They are simply properties of lines which happen to be tangents. In the context of the conics, 'tangent' is taken as a primitive concept, which is clarified by the accumulation of geometrical information concerning it.[33]

In this case it is clear that the clarity of the concept owes a lot to visual intuition. It is obvious what a tangent is – you see this. That it is such that no other line may be placed between it and the curve is then an interesting discovery. In general, the logical role of the diagram can help explain why less need is felt for verbal definitions. As usual, the visual may fulfil, for the Greeks, what we expect the verbal to do. Setting aside such possible explanations, the fact remains: in some cases (and I do not claim at all to have exhausted such cases) Greek *defined* concepts were ambiguous.

1.5 *Summary*

Most of my results are negative – the negations of what were at least my own innocent views on definitions.

The definiendum, usually, is not a single word but a short phrase. This of course is related to the central place of formulae, which I will discuss in the next chapter.

Further, most commonly, definitions do not settle linguistic usage but geometrical propriety: they set out when a property, independ-

[31] συμπίπτουσα ἐκβαλλομένη ἐφ' ἑκατέρα ἐκτὸς πίπτῃ τῆς τομῆς. See, with variations, *Conics* I.23 76.26–7, I.24 78.19–21, I.25 80.7–9, I.28 86.16–17, I.33 100.3–4, 100.23, I.34 104.16–17.

[32] εἰς τὸν μεταξὺ τόπον τῆς τε κώνου τομῆς καὶ τῆς εὐθείας ἑτέρα εὐθεῖα οὐ παρεμπεσεῖται. See, with variations, *Conics* I.32 94.23–4, I.35 104.23–5, I.36 106.28–30.

[33] Properties of intersections between lines and sections occupy the section of *Conics* I immediately following the generation of the conic sections, propositions 17–32. Just as the generation of the conic sections renders their precise definition redundant, so the accumulation of properties on tangents (and cutting lines) renders redundant their definition. Ambiguity is dispelled by accumulating information on the objects, not on the terms.

ently understood, is considered to hold. This is why definitions can function as axioms, as stressed in Mueller (1991).

Definitions do not occur as clearly marked, discrete units. They occur within larger second-order contexts. Their motivation may be within this second-order context: sometimes they may be not so much preparation for mathematics as reflection upon mathematics. They stand in such cases apart from the main work of demonstration.

Finally, Greeks did not set out systematically to disambiguate concepts with the aid of precise definitions. This, then, is not the main goal of definitions.

Why are definitions there? Partly because of the general second-order curiosity, suggested already – the wish to say something on the 'what is it?' question. Or definitions may function as axioms; or they are used to express explicitly the extension of concepts, e.g. as is done in Archimedes' *CS*. The reasons are diverse, but, as I have said, disambiguation is not one of them. Definitions are not there to regulate the lexicon – and they cover only a small part of the actual lexicon in use.

One significant contribution of definitions, however, is in their very presence. The *content* of definitions – the definiens – is perhaps less important than the very existence of a place where a definiendum is set out. The existence of a definition must strengthen, to some extent, the tendency to employ the definiendum instead of other, equivalent, expressions or words. But this is no more than a tendency. As explained above, brief formulae such as *to A* may appear in the text, while the definiendum is *sēmeion*. Or the notorious *Elements* I. Def. 22: *heteromēkēs*, there, is what Euclid himself generally refers to as *chōrion*, while *rhomboeides*, there, is what Euclid generally refers to as *parallēlogrammon*. Again and again, the lexicon is found to be governed by forces other than the definitions. These, in turn, were never meant to govern the lexicon. We should therefore move on to look at the realities of the lexicon.

2 THE SHAPE OF THE LEXICON

I have tried to argue against the view that the Greek mathematical lexicon was structured mainly through definitions. I have concentrated on the nature of Greek definitions. Now is the time to say that the emphasis on definitions is fundamentally misplaced, regardless of what definitions may do. This is because the emphasis on definitions implies an emphasis on words, piecemeal, rather than on the lexicon as a whole. It is the lexicon as a whole which is the subject of the following discussion.[34]

2.1 *Description*

As already mentioned above, I have made a census of the words of Apollonius' *Conics* I.15.

Most importantly, the text is made up of few word-types: 74 different word-types are responsible for 783 word-tokens[35] (the first 783 word-tokens of Aristotle's *Metaphysics* Λ, which I have also surveyed, are made up of 200 word-types. I will concentrate on this pair, Apollonius and Aristotle, but I have made some other surveys, whose results can be seen in table 3.1).

Table 3.1. *Word-types and word-tokens in some Greek texts.*

	Conics[36]	*Metaph.*[37]	*SC*[38]	*Elem.*[39]	*Opt.*[40]
Tokens (+ letters)	783	783	510	578	423
Types (– letters)	74	200	86	63	65
Types (+ letters)	142	200	99	98	95
hapax (– letters)	19	100	37	14	20
hapax (+ letters)	52	100	44	26	38
Article	213/1[41]	118/1	137/1	132/1	97/1
Prepositions	107/10	48/12	44/10	76/9	40/9
Letters	169/68	0/0	44/13	101/35	76/31
Other					
Closed-class	159/23	339/54	99/22	125/19	92/27
Total					
Closed-class	648/102	505/67	324/46	434/64	305/38
Open-class	135/40	278/133	186/53	144/34	118/27

[34] To clarify my own usage: it is significant that talk of *the* lexicon of Greek mathematics is at all possible. The lexicon is a constant. Yet some variations do occur from text to text, so I will speak on occasion of, say, 'the lexicon of Archimedes' *Floating Bodies*', and it is also possible to speak of lexica of specific portions of texts such as 'the lexicon of introductions'.

[35] In this calculation, all the *A,B* words are taken as a single type. If we differentiate them according to their different references (AB being a different word from BΓ, though not from BA) the number of word-types rises to 142. Both ways of counting are legitimate. Approached from the outside, the whole *A,B* phenomenon is a characteristic of Greek mathematics, best understood as a single word, highly declinable. Approached from the inside, each *A,B* word has its own individuality. The outside approach helps to characterise Greek mathematics as distinct from other genres and, for our immediate purposes, it is more useful.

[36] Apollonius' *Conics* I.15. [37] First 783 words of *Metaphysics* Λ.

[38] *SC* I.32. [39] *Elements* II.10. [40] *Optics* 34.

[41] In the form #/#, the first number is the number of tokens and the second number is that of types.

Why does Greek mathematics use so few word-types? First, it uses few *hapax legomena* (in a relative sense, that is, words occurring only once within the specific text). There are 19 such *hapax legomena* in I.15, 100 in Aristotle's text: approximately 26% of the word-types in the mathematical text versus 50% in the philosophical, clearly a statistically meaningful difference.[42] The mathematical text is strongly repetitive: there are no dead ends which are entered once but never followed up later. Further, the relative paucity of *hapax legomena* results also from another, more purely lexical feature: the text includes no synonyms or near-synonyms (i.e. words expressing close shades of meanings). I shall return to this later.

Ignoring *hapax legomena*, one is left with 63 word-types in Apollonius responsible for 763 word-tokens (approximately 12 tokens per type), against 100 word-types in Aristotle responsible for 682 word-tokens (approximately 7 tokens per type). This is still a meaningful difference. How should this be accounted for? One surprising feature is that mathematical texts employ the article more often. The article is responsible for 213 word-tokens in Apollonius' text, approximately 28% of the whole, against 118, approximately 15%, in Aristotle.[43] Adding to this the prevalence of *A,B* words in the mathematical text (169, approximately 22%), and of prepositions (107, approximately 14% in Apollonius; 36, approximately 5% in Aristotle), the face of the lexicon is at once made clear. The majority of word-tokens are made up of short phrases composed of letters, articles and prepositions (the three classes together responsible for 64% of the whole in Apollonius): phrases such as:

ἡ ὑπὸ τῶν ΑΒΓ
literally, 'the by the ΑΒΓ', meaning:
'The <angle contained> by the <lines> ΑΒΓ'.

I will discuss such phrases in detail in the next chapter.

[42] Ledger (1989) 229–34 has statistics on the first 4950 words in Plato's *Euthyphro* and *Critias*. These show that in a more literary context *hapax legomena* are the majority – approx. 70% – of the word-types used in the text. Ledger's sense of 'word', however, is that of a string of characters: his computer-searches are morphology-blind. Still, one would certainly get much higher results for 783-word-long chunks of literary works. It seems therefore that even Aristotle has relatively few *hapax legomena* compared to literary texts – which should not come as a surprise. (It is interesting to note in this context results quoted by Ledger of approx. 20% *hapax legomena* among word-*types* in Shakespearian plays, each about 20,000 words long.)

[43] Statistically more significant, the article is used 4196 times in Euclid's *Data*, whose total length is about 19,000 words: approx. 22%. The low percentage compared with Apollonius' *Conics* I.15 is due to the fact that the *Data* is made up of very short propositions, which means that the ratio *protasis/ekthesis* is higher in the *Data*. In the *protasis* context, the article+prepositions+letters combination does not feature, and more normal Greek takes its place (just a chance example: the *protasis* of *Data* 71 is 28 words long, of which 2 are articles, approx. 7%).

Greek mathematics is object-centred. It is not about circles, lines, etc., i.e. about general objects and their properties, but about concrete objects, individuated through the article and the letters and spatially organised through the prepositions. However, this 'concreteness' (stressed already in chapter 1 above) is tempered by the almost surreal nature of the text, which is due to the fact that the most common word-tokens are, in a sense, not lexical items of the Greek language at all. They are either grammatical (the article and the prepositions), or the *ad hoc* indices constituted by Greek letters. The Greek mathematical lexicon is strange – marking mathematics strongly.

The article is used often; other grammatical words are, relatively speaking, ignored. Apollonius' text has 34 closed-class word-types, responsible for 479 word-tokens; Aristotle has 67 closed-class word-types, responsible for 505 word-tokens. I will not take the difference in the number of word-tokens as statistically meaningful.[44] The difference in the numbers of word-types is qualitative and important. It means that the mathematical text is less diversely structured. To begin with, take demonstratives. Aristotle has, e.g., *ekeinos*, *hode*, *houtos*, all giving various shades of 'this' and 'that': in all, he has 11 such word-types, responsible for 67 word-tokens. Apollonius has 6 such word-types, responsible for 15 word-tokens. The mathematical text does not refer to objects in periphrastic ways; it confronts them directly.

Even more surprisingly, the mathematical text has few logical words such as conjunctions and negations. The Apollonian text has 15 such word-types, responsible for 111 word-tokens; Aristotle has 32 such word-types, responsible for 227 word-tokens. Aristotle constantly correlates his statements in ever-shifting ways. Apollonius simply states them, with monotonic *kai* ('and', 36 occurrences) or *ara* ('therefore', 15 occurrences) interspersed between the logical components.

The set of closed-class words apart from the article and the prepositions includes 23 word-types in Apollonius. In the other mathematical texts in table 3.1, the numbers are: 22 in Archimedes, 19 in the *Elements*, 27 in the *Optics*. Eleven of these words are common to all four mathematical texts,[45] a further two are common to Apollonius, the

[44] I ignore here *A,B* names. If they are included, Apollonius has approx. 61% closed-class, against approx. 65% in Aristotle; the comparative texts have: Archimedes, approx. 55%; *Elements*, approx. 60%; *Optics*, approx. 55%. But then again, even if there is a consistent difference here, it may be related to Aristotle rather than to the mathematicians, especially to Aristotle's elliptic sentences, made up of closed-class items.

[45] ἄρα, αὐτός, γάρ, δέ, δή, ἐάν, εἶναι, ἐπεί, καί, μέν, ὅτι.

Elements and the *Optics*,[46] one is shared by Apollonius, Archimedes and the *Elements*[47] and another, finally, is shared by Apollonius, Archimedes and the *Optics* (the relative pronoun, remarkably underused by mathematicians). I do not go into intersections between couples of texts: this Venn diagram is rich enough with the foursome and threesomes.

This brings us to the main observation: the Greek mathematical lexicon is invariant across several dimensions. It is invariant within portions of the same continuous text (this is the essence of the paucity of word-types). It is invariant within authors: the entire Archimedean corpus is made up of 851 words.[48] I have made a list of the 143 words most often used by Archimedes. Sampling the corpus,[49] I found that these words accounted for around 95% of the word-tokens in the Archimedean corpus (excluding the *Arenarius*).

This result should surprise the reader. Archimedes' work is, in a sense, highly heterogeneous, ranging from the purest geometry, through abstract proportion-theory (in some propositions of *SL*, *CS*), to mechanics, hydrostatics, and the delightful hybrids of the *Method* and the *Arenarius*. There is some local lexical variation, which I will discuss in subsection 2.4 below, but the overwhelming tendency is to carry over, from one discipline to another, the same, highly limited, lexical apparatus.[50]

Most of these 143 Archimedean stars could be found, for instance, in Euclid's *Data*. Some *TLG* surveys on Euclid's *Data* are helpful in order to get a feeling of the monstrous repetitiousness of the mathematical text. *Logos*, 'ratio', is used 655 times; *pros*, 'to', 869 times; the verb 'to be', in its various forms, about 1,000 times; *didōmi*, 'to be given', more than 1,500 times. To remind the reader: the text is about 19,000 words long, almost half of which are the article and A,B words.

The interesting variability is not so much between authors or works as within parts of the same work: the introduction/text distinction. I shall discuss this in subsection 2.5 below.

[46] τις, ὥστε. [47] τε.

[48] I have counted the number of entries in the index, which is apparently complete. The number of word-tokens in the Archimedean corpus must be in the order of 100,000. To compare: according to Ledger (1989) 229–34, 4,950-word-long Platonic chunks hold more than 1,000 word-types.

[49] I surveyed all the pages whose number divides by 50 (4% of the whole, since only even pages have Greek in the Teubner). I excluded the *Arenarius* (whose vocabulary is different) and substituted II.392 for II.400 (as the last is partly Latin).

[50] We have noticed already the tendency to use definitions not to create new terms but to extend the sense of old terms from one field to another, most noticeably in Archimedes' *CS*.

To sum up, then, our first description: the Greek mathematical lexicon is tiny, strongly skewed towards particular objects (whose properties and relations are only schematically given) and is invariant, within works and between authors.

2.2 *The one-concept-one-word principle*

We saw on p. 101, above, the trio *haptesthai/ephaptesthai/epipsauein*, all meaning 'to be a tangent'. Here the concept of the tangent is covered by three different words, and the single word *haptesthai* covers two different concepts ('touching' in the sense of sharing at least one point and 'touching' in the sense of being a tangent). The existence of many words for a single concept constitutes synonyms; the existence of many concepts per single word constitutes homonyms. I shall now claim that Greek mathematics has got very few synonyms, and even fewer domain-specific homonyms. A related feature is the absence of nuances. The lexicon is constituted of words clearly marked off from each other, not of a continuous spectrum of words shading into each other.

Why should this be the case? Consider the following pairs and trios:[51]

anthuphairesis/antanairesis (an operation, akin to continuous fractions)
gōnia/glōchis (angle)
grammē/kanōn (line)
diastēma/apostasis (interval)
emballō/enteinō (insert)
kentron/meson (centre)
kuklos/strongulos (circle)
kulindros/oloitrochos (cylinder)
logos/diastēma (ratio)
paraballō/parateinō (apply)
sēmeion/stigmē (point)
stereos/sōma/nastos (solid)
sphaira/sphairoeidēs (sphere)
chōrion/embadon/heteromēkēs (rectangle)

[51] For ἀνθυφαίρεσις, ἀνταναίρεσις, γωνία, γραμμή, κανών, διάστημα, ἀπόστασις, κέντρον, κύκλος, κύλινδρος, λόγος, σημεῖον, στιγμή, στερεός, σῶμα, σφαῖρα, χωρίον, ἔμβαδον, ἑτερομήκης, see Mugler (1958), s.vv. For γλωχίς see Hero, Def. 15. For μέσον see (e.g.) Plato's *Parm.* 137e, which see also for στρογγύλος. For ὀλοίτροχος see Democritus fr. 162. For (another sense of) διάστημα see also the *Sectio Canonis*. For ναστός see Democritus fr. 208. For ἐντείνω, παρατείνω see Plato's *Meno* 86e–87a. For σφαιροείδης see Plato's *Tim.* 33b, 63a.

At least some of the entries above reflect cases of synonymy in the early, formative period of Greek mathematics. Perhaps *kanōn* never meant 'line', perhaps Plato's *Meno* 86e–87a is intentionally 'wrong',[52] but some, e.g. *sēmeion/stigmē*, are excellently documented. In all of them, only one of the candidates survived into classical Greek mathematics (the one I give as the first). Some of these words were revived, in mathematical contexts, in late antiquity, and all of them were continuously used throughout antiquity in non-mathematical contexts – it is not as if general changes in the Greek language rendered some words obsolete. We see therefore a process, characterising mathematics, of brutal selective retention of words. This, rather than a fine adaptation into niches, was the process by which the classical lexicon evolved.[53]

The remaining cases for synonymy are, first, the case of tangents, discussed above; also the following:

* *eidos/schēma*, both meaning, roughly, 'figure'. This is such a general concept that it could almost be considered second-order. *Schēma* in Archimedes, as a rule, is not used in expressions such as τὸ ΑΕΖΓ σχῆμα ('the figure ΑΕΖΓ'), but in expressions such as τὰ σχήματα τὰ περιεχόμενα ὑπὸ τῶν ΑΕ, ΕΖ, ΖΓ[54] ('the figures contained by the [lines] ΑΕ, ΕΖ, ΖΓ'). *Schēma* is not the *name* of the object lying between those lines (as *tetragōnon*, say, is – witness expressions such as τὸ ΑΒΓΔ τετράγωνον). It is an element in a second-order *description* of the object. Further, both *schēma*[55] and *eidos*[56] are often used in locally conventionalised and distinct senses.

[52] For which see Lloyd (1992).

[53] Some such adaptation did occur, of course, to some degree; mainly, this is restricted to compound verbs, for which see below.

[54] Cf. Archimedes, *SC* I.12 48.21.

[55] I have already mentioned above (chapter 1, subsection 3.2.1) Euclid's use of the formula καταγεγράφθω τὸ σχῆμα, 'let the figure be drawn', with 'the figure' referring to a very specific kind of figure. This is an example of the way in which a local deviation from the lexicon immediately leads to a new, implicitly defined, local lexicon. Taken in the wider sense of σχῆμα, the expression καταγεγράφθω τὸ σχῆμα is strangely vague. Such vagueness is unacceptable within the Euclidean discourse. To make it acceptable, the reader supplies, without being told, a more precise meaning: a self-regulating convention at work.

[56] The word εἶδος is just as general as σχῆμα, and we see it undergoing a similar self-regulating process in Apollonius' *Conics* book II. In proposition 1 it first appears within τὸ τέταρτον τοῦ ὑπὸ τῶν ΑΒΖ εἴδους '$\frac{1}{4}$ of the shape [contained] by ΑΒΖ' – a deviation from the earlier Apollonian practice, where this idea is normally expressed by τὸ τέταρτον τοῦ ὑπὸ τῶν ΑΒΖ – '$\frac{1}{4}$ of *that* [understood χωρίον, area, and conventionally meaning rectangle] <contained> by ΑΒΖ'. So something is beginning to build up; εἶδος is about to be conventionalised. Already in the third proposition it will occur within τὸ τέταρτον τοῦ πρὸς τῇ ΒΔ εἴδους - '$\frac{1}{4}$ of the shape next to [the line] ΒΔ' – which, strictly speaking, is too vague to mean anything. This will be the regular expression from now on (with παρά sometimes replacing πρός, as in

So we are reminded of the role of self-regulating conventions. Against the background of the one-concept-one-word expectation, two words for a single concept are immediately taken as having some meaningful distinction between them – hence *some* specialisation, in this case that of occurring within specific formulae.

* Another couple is διάστημα/ἡ ἐκ τοῦ κέντρου. Here διάστημα is taken in the sense of a 'radius', which is otherwise expressed by the formula ἡ ἐκ τοῦ κέντρου. Of course διάστημα means 'distance' and no more; it is only within the formula, κέντρῳ μὲν τῷ Α, διαστήματι δὲ τῷ ΑΒ, γεγράφθω κύκλος ('Let a circle be drawn with the centre A, radius [distance] AB'), that διάστημα takes the meaning of 'radius'. So actually the synonymy evaporates. The Greeks have one concept, the construction of circles, to which one formula is attached; another, similar to our 'radius', expressed by another formula. We are reminded of the fact that the semantic units may be more than one word long.
* The participles *sunkeimenos*, *sunēmmenos* occur both in exactly the same position in a certain formula, that of the composition of ratios (on which much more in the next chapter). This is a long and very marked formula, but it is not so frequent that its constituent words could each memorised and regulated. This is instructive. It is an important feature of self-regulating convention that they work on masses. Only the explicit codification can touch the less common.
* A small class of near-synonyms is made up of some verbs and their composites: thus, *anagein* means almost the same as *agein*. This is almost inevitable in Greek, especially the later Greek of the mathematical texts. Many of these compounds, however, do come to have very specific meanings, for instance *anagein* is often used especially for the construction of perpendiculars (see Mugler, s.v.). More on this in subsection 2.3 below.
* In a few cases, well-differentiated senses may have overlapping references: e.g. *periphereia*, 'circumference' (the line constituting the border of a circle, or an arc) may overlap with *perimetros*, 'perimeter' (the border of any two-dimensional object, curved or

proposition 15). Again, there is a combination of two facts: the expression deviates from the regular lexicon, and it is used schematically within a given context (where we are never interested in anything but the *quarter* of the shape!).

For Pappus, this becomes one of the technical senses of the word εἶδος. This is the result of Pappus' reading in Apollonius: so we see how such conventions are transmitted (note that, of course, this convention is nowhere fixed in a definition).

rectilinear).[57] The adjective *euthugrammon*, 'rectilinear', is used for any object which is not curved;[58] the adjective *polugōnon*, 'many-angled' – 'polygonal' – is used for any five- or more-sided polygon. These polygons may also be referred to as *euthugrammon*.[59] Or the two expressions, κάθετος ἐπί, 'perpendicular on', and πρὸς ὀρθάς, 'at right angles to', may refer to the same line though, of course, from different points of view. Such cases are interesting but they have nothing to do with synonymy.

* The one major exception is a set of synonyms, or very close near-synonyms, occurring within second-order language. These include:[60]

  aitēma/hupolambanon (postulate)
  analuein/luein (to solve)
  axiōma/koinon ennoion (axiom)
  deiknumi/apodeiknumi (to prove)
  epitagma/protasis (that which the proposition sets out to obtain)
  katagraphē/diagraphē, etc. (diagram)
  horos/horismos (definition)

  One should add a couple occurring within first-order proofs, but of course essentially a second-order expression, logical rather than mathematical:

  atopon/adunaton (absurd/impossible – used interchangeably)

Brutal selection was applied in first-order language alone. We shall return to this difference between first-order and second-order in subsection 2.5 below.

To sum up on synonyms: the Greek mathematical, first-order lexicon does not accumulate words. Diachronically, it does not deepen, like a coastal shelf; it is gradually eroded. Cases of synonymy are few and far between, and usually reveal, under closer inspection, a difference in use (within constant formulae) if not in sense.

[57] There is a tendency to avoid περίμετρος for curved figures: in Archimedes this is found only in the deeply corrupt *DC*, and in a single place in *SC* (10.10, 13). But already Hero speaks about the περίμετρος of a circle without any difficulty, e.g. *Metr.* 1.25.

[58] E.g. angles: the first occurrence of this adjective in the *Elements* is in Def. 9, defining 'rectilinear angles' (those contained by straight lines only; the Greeks recognised other angles as well).

[59] An example: in *SC* 1.13, 54.12, 17, the two words are used in two consecutive sentences for the same object.

[60] For αἴτημα, ἀξίωμα, ἀναλύειν, λύειν, δείκνυμι, ἀποδείκνυμι, ἐπίταγμα, πρότασις, ὅρος, ὁρισμός see Mugler (1958) s.vv. For ὑπολάμβανον see Archimedes' index; for κοιναὶ ἔννοιαι see, of course, Euclid. For the terms for diagrams, see chapter 1, subsection 3.2.1.

Now to homonyms. The definition of homonymy is of course problematic – it is something of a question whether 'the centre of a circle' and 'the centre of a sphere', for instance, use the word 'centre' homonymously. I do not think the speakers felt this was a homonymous use of the concept.

The role of formulae, once again, should be borne in mind. Consider, for instance, *meros*: taken alone, it means a 'part', but in the formula *epi ta auta/hetera merē*, and only within it, it means 'direction' (see Mugler 1958, s.v.). Similarly, when special senses become attached to certain conjugations of the verb and no others, this should not be considered homonymous. *Hupothesis* is a hypothesis, *hupokeimenon* is a figure laid down; *suntithenai* is to solve a problem synthetically, but *sunkeisthai* is reserved for the composition of figures – but also for the composition of ratios (see Mugler 1958, s.vv.).

The last examples are interesting in two ways. First, we see that something approaching a homonym occurs at the boundary between first-order and second-order language (in both cases, the active sense is second-order, the passive first-order). The case of *sunkeisthai* is even stronger, and involves a real homonym, cutting across domains, in this case plane geometry and proportion theory. Generally, Greek mathematics allows several real homonyms to cut across domains, especially across the first-order/second-order border. *Akros* and *enallax* belong to both plane geometry and proportion theory; a large group, *apotomē*, *gnōmōn*, *dunamis*, *epipedos*, *kubos*, *homoios*, *stereos* and *tetragōnos* belong to both geometry and number theory. All three Apollonian names for conic sections also retain their meaning from the theory of the application of areas. *Anagein*, *anastrephein*, *ekkeisthai*, *epharmozein*, *horos* and *sunagein* mean different things as first- or second-order terms.[61] Two more surprising cases are *legō* and *luō*. *Legō* has the natural meaning of 'say', which is of course employed within second-order language, as in (but not only) the formula *legō hoti*, 'I say that'; it also has the more technical first-order sense of the assignment of ratios.[62] *Luō* is a second-order term, a near-synonym of *analuō*, for which see Mugler (1958) s.v.; it also has the first-order spatial significance of 'being suspended'.[63]

[61] For all of which see Mugler (1958), s.vv., and s.v. λόγος for ἀναστρέφειν. The case of συντίθημι as used with λόγος presents a further complication: the dative form of a participle, συνθέντι, is a certain operation on proportions (nothing to do with solving a problem synthetically).

[62] See Archimedes' index, s.v. λέγω, the references to the passive, finite, present forms of the verb (e.g. λέγεται, I.378.19).

[63] Archimedes' *QP* 6 274.11, 10 280.19, 12 284.2.

To conclude: within any treatise (which will concentrate – as Greek treatises do – on a single domain) the only synonymy or homonymy is between first- and second-order discourse.

2.3 *Holistic nature of the lexicon*

My theory is simple. We have seen that the lexicon is small, and operates on the principle of one-concept-one-word. As soon as such an environment is established, the users of the lexicon will easily get a grasp of the lexicon as a whole and, working with the one-concept-one-word principle, they will assume structures of meaning. This is self-regulating conventionality.

The most ubiquitous case and the best example clarifying what I mean by 'structures of meaning', is *to/hē*. These words (the neuter and feminine articles) are technical in Greek mathematics (as soon as letters are added to them): the neuter means a point, the feminine a line. These are the two pillars on which Greek mathematics was erected,[64] and they are defined simply by being set apart from each other, each with its own gender. And here clearly some historical process took place, since, in earlier times (and, in non-mathematical contexts, always), *stigmē*, feminine, meant 'point' as well. The disappearance of *stigmē* from the mathematical scene occasioned some discussion,[65] which tried to clarify what had been wrong with the word *stigmē* as such. The truth is that there is nothing bad about the word (it is, in the relevant senses, practically synonymous in everyday Greek with *sēmeion*). It is just that there is a competitor, the word *sēmeion*, and there is a niche waiting to be taken, that of the neuter (because the feminine is already firmly taken by *grammē*, a 'line'). So we see that the relevant factor is the entire ecological system, not the merits of this or that word.[66]

Such 'systemic' processes could take place when new words were introduced. Their very novelty was striking for the practitioner who was versed in the system, and therefore such a practitioner, employing the one-concept-one-word principle of interpretation, would look for a relevant *new* concept to match the new word. This could happen without any explicit definition.

[64] τὸ occurs 1001 times in the *Data*, ἡ or τὴν occur 1058 times, while ὁ or τόν occur only 92 times.

[65] See, e.g. Vita (1982). [66] I shall discuss the system of such phrases in the next chapter.

I have mentioned in the preceding section several cases of remarkable specialisation, not *between* verbs, but *within* verbs – the way in which different forms of *suntithenai*, for instance, acquire specific, consistent usages. Less spectacularly, such specialisations occur between verbs, again excluding *haptesthai/ephaptesthai*. Take, to begin with, *agein* and *graphein*.[67] Both are practically synonymous within Greek mathematics – 'to draw' – that is, if the contexts of usages are ignored. The contexts are consistent: *agein* is used when drawing straight lines, *graphein* is used when drawing curved lines. Now look further. The respective compounds, *anagein* and *anagraphein*, both mean 'to erect', and, again, their meaning is differentiated in practice, but the difference does not correspond to the difference between *agein* and *graphein*, presumably because 'to erect a curved line' is hardly a necessary concept. Thus *anagein* is to erect a line, *anagraphein* is to erect a figure.

The case of the adverb *mēkei*, 'in length', illustrates well the holistic nature of the lexicon. Generally, in the expression for the ratio '<the line> *AB* to <the line> *CD*', it is understood implicitly that the relation is between lengths. In some cases, however, the author may need to refer to ratios between (what we see as) squares of those lines, and this is expressed by '<the line> *AB* to <the line> *CD dunamei*' (the adverb *dunamei* meaning in this context – of course without a definition – something like 'in square'). In such cases, and only then, *mēkei* may be used as well, e.g. in a proportion: 'as <the line> *AB* to <the line> *CD dunamei*, so <the line> *EF* to <the line> *GH mēkei*'. The *mēkei* means nothing. In this context, however, and only here, it acquires a meaning, dependent upon that of *dunamei*: it signifies the negation of *dunamei*.

There is nothing surprising, then: words acquire their meaning through internal structural relations. This is true in general, no doubt. The fact that the mathematical lexicon is so small means that such processes are much more conspicuous. The number of components is small enough to make their internal relations immediately transparent. This will be even stronger in the smaller or local lexica, to which I now pass.

2.4 *Smaller or local lexica*

I offer here two test-cases, significant in different ways. The first, that of logical connectors, is important because of its contents, obviously

[67] All the terms discussed from now on in this section are in Mugler (1958).

relevant for deduction. The second, that of Archimedes' *Floating Bodies*, is important as a case where a lexicon is introduced for the first time.

2.4.1 *Logical connectors*

There are a few indicators of logical relations which are consistently used in Greek mathematics. 'Consistently' here means that:

* connectors are almost never missed out. If a logical move is made, a logical indicator is very often employed;[68]
* there are only a few such indicators employed;
* those which are so employed are used in a few limited ways.

Take, for instance, Apollonius' *Conics* I.41. The logical connectors are:

kai epei / alla / de / ara / kai epei / toutestin / eti de / ara / [koinos aphēirēsthō] / ara / alla men / de / ara / de / ara / [enallax] / de / gar / kai / kai ara / ara / toutestin / de / [to homoion kai homoiōs anagegrammenon] / ara / ara / epei oun / kai / de / ara / alla / ara / ara / ara

Compare this with Euclid's *Data* 64:

epei gar / kai / de / kai ara / ara / ara / kai / hōste kai / kai ara / alla / kai / kai / ara

In three cases in the text from Apollonius, the assertion was a formula of a kind having a clear relation to the logical structure of the argument, and the connector was consequently dropped. Otherwise, all the assertions in these two proofs were marked by some connector, signifying their relations to other assertions. The general principles by which the connectors are given are simple.

Most results are marked by an *ara* ('therefore'; all such translations are only rough),[69] or *hōste* ('so that').[70] When an assertion is added to a

[68] The exceptions have to do with formulae implying logical relations, as explained below.

[69] The word most typical of Greek mathematical discourse; it is used 567 times in Euclid's *Data* (total about 19,000), 3778 times in Euclid's *Elements* (total about 150,000), 72 times in Hypsicles' 'book XIV' (total about 4000) and 429 times in Hero's *Metrica* (total about 21,000). It is probably responsible for 2–3 per cent of the total word-tokens in Greek mathematics. The *Politicus*, a Platonic work roughly the same size as the *Data*, and with some interest in argumentation, has the word 32 times; and Plato is very lavish with his particles. Generally in Hellenistic Greek, the main particles are δέ, καί, ἀλλά, and γάρ (Blomquist 1969, table 21). Greek mathematics can be characterised as *Greek prose with* ἄρα – and my study can be characterised as *The Shaping of* ἄρα, a study of the new specific sense of an old emphatic particle.

[70] Much less common than ἄρα, it is still a regular feature of the style, used 44 times in the *Data* (compared with 6 in the *Politicus*. Also used 112 times – quite a lot – in Hero's *Metrica*, 6 times

previous assertion, so that in their combination they will yield some other assertion, the added assertion is marked by *de*[71] (best translated simply as 'and'), *alla*[72] ('but') or *kai*[73] ('and'). When an assertion supports a preceding assertion, the supporting assertion is marked by a *gar* ('for')[74] or, less commonly, an *epei* ('since')[75] or even (in some contexts) *dia* ('through').[76]

The above holds for continuous stretches of argumentation, where *P* leads to *Q*, leading to *R*, etc. When a stretch of argument starts, or is restarted following a hiatus in the argumentation, *epei* ('since') is used, anticipating the following *ara*.[77]

As with all rules in Greek mathematics, this system is not religiously followed, and in particular contexts there are some variations, e.g. *alla mēn* is not very common in Euclid's *Elements* – 30 times in all. But 20 of these 30 times are in the arithmetical books. A similar situation obtains with *epeidēper* and solid geometry. The use of *oukoun* in a number of treatises, and in them alone, is a similar phenomenon.[78] As ever, there are local variations.

in Hypsicles' 'book XIV'). There is a tendency to use ὥστε when a result of an activity, rather than a logical result, is intended (e.g. all of the 18 occurrences in the first hundred pages of Archimedes' *SC* are of this sort).

[71] Very popular in mathematical texts; in fact, a feature of the mathematical language is that δέ is used much more often than μέν (δέ is used 409 times in the *Data*, while μέν is used 70 times; the corresponding figures for the *Politicus* are 276/213. Also: 95/24 in Hypsicles, 610/166 in Hero). In other words, Greek mathematical texts tend, on the one hand, to use δέ absolutely, and, on the other hand, not to use it as the correlate of μέν.

[72] Not as common as δέ: used 440 times in the *Elements* (compared with 2452 times for δέ). Still, of course, this is often enough to be a normal part of the mathematical lexicon. (Compare: Hypsicles: 3; Hero: 46.)

[73] Naturally, the word is enormously common – 6143 times in the *Elements* – but many of these are not logical connectors between sentences.

[74] Very common, where applicable: it is used 964 times in Euclid's *Elements*. (Compare: Hypsicles: 14; Hero: 81). The word is generally very common in Greek argumentative discourse: e.g. 146 times in the *Politicus*, compared with 188 times in the *Data*. However, it should be noted that most often the word is simply inapplicable in Greek mathematics, where justifications of assertions by *following* assertions are rare (I shall return to this in chapter 5).

[75] In the Archimedean corpus, ἐπεί is more common in assertions *following* the assertions they support, than in assertions preceding the assertions they support. Even so, there are only 46 cases in the Archimedean corpus. The ἐπεὶ γάρ, common in Euclid's *Elements* (168 uses), appears usually at the very beginning of proofs, and I discuss this strange-looking γάρ in chapter 6.

[76] This happens 29 times in the Archimedean corpus, and 26 times in Euclid's *Elements* (I am excluding the formula διὰ τὰ αὐτά).

[77] ἐπεί is used 1110 times in Euclid's *Elements*, 1007 of which are in the contexts καὶ ἐπεί (558 uses), ἐπεὶ γάρ (168), ἐπεὶ οὖν (164) or πάλιν ἐπεί (117). ἐπεὶ γάρ and ἐπεὶ οὖν are used, generally speaking, to start an argument; καὶ ἐπεί and πάλιν ἐπεί to restart it. Other statistics for ἐπεί (without distinctions of use): Hypsicles: 15; Hero: 89.

[78] See Knorr (1994) 25–8. The treatises are Euclid's (or pseudo-Euclid's?) *Optics*, *Catoptrics* and the *Sectio Canonis*.

However, the main elements are fixed. The preceding footnotes show a considerable degree of fixity of ratios between the frequencies of the connectors in different works. The senses are also well regulated, especially compared to the general context of ancient Greek (*gar*, for instance, is very limited in scope). It is true that Greek mathematics derives from the Hellenistic period when, as is well known, a general process of simplification occurred in the Greek use of particles in general. However, comparisons between mathematical Greek and Hellenistic Greek prove the independence of mathematics.

I follow here Blomquist (1969). Chapter 1, 'Frequency and use of individual particles' begins with *mentoi*, and Blomquist notes (27) its frequency in Hellenistic Greek in general, except for its absence from mathematical authors.[79] This in fact can be repeated for each of the case-studies offered by Blomquist – except one, *alla mēn*, where mathematical authors account for about half of *all* examples from Hellenistic Greek (61). So Greek mathematicians do not follow the wider profile: they have their own, very distinctive, distribution of particles. And, as Blomquist shows (chapter 3), it is wrong to think of Hellenistic Greek as much less rich in particles than classical Greek. It uses fewer emphatic particles, no more than that.

Moreover, there is more to logical connection than just particles. For instance, the genitive absolute, ubiquitous in Greek in general, and appropriate for the description of many logical relations, is hardly ever used within mathematical proofs.[80] Perhaps most surprising is *ei*, 'if'. What is more fitting for mathematics? It is used only 16 times in the first book of Euclid's *Elements*, and there almost always in the formulaic context of the beginning of the reductio, *ei gar dunatos*, 'for if possible . . .'. As a connector, *ei* is hardly ever used.[81]

A final remark from Blomquist: to argue for his thesis – that particles did not decline sharply in use in Hellenistic times as was sometimes thought – he points out that two particles, never used elsewhere in mathematical works, are used by Apollonius in his *introductions*: *mentoi*,

[79] As well as Apollonius of Citium.

[80] I am ignoring the formula used sometimes before proofs, as an abbreviation of construction, τῶν αὐτῶν ὑποκειμένων (see especially *Conics* III, e.g. the beginning of propositions 2–3, 6–12). For an exceptional use of the genitive absolute within proofs, see *Conics* I.33, 100.18–19.

[81] In the Archimedean corpus there are a few occurrences, all of them within the *PE* (150.7, 156.25, 202.12, 206.11), always within the locally conventionalised formula, εἰ δὲ τοῦτο.

ou alla mēn.[82] Once again, we see the systematic difference between first- and second-order discourse.

2.4.2 *On floating bodies*

When Archimedes wrote *On Floating Bodies*, he was establishing a wholly new field. Of course, his treatment was geometrical, so many terms were taken over from his general mathematical lexicon, but the introduction of some specific terms was inevitable. The fact that the lexicon was new means that it stands on its own. And this is all the more remarkable, since the axiomatic apparatus of the work[83] is made up from a single hypothesis. As far as vocabulary is concerned, the work is left to fend for itself.

The source from which Archimedes builds up his local lexicon is everyday Greek. Below is a list, aiming to be complete, of such words (numbers are occurrences in *On Floating Bodies*, based on the index; 'at least' represents failings of that index):

(a) Verbs expressing motion or rest:
φέρειν (at least 41), ἀναφέρειν (9), καταφέρειν (3), ἕλκειν (1), ἀφίημι (39), καθίστημι (27), ἀποκαθίστημι (9), δύνειν (14), καταδύνειν (5), καταβαίνειν (3), μένειν (29);
(b) Verbs expressing force and other relations:
ἐξωθεῖν (5), θλίβεσθαι (18), ἀντιθλίβειν (2), βιάζειν (3), καθείργειν (1);
(c) Nouns:
ὑγρός (at least 100), γῆ (12), βάρος (at least 18), ὄγκος (17);
(d) Adjectives:
κουφότερος (20), βαρύτερος (6), ἀκίνητος (9);
(e) Adverbs of direction:
ἄνω (16), ὑπεράνω (1), ἐπάνω (1), κάτω (18), ὑποκάτω (2).

The first overall observation is that this local lexicon shares the usual leanness of the Greek mathematical lexicon. All the words are very mundane, and many synonyms would have been possible in principle. Yet the classes (b)–(d) are free of any synonyms. In the classes (a) and (e), where synonyms occur, they are rare, as if they represented an oversight.

Some precise senses are specialised, e.g. the clusters φέρειν/ἀναφέρειν/καταφέρειν, or καθίστημι/ἀποκαθίστημι. A more interesting specialisation is that of καταβαίνειν, as distinct from the δύνειν root; the

[82] Blomquist (1969) 140. [83] *CF* 318.2–8.

former, but not the latter, denotes in Archimedes a complete sinking, though both may mean this in common Greek. But some synonyms are still there, namely ἕλκειν (1), καταδύνειν (5), ὑπεράνω (1), ἐπάνω (1), ὑποκάτω (2). It will be seen that, quantitatively speaking, these are not very important. Qualitatively, however, they show that, while Archimedes tends to use the same words again and again, he is not religiously avoiding any variation.

2.5 *Compartmentalised nature of the lexicon*

To say that the lexicon is compartmentalised, following my arguments for its holistic nature, may sound paradoxical. However, the discussion concerning local lexica should make us aware of the possibility of slight differences from one part of the text to another, and it is such differences, though they occur on a regular basis, which I call 'compartmentalisation'. The texts are made up of several types of discourse.

The best way to see this is through *hapax legomena*. We saw how few of them there are locally in each individual proposition. Paradoxically, taking corpora as wholes, there are many *hapax legomena* in Greek mathematics. In the Archimedean corpus, out of a total of 851 words, 228 are *hapax legomena* – about 26%, an extraordinary result for a corpus in the range of 100,000 words. This is easily explicable. When a text is split into different registers, the number of *hapax legomena* must rise, since the chance that any given word would be repeated is smaller in such a case. The text is really a juxtaposition of different, unrelated small corpora, and the number of *hapax legomena* rises accordingly.

The Archimedean text is divided into three parts:

(a) introductions, taking the form of letters;
(b) the *Arenarius*;
(c) the remaining, ordinary mathematical text.

Those types of text vary enormously in quantity, with (a) and (b) each responsible for no more than a few per cent of the corpus as a whole, which is taken up almost entirely by (c). The results for *hapax legomena* are much more equal. Eighty-two of the *hapax legomena* occur in the introductions; 72 in the *Arenarius*; only 74 occur in the whole of the rest of the text. Here then we come to the real gulf separating registers in Archimedes. One is the letter form (the *Arenarius* being, after all, a letter), the other the mathematical form. The first talks *about* mathematics; the latter *is* mathematics. We have glimpsed this distinction often before: we now confront it directly.

It should be stressed that Archimedes does not use the letter form to commend the attractions of Syracuse or to complain about his health. The introductions are mathematical texts, as dry as any. The difference of register could not be explained by a difference of subject matter.

Or perhaps, as ever in the lexicon, a more structural approach should be taken? The important thing is not *how* the second-order lexicon is different, but *that* it is different. The two are separated. Indeed, they are sealed off from each other, literally. Second-order interludes between proofs, not to mention within proofs, are remarkably rare.[84] The two are set as opposites. And it is of course the first-order discourse which is marked by this, since the second-order discourse is simply a continuation of normal Greek prose.

I have said already that Greek mathematics focuses directly on objects. The compartmentalisation of lexica is one of the more radical ways in which this is done. When it is talking about objects, the lexicon is reduced to the barest minimum, so that any wider considerations are ruled out – because there are no words to speak with, as it were. And, in this, the main body of Greek mathematics is marked off from ordinary Greek, as no other Greek subject ever was.[85]

2.6 *Summary*

Some of the results we have obtained are significant for the question of the shaping of the lexicon. These are the small role of definitions, the persistence of exceptions to any rule, the role of the structure of the lexicon as an entirety, and the great divide between first- and second-order discourse. I shall return to such points later on.

More positively for the lexicon itself, we saw that it is dramatically small – not only in specifically mathematical words, but in any words, including the most common Greek grammatical words. It is strongly repetitive, within authors and between authors. And it follows, on the whole, a principle of one-concept-one-word.

[84] One thinks of *CF* 374.16, coming in the middle of proposition 8: τοῦτο δ' ἦν εὔχρηστον ποτὶ τὸ δεῖξαι – 'now that was useful for the proof'. I do not think there are more than a handful of such remarks in the whole of Greek mathematics (and, needless to say, εὔχρηστος is a *hapax legomenon* in the Archimedean corpus).

[85] I shall return to show this in section 3 below. Blomquist (1969) also shows that no other set of authors had similar peculiarities in their use of particles.

We saw how the lettered diagram created a small, finite, discrete system out of infinite continuous geometrical situations. The Greek mathematical lexicon did the same to Greek language (hardly less infinite or continuous than Euclidean space!) and the significance for deduction is similar. But the moment to discuss this in detail will come only after we have seen the all-important phenomenon of formulae.

3 COMPARATIVE REMARKS

At this stage readers may respond by saying that my description can, *mutatis mutandis*, serve to characterise any scientific discourse – that it is simply the nature of a scientific discourse to be brief, free from ambiguity and redundancy. I beg such readers to notice that in the third century BC *no* discourse approached the mathematical one in these terms. Perhaps the lexica of other scientific disciplines began to approach the mathematical lexicon only following the scientific revolution with its *more geometrico* bias. But leaving aside such speculations, let us take, as an example, the pre-Galenic Greek anatomical lexicon, following Lloyd's (1983) discussion.[86] Here are several important points of difference:

* *Date.* Though an anatomical vocabulary of sorts is part of any natural language, and as such is as old as dated references to 'anatomy' (in our case Homer), a technical use of anatomical terms was no more than foreshadowed in the Hippocratic treatises. It was mainly a Hellenistic creation, and stabilised only following Galen. The scientific terminology of anatomy was much younger than the science of medicine, while the science of mathematics seems practically to have been born armed with its terminology.[87]
* *Authority.* There are good reasons for ruling out the hypothesis according to which 'Euclid (or any other individual) settled the Greek mathematical lexicon'.[88] The anatomical terminology was settled – to a great extent – following Galen.
* *Ambiguity.* Ambiguities are common, especially in the Hippocratic corpus (that is, what the terms refer to is a matter of guesswork –

[86] I am referring to the chapter entitled 'The Development of Greek Anatomical Terminology', 149–67.
[87] Homer: ibid. 152. The Hippocratic corpus: 153–7. Hellenistic times: 157ff. Galen: 167 n. 106.
[88] I shall return to this in chapter 7.

for modern lexicographers, and apparently for many Greek readers as well). It is true that anatomical objects do not allow simple definitions (compared to mathematics), but this could mean that anatomists would have to use more elaborate and complex terminology. That is what they do today. As a matter of fact, they used a lexicon that was not very elaborate, and defined hardly anything. The ancient reader, lacking our hindsight, dictionaries and, probably, patience, could never have deciphered the words unless they were defined, ostensively, in front of him – which entails an altogether different communication situation from the one we meet in the mathematical context.[89]

* *Homonymy and synonymy*. Even when, in Hellenistic times, a certain degree of clarity was achieved, homonymy and synonymy were still widespread in the anatomical lexicon.[90] Note that both were present simultaneously. It is not as if the physicians had too many or too few words: they had, as it were, just the correct number, only they were unevenly spread, some objects getting too many words and some words getting too many objects.
* *Coinage*. Though possessing a rich stock of pre-technical terms to borrow from and refine, anatomists were fond of inventing new words, in marked contradistinction to the mathematical lexical conservativeness.[91] Indeed, the main feature of the anatomical vocabulary as discussed by Lloyd is its permanent flux. Lloyd shows the relation between the shape of the anatomical lexicon and a strongly competitive, partly oral communication situation, one which is inimical to the development of *any* consensus.[92]

Very briefly, I should like to add some comments on the philosophical lexicon. Here the first noticeable feature is a strong tradition of second-order discussion. The *Metaphysics* Δ is a concentrated example of what is after all one of the main traits of Aristotle, at least: a conscious, critical use of philosophical language, often made in order to solve philosophical problems by pointing to their lexical origins. That this was not only an Aristotelian game is clear, e.g. from an Epicurean fragment[93] where several terms related to 'space' and 'void' are defined and distinguished. A concern over terms is indeed probably necessary for anyone seriously interested in philosophy. However, this did

[89] For ambiguity and its modern deciphering, see especially Lloyd (1983) 154–7.
[90] Synonymy: ibid. 160–1. Homonymy: 161–5. [91] Ibid. 163. [92] Ibid. 165–7.
[93] L&S, 5D = Usener 271.

not lead to any consensus in antiquity: first, because much of this interest is descriptive rather than prescriptive; second, because the situation was no less competitive in philosophy than it was in medicine. Many of the entries in Urmson's *Greek Philosophical Vocabulary*,[94] for example, have special meanings, specific to different authors or schools.[95] If the terminology was at all settled, this was not at the level of the entire scientific community, as in mathematics, but within more local groups. Nor was such local codification ever complete. Sedley (1989) discusses the nature of intra-school controversies in Hellenistic periods. The presence of an institutional setting, together with a quasi-religious cohesion, did not make the communicative interchange any less competitive. And much of the controversy was about the meaning of words as used by the founders of the school.[96] What we see is the philosophical habit of referring to terms in a second-order approach, the tendency to problematise the lexicon. This tendency results in the fastening of controversies, whatever their origin, upon terminological issues. This, in turn, prevents the development of any established lexicon. We begin to see the importance of the historical setting, and I shall add a few comments on it below. A detailed discussion of the historical setting will wait until chapter 7.

4 THE SHAPING OF THE LEXICON

The lexicon did not emerge through some explicit codification, an 'Index of Prohibited Words'. I have made this claim already; more should be said to prove it. The *hapax legomena* in Archimedes are relevant. As mentioned above, the purely mathematical portions of the text contain 74 *hapax legomena*. Some of these are due to rare 2-order interpolations, such as εὔχρηστός, στοιχειώσις, etc.; some are simply very rare words which could not be ruled out by any 'Index', for instance specific numbers like ἑκκαίδεκα; but this still leaves us with at least 12 words:[97]

[94] Urmson (1990). Unfortunately, this is not, nor does it profess to be, the philosophical counterpart of Mugler (1958). Apparently, the relative anarchy of the Greek philosophical terminology makes such a project enormously complicated.

[95] Of the first 30 entries, 22 have such specific meanings. I think it is safe to say that most entries are like this.

[96] The specific example is the use of τέχνη by the founders of Epicureanism.

[97] See the index to Archimedes s.vv. It is easier to check *hapax legomena* but further evidence could be set forth: why, for instance, is such a natural preposition as πλήν used only twice?

ἅπλως, εἴπερ, ἔμπροσθεν, ἐνθάδε, μετάγειν, μηδέτερος, ὄλιγος, ὅποσος, ὅπου, ὅτε, οὐδέποτε, οὐκέτι.

Each of the above *could* be used often. The fact that they are used so rarely must imply some reluctance to use them. But then again, the fact that they are used at all means that the nature of this reluctance is not explicable by any prohibition. The situation is similar to that we have seen in chapter 2 above: the conventionality is not of a sacred-rule type, simply because it is sometimes broken. We have seen in *On Floating Bodies* a similar phenomenon: a strong tendency to use specific words, which, however, is not absolute. In the more established parts of the lexicon, deviations become rarer, but they do not disappear.

Words were not expelled by a fiat, nor were they similarly introduced. One should make an effort to realise how mundane Greek mathematical terms are. We translate *tomē* by 'section', *tmēma* by 'segment', *tomeus* by 'sector'. Try to imagine them as, say, 'cutting', 'cut' and 'cutter'. The Greeks had no Greeks or Romans to borrow their terms from. Greek mathematical entities are not called after persons, as modern entities are. They are not called after some out-of-the-way objects. They are called after some of the most basic spatial verbs – βάλλειν (to throw), ἄγειν (to draw), πίπτεν (to fall), τέμνειν (to cut), λαμβάνειν (to take), ἱστάναι (to stand), etc.; some everyday spatial objects – πλευρά (side), γωνία ('corner' = angle), κύκλος (circle), etc. Often, when a new name is required, an old name is extended: 3-dimensional terminology, for instance, builds upon 2-dimensional terminology, and similar extensions occur between domains such as proportion theory, the theory of numbers, and geometry (for which see subsection 3.2 above).

So was there a Style Manual? The question is rhetorical, but still, it is useful – it brings us to the heart of the *a priori* argument against the hypothesis of the ancient Style Manual. A Style Manual must be a written form, to which one can return. We know of no written discussions of style in ancient Greek mathematics, and the second-order discussions we do get make us think that no such discussion of *style* existed (apart, of course, from discussions of second-order terminology: say, whether propositions should be called 'theorems' or 'problems').[98] We cannot look for written influences in this direction. There are, however, written texts working within Greek mathematics. *These are the mathematical texts themselves*. And if their style is well regulated, this

[98] See Proclus' *In Eucl.* 77–8.

regulation must be explained through the texts themselves. The Greek mathematical lexicon must be a self-regulating organism. In the following I shall try to indicate briefly how – I believe – this could have come about.

First, it is time to remind ourselves of our main goal, namely the shaping of deduction, and for this purpose it is especially the absence of ambiguity which is of interest.

When we say that Greek mathematical texts are unambiguous we mean that the fixation of reference is distinct (the referent, taken by itself, is a well-defined object) and certain (there is no question as to the identity of the referent). The distinctness of reference is mainly due to the absolute, small size of the lexicon. This means that there is a small universe of referents, which one can learn once and for all. It also means that the process of initiation, the acquisition of the model of this universe, is short, because the items of the lexicon are repeated over and over again. By the time one has learned the first book of Euclid's *Elements*, say, a considerable subset of the Greek mathematical lexicon must have been interiorised.

The small size of the lexicon also means that it is easily marked off from other texts. One does not need a title to know that a certain work is mathematical. Even more, one does not need special headings to know that certain passages within a mathematical text are only quasi-mathematical (namely the second-order portions of the text). Thus the first-order mathematical discourse is easily identified.

Now assume an advanced student, one who has already interiorised this system – who, as I claimed above, need not be much more than a moderately good reader of some books of Euclid. Suppose him to encounter a mathematical text: he would immediately identify it as such, and expect a specific lexicon. Suppose he encounters there a word unfamiliar to him. He would immediately try to associate with it a new reference, driven by the logic of self-regulating conventions. And this means that, had no new meaning been intended, the text would fail to convey its meaning.

What we see is that the logic of self-regulating conventions is a constraint on the development of the lexicon. A small, well-defined lexicon is thus a self-perpetuating mechanism.

In the discussion above I have fastened upon the process of disambiguation in, as it were, the ontogenetic plane. It will be seen that my hypothesis concerning the phylogenetic plane is exactly similar. Once the lexicon began to take the shape described in chapter 2, it could not but go on to develop in the same direction – given, of course, a co-operative, rather than polemic, communication situation.

And what made the lexicon begin to move in such a direction? Partly, the answer involves the nature of the mathematical world. It is a small world; no surprise that it has a small lexicon. It is a well-defined world; no surprise that it has a well-defined lexicon. In other contexts, nuances in terminology could be made to correspond to undefined, barely felt nuances in the objects themselves. Thus, *haptesthai* and *ephaptesthai* are probably only a shade away in normal discourse: but the mathematical world knows only sharp black and white, either a tangent or not. Such a line of explanation surely has its validity, and to this extent we may say that the Greek mathematical lexicon is the natural result of putting mathematics into a written communication context. Not all written communication contexts, however, would necessarily have brought about a similar result. I have mentioned the need for a co-operative communication situation – a strong demand indeed in Greek settings. Add to this the following. Paradoxically, one important aid to the development of the lexicon was the *absence* of explicit codification. After all, a writer could easily say 'and, by the way, by all the words *A*, *B*, and *C* I mean exactly the same thing', thus making sure that some synonyms would enter the lexicon. The Invisible Hand would codify the discourse only under an invisible, free competition. The Greeks do not speak about their mathematical language, and thus it develops according to the rules of a self-regulating conventionality (here the comparison with the philosophical vocabulary is especially revealing). So, in this way, the desire to distinguish first- and second-order discourse is a contributing factor to the shaping of the lexicon.

More generally, without this distinction between first-order and second-order lexicon, the lexicon could never have been so small and regimented. But how can we explain this distinction? Here probably wider sociological factors are at work. I gave some background in the comparative remarks in section 3 above; I shall return to this issue in chapter 7 below.

CHAPTER 4

Formulae

INTRODUCTION AND PLAN OF THE CHAPTER

The term 'formula' is inevitable. Yet it is also very problematic. 'Greek mathematics' is at an intersection of two fields, in both of which the term has (completely independently) already been put to use. The general mathematical use is easy to deal with. Suffice to say that my 'formulae' are *not* equations. They are a (relatively) rigid way of using groups of words. And here is the second problem. 'Formulae' as groups of words are the mainstay of twentieth-century Homeric scholarship (and, derivatively, of much other folklorist and literary scholarship). Thus, they evoke a specific – if not always a precise! – connotation. This connotation is not wholly relevant to my purposes. However, it is not entirely misleading, either. While my use of 'formulae' is not the same as that of Homeric scholarship, it is related to it in some ways.

This chapter, therefore, will start from the Homeric case. In section 1, both the problematics and the definition of 'formulae' will be approached through the Homeric case. Section 2 offers a typology of Greek mathematical formulae. Besides giving the main groups, I also describe some key parameters along which different formulae may be compared. Section 3 then analyses the behaviour of formulae. In section 4 I return to the Homeric case, and summarise the possibilities concerning the emergence and the function of Greek mathematical formulae.[1]

[1] As with diagrams, so with formulae: it is often asserted that 'Greek mathematics employs a formulaic language', but hardly anyone ever studied the subject. The outstanding exception is the important article Aujac (1984), which analyses a wealth of useful material (parts of which are not covered in this chapter). I owe a lot to that article, but I must warn the reader that Aujac's main interpretative claim (that Greek mathematical education was oral and comparable to the transmission of oral poetry) is unconvincing and unsatisfactorily argued. Aujac makes it clear that this is no more than a tentative guess, and in the light of the current literature on the oral and the written this guess is seen to be very implausible.

1 THE HOMERIC CASE AS A STARTING-POINT

1.1 The problem

People have noted repetitions in the Homeric corpus from time immemorial; the term 'formula' was given its technical Homeric sense only in the context of a certain theory, to which my present study is related in several ways. The theory is roughly as follows:[2] Homeric singers were illiterate; further, they were public performers. To cope with the necessity of singing long stretches of metrical text without a script, they developed a certain tool, namely 'formulae'. These are short phrases, of given metrical shapes, and therefore fitting into specific 'slots' in the Homeric line, which are then used schematically. The point about 'schematically' is that these formulae have very little variability: on the whole, once the metrical conditions as well as the general idea are set, the formula can not vary very much.[3] Very roughly speaking, therefore, the Homeric singer who has internalised the formulaic lexicon should be able to produce metrical verse with relatively little need for creativity – in some senses of creativity. Instead of laboriously producing word after word and then checking their metrical fitness, the singer chooses, for much of the time, from among ready-made metrical sequences.

Such – again, very roughly speaking – was the form of the theory. Note that this is cognitive history. It defines a certain cognitive tool, and claims that this tool was necessary for the achievement of a certain cognitive task (namely, the creation of Homeric poems), given the absence of some other cognitive tool (namely, writing). The theory posits some universal claims: it is *impossible* to improvise Homeric poems unless one has formulae. Conversely, with the aid of such formulae, this is *possible*. Writing makes formulae *superfluous*. All italicised adjectives are supposed to be cognitive universals. The theory posits a certain historical hypothesis concerning the availability of cognitive

[2] I follow the original formulation of Parry (first offered in 1928, most easily accessible in Parry 1971), not because this is still considered valid in its detail (it is not), but because it is the clearest presentation of the theory as a *theory*, a single solution to a single problem. The main line of development since Parry has been to break down 'formula' into types, each having its own cause and function (see, e.g. Hainsworth 1993, with references to the important contributions). In basic outline, however, Parry's theory is still adhered to, so I present something which has much more than just historical interest.

[3] This aspect of the theory has been corrected. Many Homeric formulae are flexible (Hainsworth 1968). Still, the Homeric formula is much less flexible than similar structures from written literature.

tools in Homer's time, namely the absence of writing. It also implies some assumptions concerning the social background, for it is assumed that the structure of formulae evolved gradually, within a community which approved of the formulae and accumulated them.[4]

It will thus be seen that the main thesis of this book is structurally similar to the formulaic theory as developed in Homeric scholarship. This is not surprising: Homeric scholarship is at the heart of the orality/literacy issue, which is the single most important topic in the discussions of history and cognition. All the more embarrassing, therefore, that the case of mathematical formulae should seem to be so unlike that of the Homeric formulae. The illiteracy of 'Homer' is a hypothesis based on some implicit evidence in the poems as well as on external archaeological and historical data; the literacy of Greek mathematics is explicitly stated by the texts, which, after all, rely on a specific use of the alphabet in the lettered diagram. Mathematical texts were edited by their authors, quite unlike the improvisation visualised by the Homeric formulaic theory. Consider Apollonius' Introduction to book I: the text asserts itself as a corrected second *edition*, superior to an earlier one which suffered from being prepared in haste. Likewise, the text sets itself up as competing with an earlier written attempt by Euclid.[5] One therefore pictures the Greek mathematician as toiling over wax tablets and papyri: those of his fellow mathematicians and those of his own. This picture is confirmed by further evidence. Hypsicles reads Apollonius, just as Apollonius read Euclid, and Archimedes refers to his own manuscripts, in ways similar to Apollonius' self-reference.[6]

The following chapter, therefore, would not try to import the results of Homeric scholarship into the history of Greek mathematics. It is simply absurd to try to imagine Greek mathematicians as more or less illiterate improvisers. But I shall try to keep the structure of the argument of the Homeric formulaic theory, which indeed is the structure of my argument throughout: for formulae to exist, they must perform some cognitive task for the individual who uses them, and they must

[4] The process of initiation of singers and their interaction with their community are the focus of, for instance, Lord's presentation of the theory in Lord (1960), chapter 2.

[5] A corrected second edition: introduction to *Conics* I.2.7–22. The earlier edition being done in haste: 13–17. Bad marks for Euclid: 4.3–5 (implied), 4.13–16 (named).

[6] Hypsicles: see introduction to *Elements* XIV. Archimedes: see introductions to *SC* I, II, *CS*, *SL*, *Meth*. Indeed, it seems that one of the main functions of the mathematical introductions was to fix the bibliographic coordinates of the work in question: where it stood in relation to other *written* works of the same author and of his predecessors.

be somehow encouraged and maintained by the context of the communicative exchange.[7]

1.2 The problem of definition

Parry's classic definition is: '*a group of words which is regularly employed under the same metrical conditions to express a given essential idea*'.[8] Obviously the 'metrical conditions' mentioned in this definition have no counterpart in the mathematical case. This is not a trivial point, given that much Greek science – and other science – was transmitted in metrical form.[9] Greek mathematics was written in prose, which makes its formulae even more perplexing.

Two basic ideas in Parry's definition are left. One is 'regular employment', the other is 'to express a given essential idea'. Both require some analysis.

The couple of English words 'in the' has been used seven times so far in this chapter, so it has been 'regularly used': is it a formula of my style? Clearly not, and this for two reasons. First, while this couple of words was used often relative to other couples of words, it was not used often relative to its use in other comparable texts. The use of this couple of words does not mark my text from other texts. Second, this couple of words was the natural way to say a certain thing which, for independent reasons, was said over and over again. The repetitive use of the words 'in the' was parasitic upon the repetitive use of the syntactic structure represented by 'in the'. There was nothing repetitive about the specific choice of *words*.

So we can begin to refine our terms.

A group of words is *regularly used* if it is used more often than other groups of words.

[7] The point can be stated in much more general terms. Once you look for them, formulae are everywhere: in law as in poetry, in everyday speech as in literature. You can find their parallels in music and, if so, why not in painting? (Think, for instance, of the emblems attached to medieval saints as the visual counterparts of the Homeric ornamental epithet.) Or look around you, at the buildings, the furniture: everywhere, you can find *repeated patterns in artefacts*. Is this then the relevant unit of description? It is useful to have this level of generality in mind – but mainly in order to avoid the tendency to transfer results hastily from one field to another. Repeated patterns are everywhere, but this means that their causes and functions would vary.

[8] Parry (1971) 272.

[9] Concentrating on the Greek case, there are well-known philosophical examples (e.g. Parmenides, Empedocles), but also an important genre closely related to mathematics: descriptive astronomy, often rendered poetically (Aratus is the foremost example, but the tradition begins with Hesiod, and gets even closer to mathematics proper with the poetic (lost) original of the *Ars Eudoxi*, for which see Blass 1887). One of the 'modern' features of the *Almagest* is its use of a specifically written form (the table) for catalogues of stars which were traditionally represented in metrical poems.

It is *markedly repetitive* if it is used in the text more often than it is used in comparable texts.[10] I take Parry's 'regular employment' to refer to this.

Finally, a group of words may be *semantically marked*. This, in its most general form, can be defined as follows. For the expression E, find the set of *equivalent expressions* (a fuzzy concept, depending on how wide your sense of 'equivalence' is). Find the frequency in which E is used relative to its equivalent expressions in a group of texts. If, in a particular text, the frequency of E relative to its equivalent expressions is higher than in the wider group of texts, then E is *semantically marked* in that text (thus, 'in the' was not semantically marked in the above, since it was not used more often than is necessary for saying 'in the'). Sometimes, however, a group of words may be semantically marked *per se*, even if it is not comparatively frequent. This happens through non-compositionality.[11] Sometimes the non-compositionality may be the result of ellipsis only (see n. 11); sometimes it will involve a metaphorical use of language. Whatever the route leading to non-compositionality, it yields semantic markedness. There are two routes to semantic markedness: quantitative, through the relative frequency defined above, and qualitative, through non-compositionality. This semantic markedness is, I think, behind Parry's 'to express a given essential idea'.

Each of the three concepts (*regularly used*, *markedly repetitive*, *semantically marked*) amounts to 'being used more often than' – relative to different contexts. Regularly used groups of words are used more often than other groups of words. Markedly repetitive groups are used more often than the same groups of words in other relevant texts. Semantically marked groups (in the quantitative sense) are used more often than their semantic counterparts, relative to the distribution in other relevant texts. So I pause for a moment. The definition proves complex – and still it leaves the definienda fuzzy, for the concepts are contextual, and the contexts cannot be set precisely and are not accessible.[12] Further, the definitions use the word 'more' – which is fuzzy. My goal is not to eliminate fuzziness, but to obtain a clear understanding of fuzzy objects. This clarity is worth our trouble; let us continue.

[10] Notice that groups of words may be regularly used but not markedly repetitive, as are 'in the' in my text; or markedly repetitive but not regularly used, as are in fact *most* Homeric formulae.

[11] A non-compositional phrase is one whose content can not be reduced to the contents of its constituents and their syntax. For example: I once went to a Cambridge feast wearing what I thought was a 'black tie' – i.e. a tie which was also black. I analysed the expression into its constituents and its syntax. Only there I discovered that 'black tie' is a formula for 'a black bow-tie' (what I will call non-compositionality through ellipsis).

[12] Either in the Greek case or in any other case, since the relevant contexts can always be extended beyond written corpora.

Of the three concepts, only the last two lead to formulaic status. An unmarked group of words can not have any special status – and a regularly used group of words, which is neither markedly repetitive nor semantically marked, is certainly unmarked.

Homeric formulae are both markedly repetitive and semantically marked. Now let us think of other possible cases. What about the case of being markedly repetitive, but hardly, if at all, semantically marked? This means that the text repeatedly uses some groups of words, referring to a specific content. The groups are used much more often than they are used elsewhere. But they are just as interchangeable as they are elsewhere; none of them is semantically marked. So perhaps the text may be said to be non-formulaic? One may say, perhaps, that it does not use language in a limited, constricted fashion, that it simply talks over and over again about the same topic. But this first reaction is misleading, and again some fuzziness must be admitted. For here quantity becomes quality. By repeatedly using the same groups of words *much* more (a doubly fuzzy relation) than elsewhere, at least a quasi-formulaic effect is created. The relevant groups of words become all too well known. They are glanced over, processed away through sheer familiarity. They numb our semantic apparatus: we no longer reconstruct them from their constituents, we read them off as unanalysed wholes. While originally compositional, they are read *as if* non-compositional – and therefore they are not unlike semantically marked formulae. This does not fall under Parry's definition – but then the Homeric text (while repetitive in its way, with repeated themes) is nowhere as repetitive in subject matter as Greek mathematics is.

Here is my definition, then: I count a group of words as formulaic if it is semantically marked OR it is *very* markedly repeated – a non-exclusive disjunction. One corollary of the definition should be pointed out immediately: it allows one-word-long formulae – so we have come a long way indeed from the Homeric starting-point.

The definition offers symptoms, means for diagnosis.[13] The essence behind them is different, and I will not attempt to define it here. This is a cognitive entity, having mostly to do with the non-compositional, holistic parsing of certain groups of words: it will emerge as we proceed.

[13] And it offers them in a particular context. In the general field of 'repeated patterns', different means for diagnosis will be useful with different patterns. The definition is not meant to be universal.

2 GREEK MATHEMATICAL FORMULAE: A TYPOLOGY

I have already said that one of the lessons of the Homeric scholarship since Parry is the need to look carefully at the different types of formulae. Mathematical formulae can be analysed into five distinct, each fairly homogeneous types – in itself a mark of how simple the system is. This is the taxonomy. To this I add a typology based on 'parameters' – no longer discrete types, but dimensions expressing features which formulae may have to varying degrees.

2.1 *The taxonomy*

2.1.1 *Object formulae*

The main group of formulae is that of objects: the diagrammatic 'things'. In the most important cases, these are made up mainly of letters.

As explained in the preceding chapter, the common expression for 'a point' in Greek mathematics is not *sēmeion* – what the dictionary offers for 'point' – but rather τὸ Α, 'the (neuter)[14] + {letter}'. 'τὸ Α' is a minimal formula. It is semantically marked (in the wider context – i.e. anywhere outside mathematical works – the noun will not be elided, and in this wider context it may be interchanged with *stigmē*, for instance). By its ellipsis, it is non-compositional (the neuter hints at the noun, but does not supply it fully. Note that this is a very mild non-compositionality). It is very markedly repetitive (though the repetitiveness results from subject matter).

The Homeric system fits an epithet for many combinations of character and metrical position,[15] and the Greek mathematical system fits an object formula for the most important 'characters' of Greek mathematics. Naturally, articles run out quickly, but there are prepositions as well, and thus a rich system is constructed:

1. τὸ Α – 'the (neut.) {letter}' – point
2. ἡ ΑΒ – 'the (fem.) {2 letters or more}' – line
3. ἡ ὑπὸ τῶν ΑΒΓ – 'the (fem.) by the (gen. pl. fem.) {3 letters}' – angle
4. ἡ πρὸς τὸ Α – 'the (fem.) next to the (neut. acc.) {letter}' – angle
5. ὁ ΑΒΓ – 'the (mas.) {3 letters or more}' – circle

[14] The Greek article has different forms for the three genders (feminine, masculine and neuter).
[15] See e.g. Hainsworth (1968) 8–9.

6. τὸ ΑΒ – 'the (neut.) {2 letters/4 letters or more}'[16] – area
7. τὸ ΑΒΓ – 'the (neut.) {3 letters}' – triangle
8. τὸ ἀπὸ τῆς ΑΒ – 'the (neut.) on the (gen. fem.) {2 letters}' – square
9. τὸ ὑπὸ τῶν ΑΒΓ – 'the (neut.) <contained> by the (gen. fem.) {3 letters}' – rectangle
10. ὁ Α/ΑΒ – 'the (mas.) {1/2 letters}' – number

These are the most important object formulae. They are characterised by the fact that *everything* save article, preposition and letter may be elided. Sometimes fuller versions are given, and then the same formulae become fully compositional.

The first striking fact about this system is its generative nature: 3, 8, 9 use, as one of their building-blocks, the formula 2. For instance, 3, 'the (fem.) by the (fem. gen. pl.) {3 letters}' is, when fully unpacked, 'the angle contained by *the lines AB, BC*'. The italicised element is formula 2. Indeed, in a sense, all geometric formulae of this sort build upon the formula for a point. All these formulae include a 'variable' element, letters standing for points.

The generative structure explains the slight extent to which the system is uneconomic. 'Angle' is covered by two formulae, 3 and 4. Formula 3 is generated from the formula for 'line', 4 is generated from the formula for 'point'. Whatever the origins of the system, it is not motivated by a conscious effort to reduce to one-formula-per-concept.[17] However, it is quite economical and, especially, it avoids 'homonyms'. To explain: 3 and 4 are 'synonyms' – two different formulae expressing the same object. Homonyms will be the same formula referring to two different objects. These are avoided *throughout the literature*. This can be seen as follows. In book II of the *Elements*, the word '*gnōmōn*' is unabbreviable. The expression is throughout

11. ὁ {3 letters} γνώμων 'the gnomon {3 letters}',[18]

not

*11. 'The (mas.) {3 letters}'.

[16] Areas may be conceptualised either through (in the quadrilateral case) two opposite vertices, or through their entire vertical circuit (the distinction from triangles should be kept clear: the elided noun in areas is χωρίον, while in triangles it is τρίγωνον).

[17] Note also that formulae 3 and 4 coexist in the same works: e.g. *Elements* II, 148.25–6: formula 3; 154.1–2: formula 4.

[18] This is semantically unmarked, but in its own context it is sufficiently markedly repetitive to count as a formula.

This is because *11 would result in a homonym, for *11 is the same as 5 above, circle. But the crucial thing to notice is the context in which 11 occurs, *Elements* II. This is interesting, because the context of *Elements* II *makes no reference to circles*. It is the system which is universally felt, not its individual constituents.[19]

As we begin to see, there are more specialised object formulae, and some of these do not involve drastic ellipsis. One of the more elliptic is

13. ἡ ἐκ τοῦ κέντρου τοῦ {3 letters or more} κύκλου, 'the <line drawn> from the centre of the {3 letters or more} circle <to its circumference>'.

This is the common expression for 'radius'. Here ellipsis yields strong non-compositionality, and this is a relatively rare case, where a Greek mathematical formula would have been identifiable even in isolation. It is qualitatively semantically marked, in a strong way.

Other cases are almost compositional: for instance, the expression

14. τὸ διὰ τοῦ ἄξονος τρίγωνον, 'the triangle through the axis'.

The verb is elided, and should be, perhaps surprisingly, 'cut'.[20] In the context of the beginning of Apollonius' *Conics* I, where axes of cones and planes cutting them are constantly referred to, the sense is clear, and the expression feels natural: repetitive, no doubt,[21] but clear. In Archimedes' *SC* I.16, however, the setting-out starts with 'let there be a cone, whose triangle through the axis is equal to ΑΒΓ',[22] and here the sense can not be given by the context. In this more alien context, the formula is revealed much more sharply. This is the general rule: idioms which the native speaker does not notice strike the alien, and formulae are most noticeable when they are 'alien', when they appear outside their most natural context.

[19] Similarly, in Archimedes' *SL*, '*chronos*', 'time' is unabbreviable: the expression is throughout

12. ὁ {2 letters} χρόνος, 'the time {2 letters}',

not

*12. 'The (mas.) {2 letters}'.

This is meant to prevent homonymy with 'number', which again is not a formula used at all in *SL*.

[20] This is supplied from the context in which the term is introduced by Apollonius, *Conics* I.5 16.25–7.

[21] I have counted 15 occurrences in *Conics* I. pp. 16–48.

[22] *SC* I.70.9–10.

In *Elements* II, which I have selected as a case-study (admittedly, as a simple, short text), there are 15 object-formulae, which include formulae 1, 2, 3, 4, 6, 7, 8, 9 and 11 above, as well as 6 others.[23]

2.1.2 *Construction formulae*

Objects are constructed, and they are constructed in specific ways, with specific verbs, in specific tenses and moods:[24] in a word, constructions have their formulae. Obviously, these formulae have a structure which consists, among other things, of specific object-formulae.

Parallels, for instance, are almost always drawn with the following matrix:

> 21. διὰ {formula 1: point, in the genitive} {formula 2: line, in the dative} παράλληλος ἤχθω {formula 2: line, in the nominative}, 'let {formula 2: line} a parallel to {formula 2: line} be drawn through {formula 1: point}'.

The verb 'to draw' may occur in other structures as well, e.g.

> 22. ἤχθω {formula 2: line} ἐπιψαύουσα, 'let the tangent {formula 2: line} be drawn'.

In a case such as 22, the formula is too brief to allow semantical markedness, and in some cases there is no semantical markedness. This is most clear in

[23] As there may be an interest in the list of formulae in *Elements* II, I shall give in a footnote, at the end of each description of type, the formulae belonging to this type. To save space, I shall not translate the added formulae in these footnotes.

Object formulae of *Elements* II are:

A. Specific to the universe of book II:

15. τὰ περὶ τὴν διάμετρον παραλληλόγραμμα
16. τὰ δύο παραπληρώματα.

B. Types of angles:

17. ἡ ἐκτὸς γωνία
18. ἡ ἐντὸς καὶ ἀπεναντίον γωνία
19. αἱ ἀπεναντίον αἱ {either of formulae 3/4}.

C. Finally,

20. ἡ ὅλη, 'the whole'

is repeatedly used instead of 'the whole line'. I see this systematic abbreviation as a basis for granting formulahood, but this brings us face to face with the fuzziness of the definition.

[24] The distinction between subjunctive (in the general enunciation, 'if a line *is cut* . . .') and imperative (in the particular construction and setting-out, '*let* a line *be cut*') does not, however, change the structure and function of such formulae, and therefore I see the two as two different morphological expressions of the same formula. But this is a delicate decision (and I will deal differently with the indicative, as we shall see below with predicate formulae, e.g. 71, 31).

23. ἔστω {any unlettered object formula} {a lettered equivalent}, 'let there be {some object} <namely> {some lettered object}'.

This is certainly very markedly repetitive, but is also a very transparent piece of Greek.

The construction formulae in *Elements* II include 21, 23 above, as well as 16 others.[25]

2.1.3 Second-order formulae

So far one distinctive feature of the types of formulae has been their form: noun phrases, often lettered, in object formulae; verb phrases, often in the imperative, in construction formulae. The remaining groups are defined by subject matter. The group of second-order formulae is very distinct. This is to be expected, since, as we saw in chapter 3 above, the first-order/second-order distinction is central to the Greek mathematical language. However, these formulae occur within the main, first-order discourse, not in the second-order introductions to works (such introductions are relatively speaking unformulaic). Thus, second-order formulae, unlike second-order *words*, are as regimented as their first-order counterparts.

[25] Very markedly repetitive but not semantically marked:

24. κείσθω {any object formula in the dative} ἴσος {any object formula}
25. ἐκβεβλήσθω {formula 2: line}
26. ἐπεζεύχθω {formula 2: line}
27. συνεστάτω {any figure}.

A particular development of 25:

28. ἐκβεβλήσθωσαν καὶ συμπιπτέτωσαν κατὰ {formula 1: point} (II.10 148.10–11).

Less compositional are:

29. ἀναγεγράφθω ἀπὸ {formula 2: line, in the genitive} τετράγωνον {formula 6: area}
30. κέντρῳ μὲν {formula 1: point} διαστήματι δὲ {formula 2: line} γεγράφθω {formula 5: circle}
31. προσκείσθω αὐτῃ εὐθεῖα ἐπ' εὐθείας (146.24)
32. καταγεγράφθω τὸ σχῆμα (136.5).

The 'cutting' family, especially important in this book:

33. τετμήσθω {formula 2: line} κατὰ {formula 1: point}
34. τετμήσθω {formula 2: line} δίχα
35. τετμήσθω {formula 2: line} εἰς μὲν ἴσα κατὰ {formula 1: point}, εἰς δὲ ἄνισα κατὰ {formula 1: point} (II.5 128.23–130.1)
36. τμηθῇ εἰς ὁσαδηποτοῦν τμήματα (II.1 118.10–11).

The 'drawing' family, which includes formula 21 above and:

37. διήχθω {formula 2: line} ἐπὶ {formula 1: point}
38. ἤχθω {formula 2: line} πρὸς ὀρθὰς {formula 2: line, in the dative}
39. ἤχθω ἀπὸ {formula 2: line, in the genitive} ἐπὶ {formula 1: point} κάθετος {formula 2: line}.

Second-order formulae occur as signposts within arguments. For instance, in the structure known as analysis and synthesis, the mathematician does the analysis, and then most often introduces the synthesis by:

40. συντεθήσεται δὴ τὸ πρόβλημα οὕτως, 'so the problem will be "synthesised" as follows'.[26]

This formula occurs within a specific structure. Freer, 'floating' second-order formulae include:

41. ὁμοίως δὴ δείξομεν, 'so similarly we will prove . . .',
42. διὰ τὰ αὐτὰ δή, 'so through the same'.

Both, of course, can occur in many contexts.

It is typical that second-order formulae come armed with specific particles. This reflects the technical use of particles, described in the preceding chapter, and it is such features which make these expressions (in themselves quite natural) semantically marked.

So far we saw simple cases: the second-order expression demanded some complement, but it consisted of a continuous stretch of Greek. Other, more complicated cases are second-order formulae constituted mainly by their complement, which is *infixed* rather than suffixed to them. One such case is the Euclidean

43. . . . ἄρα . . . (the enunciation, repeated verbatim). ὅπερ ἔδει δεῖξαι: 'Therefore . . . QED'.

The formula is a matrix: a syntactic, no less than semantic, unit. It is in such a context that the famous

44. λέγω ὅτι, 'I say that'

should be understood. The formula is constituted not only by its two words, but also by what precedes it (setting-out) and what follows it (definition of goal). On a higher level still, it is possible to see the entire proposition structure as such a matrix, with 43 and 44 as two of its constituents. It is through formulae that the structure of the proposition should be approached. I shall return to this point later on.

[26] E.g. *SC* II.1, 172.7.

Elements II contains 12 second-order formulae, including 42, 43 and 44 above.[27]

2.1.4 Argumentation formulae

These are expressions validating an argument. Their essence is that they combine assertions in a fixed matrix in which the result is known to derive from the premisses.

A central group of such formulae covers arguments in proportion theory. All these formulae are generated from the basic formula for proportion, which in turn is generated from the formula for ratio. As an introduction, therefore, I put forward the following two formulae:

54. ὁ λόγος {any object in the genitive} πρὸς {any object in the accusative}, 'the ratio of {any object} to {any object}'.

54 is an object formula.

55. ὡς {formula 54: ratio} οὕτως {formula 54: ratio}, 'as {ratio} so {ratio}'.

55 belongs to another type of formula, which I will describe below. It will also be seen that the words ὁ λόγος are regularly elided in this formula.

We can now move on to the basic group of argumentation formulae in proportion theory:

56. {formula 55: proportion} καὶ ἐναλλάξ {formula 55: proportion} – 'alternately',

[27] Three formulae tied to specific tasks:

45. ἐδείχθη δὲ {clause in the indicative}
46. ἐκ δὴ τούτου φανερόν, ὅτι {clause in the indicative}
47. γέγονος ἂν εἴη τὸ ἐπιταχθέν (160.15).

Two formulae are among the main matrices of definition:

48. {noun phrase} λέγεται {verb phrase} (Def. 1)
49. {noun phrase} {noun phrase} καλείσθω (Def. 2).

The two main matrices for the enunciation are:

50. ἐὰν {clause in subjunctive} {clause in indicative}
51. ἐν {noun phrase} {clause in the indicative}.

Problems appear here always with the same structure, which involves two matrices:

52. The matrix for the enunciation is: {a clause in the genitive absolute} {a clause governed by an infinitive}
53. The matrix for setting-out is: {a series of verb formulae} δεῖ δὴ {a clause governed by an infinitive}.

57. {formula 55: proportion} καὶ ἀνάπαλιν {formula 55: proportion} – 'inversely',
58. {formula 55: proportion} καὶ συνθέντι {formula 55: proportion} – 'in composition',
59. {formula 55: proportion} καὶ διέλοντι {formula 55: proportion} – 'separately',
60. {formula 55: proportion} καὶ ἀναστρέψαντι {formula 55: proportion} – 'conversely'.

All these formulae above involve also a specific relation between the their constituent formulae. Let us analyse, for instance, 56.

56 =
ἐπεὶ 55 καὶ ἐναλλάξ 55 =
ἐπεὶ ὡς 54 οὕτως 54 καὶ ἐναλλάξ, ὡς 54 οὕτως 54 =
ἐπεὶ ὡς {first object} πρὸς {second object} οὕτως {third object} πρὸς {fourth object} καὶ ἐναλλάξ, ὡς {first object} πρὸς {third object} οὕτως {second object} πρὸς {fourth object}

The whole group 56–60 can be analysed similarly, though only 56–57 can be described 'syntactically', without reference to the *contents* of the objects.

Besides their rich internal structure, such formulae stand in a structural relation to each other. It is possible to conceptualise this group as a single formula, with five different manifestations, or to see the five formulae as equivalent transformations of each other. I will return to this later on.

Argumentation formulae are often richly structured: they correlate rich contents. They are thus not unlike some of the matrices described for second-order formulae. And in fact the following argumentation formula seems almost like a second-order matrix, so fundamental is it:

61. εἰ γὰρ δύνατον, ἔστω {some property}. {some argument} {some property} ὅπερ ἔστιν ἀδύνατον/ἄτοπον. οὐκ ἄρα {the first property}. ἄρα {the negation of the first property}, 'For if possible, let {some property} be. {some argument} {some property, considered impossible/absurd} which is impossible/absurd. Therefore not {the first property}. Therefore {the negation of the first property}'.

This is the *reductio*. To understand this form of argument, the first thing is to put it in its wider context of Greek mathematical argumentative formulae.

We see that argumentation formulae may be more or less general, more or less content-specific. A formula covering a relatively wide range of contents is:

62. {context: equality of objects 1 and 2}. κοινὸν προσκείσθω/ἀφῃρέσθω {some object 3}. ὅλον/λοιπὸν ἄρα {object made of 1 plus/minus 3} ἴσον ἐστί {object made of 2 plus/minus 3 in the dative}: {1 equals 2} 'let 3 be added <as> common, therefore the whole/remainder {1 plus/minus 3} is equal to {2 plus/minus 3}'.[28]

Some argumentation formulae, on the other hand, are very content-specific. Therefore they can not be, in general, very markedly repetitive. But they are semantically marked: they repeat the grounds for the move in a very specific way, from which there is little deviation. In other words, they may be compared to quotations of the grounds for the move. For instance, consider the following:

63. ἐπεὶ {formula 35: cutting a line into equal and unequal segments, expressed in the indicative instead of the imperative} ἄρα {formula 9: rectangle, its two lines being the two unequal segments} μετὰ τοῦ {formula 8: square, its line being the difference between the unequal and the equal cut; in the genitive} ἴσον ἐστι {formula 8: square, its line being the equal section; in the dative}.

This is what we express by $(a + b)(a - b) = a^2 - b^2$. Our neat typographic symbol is expressed, in the Greek, by a baroque structure of formulae.[29]

What makes the formula most strongly felt is the fact that it is not the general result, proved elsewhere, which is referred to. It is the general result expressed in the particular terms of the case at hand. The Greek mathematician does not pause to say 'and when a line is cut into equal and unequal segments, etc.'. Rather, he says 'and since the line *AB* was cut into equal segments at *D*, into unequal segments at *C*...'.

[28] In this case I did what was only suggested for formulae 56–60 above: I collapsed the two formulae, for addition and for subtraction. I saw the two as manifestations of a single, higher formula.

[29] Bear in mind that each of the formulae 35, 9, 8 is itself a structure of formulae. To express this in term of brackets (a representation I shall explain below), the formula is:

63[35[2,1,1],9[2],8[2],8[2]]

Even this is a simplification, however, since it does not list the matrix for equality as a formula.

Thus, the use of the general result is not a quotation: it is a formula. When a Greek proves a general result, what he does is to validate a *matrix*, in which particular objects are, from that moment onwards, allowed to be fitted.

Book II of the *Elements* contains nine argumentation formulae, including 62 and 63 above.[30]

2.1.5 *Predicate formulae*

The remaining group is formulae denoting predicates (in the wide sense, including relations).

I have already given above a few of these formulae (which, naturally, are often constituents in larger formulae): first, 55, the all-important formula for proportion. This signifies a relation between two ratios, and we saw it as a constituent in formulae 56–60.

I have also mentioned, without numbering it:

71. εἰς αὐτὰς ἐμπέπτωκεν {formula 2: line}, '{formula 2: line} meets them'.

[30] Three other formulae of the same type as 63:

64. What we express by $(a + b)^2 = a^2 + 2ab + b^2$. ἐπεὶ {formula 33: cutting a line} ὡς ἔτυχεν, ἄρα {formula 8: square} ἴσον ἐστὶ {formula 8: square} καὶ δὶς {formula 9: rectangle} (based on prop. II.4, used in 156.6–9).
65. What we express by $b(2a + b) + a^2 = (a + b)^2$. ἐπεὶ {formula 34: bisecting a line}, {formula 31: adding a line} ἄρα {formula 9: rectangle} μετὰ {formula 8: square} ἴσον ἐστὶ {formula 8: square} (based on prop. II.6, used in 152.21–4).
66. What we express by $(a + b)^2 + b^2 = 2b(a + b) + a^2$. ἐπεὶ {formula 33: cutting a line} ὡς ἔτυχεν, ἄρα {formula 8: square} ἴσα ἐστὶ τῷ τε δὶς {formula 9: rectangle} καὶ {formula 8: square}.

A similar formula is:

67. ἐπεὶ παράλληλός ἐστιν {formula 2: line} {formula 2: line}, καὶ εἰς αὐτὰς ἐμπέπτωκεν {formula 2: line}, {formula 17: external angle} {formula 3: angle} ἴση ἐστὶ {formula 18: internal angle} {formula 3: angle}. This is based on I.29, and the formula occurs in 126.4–6.

What is in fact the other half of the same formula is:

68. ἐπεὶ παράλληλός ἐστιν {formula 2: line} {formula 2: line}, καὶ εἰς αὐτὰς ἐμπέπτωκεν {formula 2: line}, {formula 3: angle} ἄρα δύσιν ὀρθαῖς ἴσαι εἰσιν.

This occurs in 148.4–6 with a somewhat different phonological form. I shall discuss the variability of formulae below. 67 and 68 show that phrases such as εἰς αὐτὰς ἐμπέπτωκεν {formula 2: line} are themselves formulae: they belong to the next and final type.

Two simpler formulae (involving no such matrices, at least in the sense that they do not infix so much) are:

69. παραπληρώματα γὰρ τοῦ {formula 6: area} παραλληλογράμμου (e.g. 140.7–8, arguing for the equality of the complements – a very local formula)
70. ἐναλλὰξ γάρ (used for the equality of alternate angles, 148.19).

This formula signifies a relation between three lines. We saw it as a constituent of 69–70 above. We also saw:

72. πρὸς ὀρθάς, 'at right angles',

a constituent of 38.

Moving from relations to predicates in the strict sense, we saw

73. ὡς ἔτυχεν, 'as it chances'.

This formula signifies a manner for an operation; it is thus naturally a constituent in construction formulae.

Another adverbial formula which we have seen already is:

74. ἐπ' εὐθείας {formula 2: line, in the dative}.

This is a constituent of formula 31 above.

We have also seen the following relational matrices:

75. {object} ἴσον ἐστὶ {object} μετὰ {object}
76. {object} ἴσον ἐστὶ {object} καὶ {object}
77. {object} ἴσον ἐστὶ {object} τε καὶ {object},

which are three ways of representing $a + b = c$, and also

78. {object} ἴσον ἐστὶ {object},

which is simply $a = b$.

Here it will be objected that I describe what is a normal distribution of a Greek semantic range. But my definition takes account of this. Formulae 75–78 are only slightly semantically marked (see below), but they are extremely markedly repetitive in Greek mathematics. They are responsible for a large proportion of the text of the proofs. For instance, of the 104 words of the proof of *Elements* II.1, 61 occur within matrices of equality.

Another important set of matrices is that of identity, indeed very simple in its formulation:

79. {object} {object} (used as a full clause)
80. {object} τουτέστιν {object} (used parenthetically within a clause).

Matrices for equality and identity are what Greek mathematics mostly deal with. They are transparent, but they are so repetitive as to be formulaic – and they are felt as formulaic within the larger formulaic context.

Elements II contains 17 predicate formulae.[31]

To sum up, then, we have seen five groups: object formulae, construction formulae, second-order formulae, argumentation formulae and predicate formulae. Very occasionally, I will call these O, C, S, A or P. This allows the use of a simple code: for instance, predicate formula 80 may be represented as P_{80}.

We have repeatedly referred to various parameters. Before moving onwards, I will recapitulate these parameters as well.

2.2 *Parameters for formulae*

2.2.1 *Markedly repetitive? semantically marked? non-compositional?*

We saw that formulaic status may derive both from semantical markedness and from marked repetitiveness. But we also saw that the semantic markedness tends to be quite weak. When formulae are non-compositional, this is almost always due to ellipsis. The degree of non-compositionality is the degree of ellipsis. Formula 13, ἡ ἐκ τοῦ κέντρου τοῦ {3 letters or more}/κύκλου, 'the <line drawn> from the centre of the {3 letters or more}/circle <to its circumference>', is a highly elliptic formula. It is also heavily semantically marked. But even here, the general practice makes it very easy to supply the noun 'line'

[31] These include a matrix, almost as important as that of equality, namely inequality:

81. {object} μείζων {object in the genitive} {object in the dative}.

A local, illuminating variation on the matrices of equality is:

82. {object} σὺν {object}.

This is used in II.6 132.8–9 and elsewhere to signify precisely that the addition is *not* intended within a matrix of equality; that the addition is simply a composition, creating a new object. The formula works by being different, by using σύν instead of μετά. If one is meant to distinguish the two concepts, these two Greek prepositions are a happy choice. But they do not signify it in themselves. It is the entire semantic structure which signifies the difference – and, in such ways, innocent prepositions such as these do begin to be somewhat semantically marked.

A local development of the matrix of equality is:

83. πλευρὰ {formula 2: line} πλευρᾷ {formula 2: line} ἴση.

Simpler cases are the relation:

84. ἐπὶ τὰ {optional: αὐτὰ/ἕτερα, or some object} μέρη (148.9–10),

the adverbial predicate:

85. ὡς ἀπὸ μίας (138.11–12)

and these widely important predicates:

86. ὀρθή
87. δοθείς.

These two are very markedly repetitive in Greek mathematics; also, through ellipsis, 86 is mildly non-compositional.

for the feminine article (after all, formula 2, for line, is easily the most common one in Greek mathematics). So the formula may perhaps be translated as 'the line from the centre of the circle', which after all is not so opaque as a 'radius'.

Very rarely, ellipsis is joined by a metaphor of kind. Perhaps the most prominent case of 'metaphor' is

88. {formula 2: line} δυνάμει, literally 'line, in potential'.

This is a predicate formula, roughly speaking the square on a line. How 'potentiality' denotes 'squareness' is an open question, which I will not tackle here.[32] It has attracted a lot of attention just because it is such a rare case of what seems like metaphor in Greek mathematics. In general, formulae – just like the lexicon itself – do not work through coinage, i.e. through the creation of original metaphorical concepts. They work through the twin processes (paradoxically twinned!) of repetition, on the one hand, and ellipsis, on the other. They start from natural language, and then employ a small subset of it. A glaring contrast to the Homeric system, then. Not for us rose-fingers, the riches of the Homeric language. Mathematical formulae are the children of poverty.

2.2.2 *Hierarchic structure*

Almost all formulae we saw included an element of 'variables', represented by {}. In a few simple cases, these variables are diagrammatic letters. In other cases, the variables are very general, such as {object}. Most often, at least some of these variables are formulae themselves. This results in a hierarchic structure. It is most naturally described by a 'phrase-structure' tree.[33] To take a very simple case, involving formulae 2, ἡ AB (the (fem.) {2 letters or more} = line), and 3, ἡ ὑπὸ τῶν ΑΒΓ – (the (fem.) by the (gen. pl. fem.) {3 letters} = angle):

or a more complicated case, that of formula 21. Its structure is:

21. διὰ {formula 1: point, in the genitive} {formula 2: line, in the dative} παράλληλος ἤχθω {formula 2: line, in the nominative}, 'let {formula 2: line} a parallel to {formula 2: line} be drawn through {formula 1: point}'.

[32] See, e.g. Szabo (1969), I.2; Knorr (1975), III.1; Hoyrup (1990b).
[33] The term – as well as the tree itself – is borrowed from linguistics.

This corresponds to the tree:

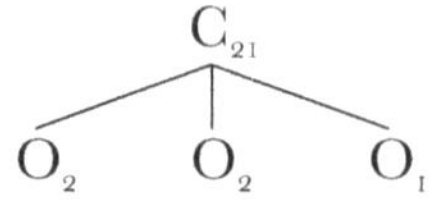

To save space, linguists often use an equivalent non-graphic, linear bracketing presentation, e.g. $C_{21}[O_2,O_2,O_1]$ (what is immediately outside the brackets governs the contents of the brackets).

Fig. 4.1 is the tree of *Elements* II.2, an entire proposition. The typical branching involves no more than two or three branches. In other words, a hierarchic formula correlates two or three simpler formulae. With the set of formulae given above, it is possible to make similar trees for the entirety of book II. It is all governed by the same limited number of recurring, hierarchically structured formulae.

2.2.3 *Contextual constraints*

In some cases formulae have formulae as constituents because these constituent formulae are written into them (thus, the formulae for line and point are written into the formula of drawing a parallel); or they may have formulae as constituents by chance, as it were: the formula demands some complement (not necessarily a given formula), which happens to be a certain formula. This is for instance the case in the matrices for equality, etc. Since expressions representing objects in Greek mathematics are formulaic, these matrices will govern formulae. So, besides the parameter of hierarchic structure, another important parameter is the strength of contextual constraints. The contextual constraints of 'draw a parallel' are strong: it occurs only with 'line, line, point'.[34] The contextual constraints of 'is equal' are minimal. These parameters have little resemblance to the Homeric case. In a limited way, it is possible to use phrase-structure trees to analyse Homeric patterns of formulae.[35] But this has not been pursued far, not even by those Homerists who set out to apply modern linguistics.[36] Why is

[34] Even this does not *determine* the context, of course: the lines and points may vary. Most significantly, they may, or may not, be lettered.

[35] See, e.g. the sentence-pattern discussed in Hainsworth (1968) 15.

[36] Nagler (1967, 1974) uses generative grammar mainly in the distinction between deep-structure and surface-structure – which, incidentally, has meanwhile been rejected by Chomsky. May this serve as a warning: never apply the most recent theories! More recently, however, some work was done on Homeric *patterns*, e.g. Visser (1987) and (following him) Riggsby (1992). (Kiparsky 1974 – the most competent discussion of the subject from the point of view of generative linguistics – is really an analysis of idioms, not of formulae, and has not been taken up in Homeric scholarship.)

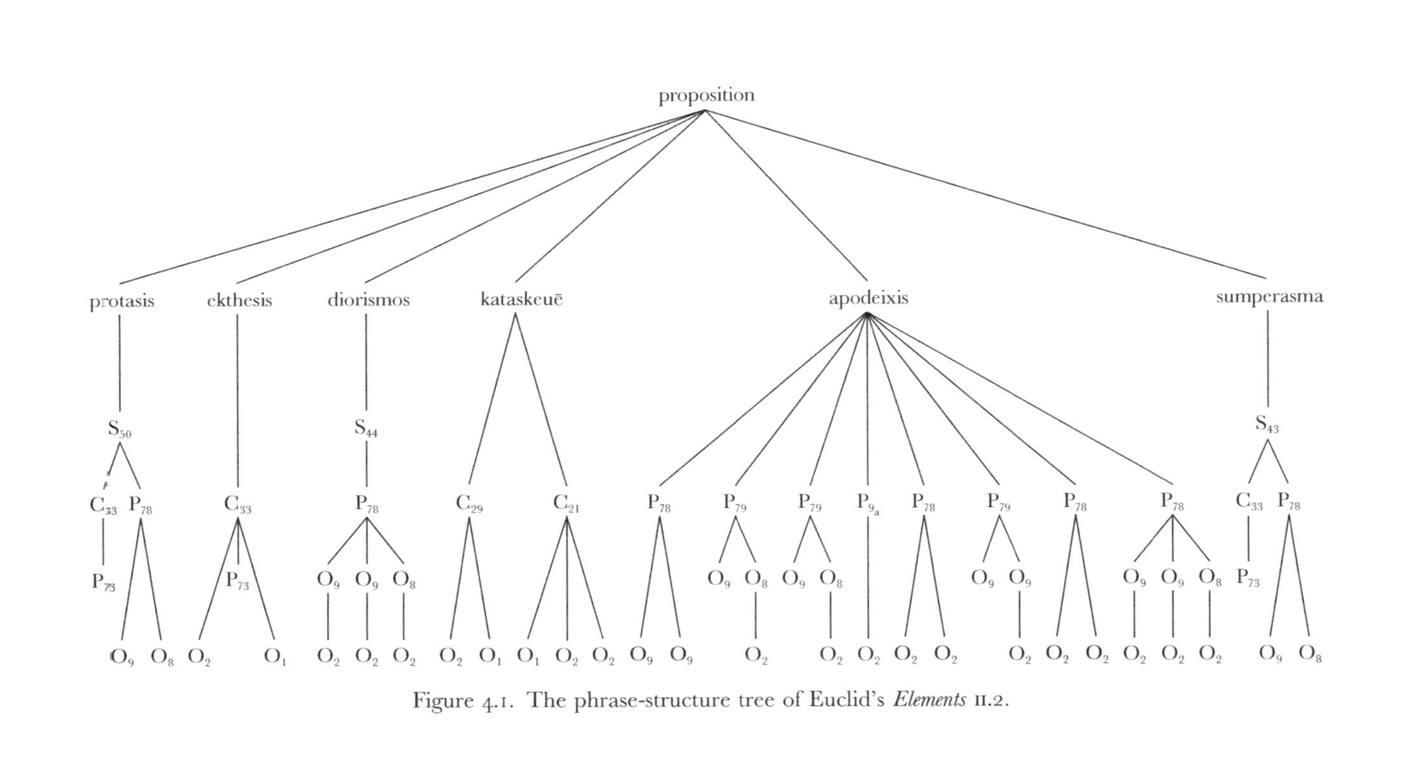

Figure 4.1. The phrase-structure tree of Euclid's *Elements* II.2.

that? It should be clear that whatever humans produce has structure which can be represented by such phrase-structure trees. People do not stop after every act (or word) to consider what their next one will be: they proceed by following structures which they fill up with acts (or words). The point is that the Homeric structure is *prosodic*. The Homeric author thinks in lines which break down into prosodic units (known as *cola*). The only relevant 'tree', therefore, is that of the line governing the cola, each colon governing (optionally) a formula (Homeric formulae are generally colon-sized). Such a representation tells us nothing beyond the fact, which we knew all along, that the text consists of lines and cola.

But the mathematical case is different. Because mathematics is written as prose, the relevant units are syntactic, and not prosodic. Therefore much greater 'depth' is possible in the trees, and the hierarchical structure becomes much more important. So some order begins to emerge: the Homeric formulae are based on prosodic form, whereas the mathematical formulae are based on syntactic form. But we have already moved from typology to analysis of behaviour.

3 THE BEHAVIOUR OF FORMULAE

3.1 The flexibility of formulae

I have made a survey of 50 occurrences of formula 89 (all the cases in Archimedes, and in Apollonius' *Conics* I. The round number is accidental). I explain the survey and the formula in a footnote.[37]

I have found 11 variants (out of the possible 64). The canonical, full version, while by far the most common, is responsible for no more than about a third of the occurrences.

But is there any regularity? Does the term 'canonical form' mean anything? My answer is positive. The variations, in the great majority of cases, take the form of ellipsis, not of substitution. In most occur-

[37] 89. (ὁ λόγος σύγκειται) ἐκ [τε] [τοῦ] [ὃν ἔχει] {X} πρὸς {Y} καὶ [ἐκ] [τοῦ] [ὃν ἔχει] {Z} πρὸς {W}.

The phrase-structure is $P_{89}[O_{54},O_{54},O_{54}]$. I collapse the constituent formulae of ratios into the formula itself, which is useful for the description of variability.

The formula has six 'degrees of freedom': each square-bracketed element is optional (so there are 64 possibilities). This is like a binary number: I attached to each occurrence its binary number, and checked, of course, for other possible variations (which turn out to be much rarer).

rences, some of the elements of the formulae are elided. But a substitution of one element for another is much rarer, occurring only 7 times out of my 50. Most commonly, therefore, the mathematician deviates from the formula in ellipsis only. He does not go through all the motions, but only through some. If indeed formulae are parsed as wholes, this is natural: why say more when you can get the same effect more economically?

Significantly, however, the Greek mathematician almost always goes through at least *some* of the motions. Of my 50 formulae, only one is a zero formula, one in which all the optional items have been left out. All the cases where the mathematician chose a more-than-zero formula are cases of redundancy. Once the motivation for ellipsis is understood, one is rather surprised by the tendency *not* to be brief. And why are full formulae so common?

The main distinction inside propositions is between the letterless, general enunciation and the rest of the proposition, structured around the proof. The two foci of the propositions are the general and the particular. These foci extend their fields of gravity. Thus, it is certainly no accident that Apollonius' *Conics* 1.39–41, where the full form of the formula is so common (10 out of 15, with 2 almost-full occurrences) are also the only cases in my survey where the formula occurs in the enunciation. The enunciation is where language is most self-consciously used, where it is not yet lost in the heat of the discursive argument, so formulae are more strongly adhered to there.

What, then, is the identity of the formula? We should separate here two questions. One is, what the mathematician *thought* that the formula was. Another is, what the formula actually was. Very probably, the Greek mathematician would say that the formula was the full version. I even imagine Apollonius would have been surprised to know that he used it so little. When the formula is being deliberately, self-consciously written, the full version is employed. But in reality, the formula is not a fixed sequence of words, it is not a phonological unit. It is a matrix within which words are fitted, and very often some of the slots are not phonologically fulfilled. Indeed, in some cases, some phonological values in the slots may be substituted for others. And this, most probably, happens without the mathematician even noticing it. The presence of such small, hardly noticed variations is a fact which must qualify all my description so far. Again: we do not see the strict rigidity of the Style Manual, of the proofread text.

3.2 *The productivity of formulae*

Formulae are productive in more than one way. First, formulae beget formulae: a formula may transform into another formula. Second, formulae may be created from non-formulaic materials. The two main practices through which such formulae are created – indeed, from which all formulae are ultimately created – are ellipsis, on the one hand, and repetition, on the other.

3.2.1 *Transformations on formulae*

This of course is related to the flexibility of formulae which was described above.

Formulae are syntax-sensitive. When the formula for 'line' is used in the genitive position, it changes from ἡ {2 letters} to τῆς {2 letters}. This is a different production of the same formula: there is nothing distinctive about the genitive form: it is not a specific form, having its own separate meaning. This is similar to some transformations from verbal to nominal form. For instance, the simplest construction formula in the 'draw' family is:

90. ἀπὸ {formula 1: point} ἐπὶ {formula 1: point} ἤχθω {formula 2: line}, 'let a {line} be drawn from {point} to {point}'.

This may transform into:

90a. {formula 2: line} ἀπὸ {formula 1: point} ἐπὶ {formula 1: point} ἀγομένη, 'the {line} drawn from {point} to {point}'.

90a is a transformation of 90: it is probably best seen as a distinct formula, since it already belongs to the clearly defined group of object formulae.

Similarly, construction formulae may transform into predicate formulae. Transforming the imperative of the construction into the indicative yields a new formulae, one describing a property. We have seen:

31. προσκείσθω αὐτῇ εὐθεῖα ἐπ' εὐθείας, 'and let a line be added to it in a straight line' (*Elements* II.146.24).

This may transform into:

31a. πρόσκειται αὐτῇ {formula 2: line}, 'a line has been added to it' (*Elements* II.152.22).

Formula 31a is a constituent in the argumentation formula:

> 65. ἐπεὶ {formula 34: bisecting a line}, {formula 31: adding a line} ἄρα {formula 9: rectangle} μετὰ {formula 8: square} ἴσον ἐστὶ {formula 8: square} (*Elements* II.152.21–4)

which I have given above: it is not 31 itself, but its transformation, 31a, however, which is used there. And the same is true for other argumentation formulae. Argumentation formulae state the legitimate transition from one predicate formula to another, so when they argue from an action to its result, they transform the construction formula involved in the action into a predicate formula.

Another possible transformation is from object formulae to predicate formulae. For instance, the formula

> 9. τὸ ὑπὸ {formula 2: line}, 'the (neut.) <contained> by {line}'

may transform into

> 9a. περιέχεται ὑπὸ {formula 2: line}, 'it is contained by {formula 2: line}' (*Elements* II.120.6–7).

The hidden participle of 9, περιεχόμενον, 'contained', becomes visible in 9a.

In the cases above, it seems reasonable to view one of the forms as more basic than the other. However, we have already seen some cases where formulae are reciprocally related, in the case of, for example:

> 62. {context: 1 = 2}. κοινὸν προσκείσθω/ἀφῃρέσθω {3}. ὅλον/λοιπὸν ἄρα {1 plus/minus 3} ἴσον ἐστὶ {2 plus/minus 3}, '{1 = 2}, let 3 be added <as> common, therefore the whole/remainder {1 plus/minus 3} is equal to {2 plus/minus 3}'.

I have chosen to see two formulae as a single, variable formula. In the case of formulae 56–60 I have done otherwise:

> 56. {formula 55: proportion} καὶ ἐναλλὰξ {formula 55: proportion} – 'alternately'

and the following with the same structure but with a different adverb:

> 57. ἀνάπαλιν – 'inversely'
> 58. συνθέντι – 'in composition'
> 59. διέλοντι – 'separately'
> 60. ἀναστρέψαντι – 'conversely'.

These formulae may be seen as a single, variable formula; or they may be seen as a set of formulae, reciprocally transformable. And it is natural to assume that, diachronically, one of these formulae came first, the rest being created by analogy from the first.

3.2.2 Ellipsis

We saw that formulae may undergo ellipsis, and yet retain their formulaic status. Paradoxically, the process of ellipsis may also be the source of the formulaic status itself. For a test-case, I take the two Autolycean treatises, and the development of the concept of the horizon. For the purpose of the discussion I assume that the *Moving Sphere* is in some sense preliminary to the *Risings and Settings*.

The horizon first appears in *Moving Sphere* 4, where, in the enunciation, it is described as a

91. μέγιστος κύκλος, 'greatest circle'

(what we call 'great circle').[38] This is an object formula. To this is added the construction formula:

92. ὁρίζῃ τό τε ἀφανὲς καὶ τὸ φανερὸν ἡμισφαίριον τῆς σφαίρας, 'divides the invisible and the visible hemispheres of the sphere'.

Immediately afterwards in the same proposition the formula transforms, in the setting-out, to:

ὁριζέτω τό τε φανερὸν τῆς σφαίρας καὶ τὸ ἀφανές – 'let it divide the visible of the sphere and the invisible'.

Something has already been omitted – the reference to hemispheres.

In the enunciation of the next proposition the horizon is that which ὁρίζῃ τό τε φανερὸν καὶ τὸ ἀφανές – 'divides the visible and the invisible' – now the word 'sphere' has been left out; and in the very same sentence it is already called, simply,

93. ὁ ὁρίζων, 'the divider'.

Hence, of course, our 'horizon'. 93 is already an object formula and, through ellipsis, it has become non-compositional.

In the following propositions, 6–11, the horizon is consistently mentioned in the enunciation. It is usually used in a relatively full form ('sphere' included, 'hemispheres' not). It is also interesting to note that

[38] Already elliptic for μέγιστος κύκλος τῶν ἐν τῇ σφαίρᾳ κύκλων.

the sequence has by now been settled ('visible' first, which makes sense). In the process of each proposition, the expression is commonly abbreviated to the plain 'the divider'. In a single case, proposition 8, 'the divider' is used throughout.

In none of the enunciations of the *Risings and Settings* is the horizon mentioned. However, it is mentioned in practically all the propositions, and almost always as 'the divider'. Twice, in II.4 and in II.6, fuller forms are used. It is not as if the group of words 'the divider' has ousted the larger group. It is felt to be an abbreviation of a fuller expression, and I guess that, had it appeared in enunciations, expanded versions would have been more common.

The general rule is as follows: repetition creates formulae of marked repetitiveness. In such formulae of marked repetitiveness, ellipsis is natural – the unit is parsed as a whole, and therefore parts of it may become redundant. Ellipsis then leads to non-compositionality and to semantic markedness. This may form not only a new source of formulaicity, but a veritable new formula.[39]

3.2.3 The variability of formulae – and their origin

Above all, of course, formulae are produced through repetition. And so, a paradox seems to confront us: we have insisted on the flexibility of formulae; we saw the transformability of formulae; we have now shown the significance of ellipsis for the very production of formulae – and ellipsis is the opposite of rigid repetition. And yet repetition is the essence of formulae. What makes a formula is the fact that an author chooses to use the same expression again and again, and then another author comes and uses it yet again.

And indeed formulae are repetitive, and are originally introduced as such. Again, Archimedes' *On Floating Bodies* is useful to show what happens where the language is formed. So take the following predicate formula:

96. ἀφεθεὶς εἰς τὸ ὑγρόν, 'immersed in the water'

[39] Compare the predicate formula:

94. παρ' ἣν δύνανται αἱ κατηγομέναι τεταγμένως (e.g. *Conics* I.11 42.2),

leading to the extreme ellipsis

95. {formula 2: line} παρ' ἣν δύνανται (e.g. *Conics* I.13 48.20).

This is already an object formula, describing the object having the property of 94.

This is repeated 29 times – put together, 116 word-tokens, a significant proportion of this short work.[40] It is Archimedes' repetitiveness which makes us see this as formulaic.

But this is not a paradox: it is another reminder of the fact that the conventions we study are self-regulating conventions. They are not the result of external, explicit codification. In chapter 2 above we saw the '15 per cent' tendency: both baptisms and many-lettered names switches followed a rule but broke it about 15 per cent of the time. The repetition of formulae cannot be quantified in the same way, but it is essentially similar. The rule is to repeat; the rule is not always followed; and when it is broken, this may yield new, meaningful structures.

3.3 *The generative grammar of formulae*

The setting-out of Archimedes' *SL* 13 contains the following:

> ἔστω . . . ἀρχὰ . . . τᾶς ἕλικος τὸ Α σαμεῖον, 'let the point A be the origin of the spiral'.

This expression is the formula

> 23. ἔστω {object} {object}

with the first object filled by the formula

> 97. ἀρχὰ τᾶς ἕλικος[41]

and the second filled by formula 1: point.

This is the first occurrence of 97 inside 23 in this work, and hence (since Archimedes discovered spirals), this is the first occurrence of 97 inside 23 in Greek mathematics.

The example is simple. Yet no further examples need to be given (though thousands could). This example proves the fundamental characteristic of formulae: they are generative. When Archimedes produces the expression above, he does not consult any manual. He follows his interiorised grammar of formulae, and this grammar allows him –

[40] This includes a transformation into a construction formula:

> 97. ἀφείσθω ἐς τὸ ὑγρόν (e.g., 322.3), 'let it be immersed in the water'.

[41] As the formula is distinctively Archimedean, I give its Doric form. Naturally, formulae can cross dialects – which in itself can be used to argue for their generative, non-mechanical nature.

certainly without any need for thought at all[42] – to generate this complex formula. Thus the different aspects of the behaviour fall together: the hierarchic structure (it is through such structures that formulae are generated); the flexibility (explained by generative, as opposed to mechanical, uninteriorised production); the transformability and the creativity (both expressions of generativity). Instead of thinking in terms of a set, once-and-for-all list of formulae, we should understand Greek mathematical formulae as a rule-governed, open-ended system.

This can be viewed from several perspectives. One is the perspective of the individual who uses the system: we see him as a creative, independent writer. Another is the perspective of the individual who acquires the system: and now we see why the explicit codification is not an option at all. Grammar can not be taught just by external dictation. It must be internalised. *How* the internalisation takes place is then another question, not so simple to answer (for indeed, do we quite understand how natural language, let alone natural second language, is acquired?). But the analogy itself is clear: the acquisition of a technical language is like the acquisition of a second language – but not quite: the acquisition of a technical language *is* an acquisition of a second language.

Finally, this may be viewed from the perspective of the system itself. Once formulae are viewed as generated according to a grammar, their systematic nature is emphasised. I have stressed the holistic nature of the lexicon. The lexicon can not be reduced to its constituents. The same is true for formulae.

The holistic nature of the lexicon had two main aspects:

(a) Economy – the one-concept-one-term principle;
(b) Recognisability – the text is manifestly 'technical', through its global lexical features.

These two features are present in the case of formulae. For instance, we saw how object formulae formed by ellipsis constitute a system, and are perceived as such (see formulae 1–10 above). This is a case of economy – there is generally a single formula for a single concept, and certainly no two concepts are referred to by the same formula (even though contextual considerations could, in principle, differentiate the

[42] A crucial conceptual point: 'generative' does not exclude 'automatic'. On the contrary: the prime example of generative behaviour, namely language, is also the prime example of automated behaviour. When I say that formulae are generated, I mean that they are internalised (and thus to some extent automated), just as a second language is.

two uses of the single formula). As for recognisability, no argument is required. The formulaic nature of the text is its most striking feature.[43]

Formulae constitute a system in a way going beyond that of the lexicon. The lexicon is atomic – it is made up of unanalysable words. Formulae are molecules. They have an internal structure, and therefore they may resemble each other, or they may result from each other through transformations, or they may be embedded in each other. I gave a number of examples for each of these procedures. I will now concentrate on a single system of formulae. An especially important system (as will be explained in the next chapter) is that of proportion theory. I have already given a number of formulae from this system. First, the building-block for all the rest, the object formula:

54. ὁ λόγος {any object in the genitive} πρὸς {any object in the accusative}, 'the ratio of {any object} to {any object}.

On the basis of this, another important predicate formula is:

55. ὡς {formula 54: ratio} οὕτως {formula 54: ratio}, 'as {ratio} so {ratio}'.

A variation on this is:

98. {formula 54: ratio} τὸν αὐτὸν λόγον ἔχει, ὃν {formula 54: ratio}, '{object to object} has the same ratio which {object to object}'.[44]

This has a number of cognates:

99. {formula 54: ratio} μείζονα λόγον ἔχει, ἤπερ {formula 54: ratio}, '{object to object} has a greater ratio than {object to object}';
100. {formula 54: ratio} ἐλάσσονα λόγον ἔχει, ἤπερ {formula 54: ratio} (the same as 99, with 'smaller' for 'greater');
101. {formula 54: ratio} διπλασίονα λόγον ἔχει, ἤπερ {formula 54: ratio} (the same as 99–100, with 'twice' instead of 'greater/smaller'. The reference is to the 'square', not to the 'double').

An even more complex predicate formula is:

89. {formula 54: ratio} σύγκειται ἔκ τε {formula 54: ratio} καὶ [ἐκ] {formula 54: ratio}, '{ratio} is composed of {ratio} and {ratio}'.

[43] Even though, paradoxically, the structure behind the system of formulae is not so obvious. The amateur reader sees an artificial, unstructured system of *ad hoc* formulae, while the professional internalises the system and no longer articulates it to himself explicitly.

[44] In 98, the elements of 54 are used directly, whereas, in 55, 54 is used as a composite unit.

And, as already explained above, a number of argumentation formulae are based on these formulae, especially on 55 (but also on 99–100, with obvious modifications):

56. {formula 55: proportion} καὶ ἐναλλὰξ {formula 55: proportion} – 'alternately',
57. {formula 55: proportion} καὶ ἀνάπαλιν {formula 55: proportion} – 'inversely',
58. {formula 55: proportion} καὶ συνθέντι {formula 55: proportion} – 'in composition',
59. {formula 55: proportion} καὶ διέλοντι {formula 55: proportion} – 'separately',
60. {formula 55: proportion} καὶ ἀναστρέψαντι {formula 55: proportion} – 'conversely'.

And another similar formula is:

102. {formula 55: proportion} δὲ {formula 55: proportion} δι' ἴσου ἄρα {formula 55: proportion} – *'ex aequali'*.

Another argumentation formula, much more complex in internal structure, is:

103. ὡς ὅλον {formula 54: ratio ὅλον} οὕτως ἀφαιρεθέν {formula 54: ratio ἀφαιρεθέν}, 'as whole X is to whole Y, so remainder Z is to remainder ω'.[45]

This, then, is a system made up of 14 formulae.[46] As explained above, it is possible to represent linearly the trees of these formulae, and such a presentation makes the interrelations immediately obvious:

[45] Here, again, the combination of the elements, in constituting a new formula, can not be reduced to simple additions. I do not go here into these difficulties, which represent not so much the complexities of Greek mathematical formulae as the complexity of Greek (or of most other languages).

[46] The list does not give all the formulae related to proportion: what it does is to give all the important formulae based on formula 54. There are a few other groups, more specialised. The most noteworthy is the system of two formulae for the mean proportional, a construction formula (and, as such, transformable into a predicate formula):

104. τετμήσθω {formula 2: line} ἄκρον καὶ μέσον λόγον, 'let {line} be cut in extreme and mean proportion'

and a predicate formula, shading into an object formula:

105. μέση ἀνάλογον (I take the feminine as representative), 'a mean proportional'.

The interest in this system is the specialisation of the two formulae. The two approaches – the construction/predicate and the predicate/object – use completely different morphological forms. This is directly comparable with the system for the radius, with the completely unrelated object formula 13 and the construction formula 30. Also, in both systems, the formulae are strongly semantically marked. Such *repeated* structures show the necessity for the structural approach.

$$O_{54}[O_x,O_x]$$
$$P_{55}[O_{54}[O_x,O_x],O_{54}[O_x,O_x]]$$
$$P_{98}[O_{54}[O_x,O_x],O_{54}[O_x,O_x]]$$
$$P_{99}[O_{54}[O_x,O_x],O_{54}[O_x,O_x]]$$
$$P_{100}[O_{54}[O_x,O_x],O_{54}[O_x,O_x]]$$
$$P_{101}[O_{54}[O_x,O_x],O_{54}[O_x,O_x]]$$
$$P_{89}[O_{54}[O_x,O_x],O_{54}[O_x,O_x],O_{54}[O_x,O_x]]$$
$$A_{56}[P_{55}[O_{54}[O_x,O_x],O_{54}[O_x,O_x]],P_{55}[O_{54}[O_x,O_x],O_{54}[O_x,O_x]]]$$
$$A_{57}[P_{55}[O_{54}[O_x,O_x],O_{54}[O_x,O_x]],P_{55}[O_{54}[O_x,O_x],O_{54}[O_x,O_x]]]$$
$$A_{58}[P_{55}[O_{54}[O_x,O_x],O_{54}[O_x,O_x]],P_{55}[O_{54}[O_x,O_x],O_{54}[O_x,O_x]]]$$
$$A_{59}[P_{55}[O_{54}[O_x,O_x],O_{54}[O_x,O_x]],P_{55}[O_{54}[O_x,O_x],O_{54}[O_x,O_x]]]$$
$$A_{60}[P_{55}[O_{54}[O_x,O_x],O_{54}[O_x,O_x]],P_{55}[O_{54}[O_x,O_x],O_{54}[O_x,O_x]]]$$
$$A_{102}[P_{55}[O_{54}[O_x,O_x],O_{54}[O_x,O_x]],P_{55}[O_{54}[O_x,O_x],O_{54}[O_x,O_x]],P_{55}[O_{54}[O_x,O_x],O_{54}[O_x,O_x]]]$$
$$A_{103}[O_{54}[O_x,O_x],O_{54}[O_x,O_x],O_{54}[O_x,O_x]]$$

Two features of this system are most important:

(a) The entire system is produced through combinations and transformations on 54–55.
(b) The system in which the formulae are interrelated mirrors the way in which the concepts themselves are related (e.g. the mathematical cognates are identical in form). In argumentation formulae, logic and form are identical. I shall return to this in the next chapter.

To sum up: the generative nature of formulae helps to explain both their accessibility and their deductive function. Formulae are accessible to the user because they are produced from a few simple building-blocks. They are deductively functional because their form mirrors logical relations.[47]

The generative grammar of formulae is their most important qualitative feature, the one which encapsulates most of the rest. But qualitative features are not enough: a quantitative detour is necessary before we move on.

3.4 *Quantitative remarks*

It is thus necessary to try to quantify the phenomena – but it is also very difficult. One would like to know, for instance, how many

[47] At this stage it should be noticed that second-order formulae form their own separate system: they are not transformable into or from other formulae, they are less flexible, and their hierarchic structures do not specify other formulae. This is typical of the way in which second-order language, in general, differs from first-order language.

formulae are used in Greek mathematics. But, the limitations of the corpus aside, how can you count an open-ended system? Or another important question is that of the percentage of the text which is taken up by formulae. In a sense the answer is an immediate (and therefore unhelpful) 'everything'. Looked at more carefully, the question becomes very difficult: if a formula is a matrix, a form of words, are we to count all the words occurring inside the matrix as part of that formula? Or only the 'topmost', fixed words? For instance, in *to A*, 'the <point> *A*', what is formulaic? The article? Probably yes: this is the fixed part of the formula. The letter *A*? Perhaps not: it may be replaced by other letters. But then what is left of the formula? It feels most formulaic with this strange-looking pseudo-word '*A*'! And, in this way, it becomes very difficult to approach another central question, that of the average size of formulae: how are we to measure such sizes? What we see is that formulae are form, not matter – and form is much more difficult to measure.

Parry could approach the question of the quantitative role of formulae in Homer in a straightforward way. The main example in Parry (1971) was two samples taken from anything but a neutral context: the first 25 lines of each of the epics. There, he simply underlined formulae. Fortunately, it took me some time to realise the complexity of the questions and, when I started working on the problem of formulae in mathematics, I followed Parry for a number of propositions.[48] Looking again at my survey, I can see that I underlined the constituents of the more strongly semantically marked formulae. This is not a meaningless survey: it gives us the portions of the text which immediately strike us as formulaic. I have discovered considerable variability. The limits, however, are clear. No proposition contains no semantically marked formulae at all. No proposition consists of such formulae alone. The rule is a roughly equal distribution of the text between semantically marked and non-semantically marked formulae. In Parry's examples, around 70% of the text is formulaic,[49] significantly above my results (though Parry's result is approached in more formulaic propositions, such as those of the Euclidean *Data*). In the mathematical text, but not in Homer, significant chunks of text may occur without any 'abnormal words' at all: for instance, the first 35 words of Apollonius' *Conics* I.I. The mathematician may speak, for a few lines, a language which is

[48] These are: Autolycus, *Moving Sphere* 11; Euclid, *Data* 57; Apollonius, *Conics* I.1–4, 12, 25; Archimedes, *CS* 20, *Meth.* 1. I also counted numbers of formulae per proposition.

[49] Parry (1971) 118, 120.

essentially natural (though repetitive); or he may use a language in which semantically marked and 'normal' words are roughly equally and evenly distributed; or, finally, he may speak a language which is composed of formulae-within-formulae and practically nothing else. Propositions, as a rule, are made up of all three types of language, but exact proportions may vary considerably from one proposition to another.

The numbers of semantically marked formulae in the propositions surveyed range between 7 and 21. This should be compared with the sizes of the propositions, which range between 100 and 600 words. The number of such formulae in propositions, while dependent to some extent on the absolute size of the proposition, is more constant: a standard proposition has between 10 and 20 formulae. The number of semantically marked formulae is a good measure of the conceptual size of a proposition. Whatever becomes a focus of interest for the mathematician tends to get a semantically marked formula. The number of semantically marked formulae is comparable to the number of objects and situations of mathematical interest. Between 10 and 20 such objects and situations are the size for a deductively interesting proposition.

Finally, the total number of formulae in Greek mathematics: we have seen the difficulty of even defining it. Could we nevertheless say anything useful about it? Mugler's dictionary offers, sometimes, the formulae within which the words surveyed by him are being used. In all, he gives 76 formulae. (Clearly, Mugler does not aim at completeness, but gives what he sees as the most interesting or common formulae.) Archimedes' index, which is much more complete (but is still conservative in its conception of what a formula is), supplies at least about 150 formulae, which include most of Mugler's formulae. In my (definitely not exhaustive) survey, I listed 105 formulae, of which 71 are the *complete* set of formulae in *Elements* II alone. Most of the 71 formulae of *Elements* II, however, appear in many other contexts as well.

The numbers have a certain coherence. A large sample (Mugler's dictionary; *Elements* II) gives fewer than 100 formulae; a huge sample (Archimedes' corpus) gives fewer than 200 formulae. The range of magnitude is therefore clear: hundreds. Of these, the 'mainstream' formulae, those commonly used, can not number more than a few hundred.

We get two sets of numbers, then. One set, that of formulae within a single proposition, are in the range 10–20. Another set, that of the total number of 'mainstream' formulae, is in the range of a few hundred. We shall meet these two kinds of numbers again, in the next chapter.[50]

3.5 *The Greek mathematical language: recapitulation*

It is possible now to put together the results of chapters 3 and 4.

A large Greek mathematical corpus – say, something like all the works in a given discipline – will have the following characteristics:

- Around 100–200 words used repetitively, responsible for 95% or more of the corpus (most often, the article, prepositions and the pseudo-word 'letters').
- A similar number of formulae – structures of words – within which an even greater proportion of the text is written (most often, lettered object-formulae). These formulae are extremely repetitive.
- Both words and formulae are an economical system (tending, especially with words but also with formulae, to the principle of one lexical item per concept).
- The formulae are flexible, without losing their clear identity. The flexibility usually takes the form of gradual ellipsis, which in turn makes the semantics of the text 'abnormal'.
- Further, about half of the text is made up of strongly semantically marked formulae, which serves further to mark the text as a whole.
- The flexibility sometimes takes the form of transformations of one formula into another and, more generally, formulae are structurally related (either vertically – one formula is a constituent in another – or horizontally – the two formulae are cognate).
- Thus a web of formulae is cast over the corpus. Alongside and above the linear structure of the text, we will uncover a structure constituted by repeated, transformed, cognate or dependent formulae.

Since all the text is formulaic, and formulae are repetitive and hierarchically structured, the text can be seen as a structured system of recurrences.

This is most strongly felt within individual propositions. In fig. 4.1, the analysis of *Elements* II.2, the proposition can be seen as a single tree, made up of 14 formulae. Given this analysis, the structural relations of

[50] We have already met the second number in the preceding chapter, on the lexicon: 143 Archimedean words account for 95% or more of the Archimedean corpus – remarkably similar to the number of formulae in the corpus!

the propositions become apparent: the enunciation is a pair construction formula/predicate formula; the same sequence is repeated in the pairs setting-out/definition of goal; construction/proof. Finally, the conclusion reverts to the original sequence. A limited group of object formulae is governed by those construction and predicate formulae: the construction and predicate formulae operate on the small group of object formulae, rearrange them and yield the necessary results.

The global relation between construction and predicate formulae – the sequence of four repetitions of these sorts of formulae – is the key for generality. I will discuss this in detail in chapter 6 below.

The more local relation between object formulae, governed by construction and predicate formulae (and, elsewhere, argumentation formulae) is one of the keys for necessity. I shall return to this in the next chapter.

4 SUMMARY: BACK TO THE HOMERIC CASE

4.1 Contexts of formulae

First, an obvious point: the use of a system of formulae (especially non-compositional formulae) implies a professionalised, inward-looking and surprisingly homogeneous group. This enhances the results of chapter 3, and I shall return to this issue in chapter 7 below.

I shall move straight to the main issue, that of orality and literacy – returning to Homer. In the Homeric case, formulae are seen by most of the scholars as signs of orality. As explained above, this can not be imported directly into the mathematical case. But the problem remains: what sort of language use is implied by the use of formulae?

The word 'oral' is sometimes used to mean 'illiterate': in this sense certainly it is inapplicable to the mathematical case. But this use of 'oral' is misleading (as is now well known, e.g. following Finnegan 1977). This is because there is no sharp dichotomy, 'oral' and 'written'. The cognitive reality is much more complex. The presence of writing may be influential to varying degrees, in varying ways. So consider, for instance, the flexibility of formulae. The text is repetitive – but not verbatim repetitive. This is comparable to the result in chapter 2 above, concerning the behaviour of many-lettered names. And we shall return to the same result in the next chapter, with the quotation of earlier results. Whether the Greek looked up his text or not, at any rate he did not operate under the expectation that one *should* look up one's text. This expectation *may* be an impact of the use of writing, but there was no such impact of writing on Greek mathematics.

Another point is more difficult. What constitutes the unity of a formula? What makes it perceived as a single, organic object? In the Homeric case, this is the prosodic form of the formulae. In the mathematical case, this is the internal structure: in general, mathematical formulae are structures. Now language in general is structural (just as language in general is prosodic – and, incidentally, prosody in general is structural). The generative structure of Greek mathematical formulae is a reflection of language in general. Formulae import the properties of natural language into a sub-language.

Both awareness of prosodic form and awareness of structure are built into the human linguistic capacity. The Homeric formulae and the mathematical formulae are both based on (different aspects of) this natural capacity. The Homeric case is specifically aural: it is based on the specifically heard properties of language. It is pre-written in a strong sense. The mathematical formula is not based on this phonological level. It is more abstract. But still, it is pre-written in a more limited sense. It is based on a capacity which antedates writing and which is independent of it. The awareness of linguistic form is a feature of natural language as such, which does not require writing. And it is this feature of natural language which constitutes the essence of mathematical formulae.

Most importantly: the linear presentation of texts in writing, unspaced, unpunctuated, unparagraphed, aided by no symbolism related to layout, was no help for the hierarchical structures behind language. It obscures such structures. In:

$$A + B = C + D$$

for instance, the hierarchical relations between the objects are strongly suggested by the symbolism. But consider this:

THEAANDTHEBTAKENTOGETHERAREEQUALTOTHECANDTHED

This is how the Greeks would write '$A + B = C + D$', had they written in English. And it becomes clear that only by going beyond the written form can the reader realise the structural core of the expressions. Script must be transformed into pre-written language, and then be interpreted through the natural capacity for seeing form in language.

Greek mathematical formulae are post-oral, but pre-written. They no longer rely on the aural; they do not yet rely on the layout. They are neutral: rather than oral or written, they are simply artefacts of language.

Yet could these formulae take root without writing? In a sense, the answer is positive. Clearly here (as in the lexicon in general) explicit definitions are of minor significance. True, some sets of definitions deal with groups of formulae. For instance, the definitions of the *Data* cover the various data formulae, and the definitions in *Elements* V.12–17 cover the main argumentation formulae of proportion theory. But even those definitions do not cover the actually used form: the *Data* defines the formulae with the verb in the infinitive ('*X* is said *to be* given . . .'), while the actual use of the formulae is in the participle ('*X* is *given* . . .'); book V defines the formulae with an adjective ('an *inverted* ratio . . .'), while the actual use of the formulae is with an adverb ('. . . *inversely* . . .'). Of the 71 formulae of *Elements* II, three are defined (though only one is defined in exactly the way it is used – and this is also a one-word-long formula). These are 9 (for rectangle, defined in II.1), 11 (for gnomon, defined in II.2) and 86 ('right' as an adjective of angle, defined in I.10).

So definitions – those written introductions to works – do not govern the behaviour of formulae. The formulae are governed by the texts themselves, not by their introductions – as is true for the lexicon in general. These texts, however, were written. Is it possible to imagine the conservation and transmission of formulae in the Greek mathematical case, without the presence of writing? Now, finally, the answer must be negative.

I will approach this through the argument that follows, returning to oral poetry. I do not mean to say anything dogmatic about oral poetry, just to remind ourselves of the possibilities by going through the better understood data of oral poetry. There is a certain paradox about oral poetry. As noted by Goody, poetry deriving from strictly oral cultures is not more formulaic than Homer's, but *less* formulaic.[51] The presence of writing (which in one way or another must have influenced the Homeric tradition) could help to conserve and transmit formulae. Even if there is no expectation of looking up the text, there is certainly the *possibility* of doing so in the written context – and this will help the conservation and transmission of formulae. But does this argument not run counter to the main oral hypothesis? After all, is the Homeric text not formulaic *just because* it is oral?

[51] Goody (1987) 99.

The paradox is illusory. The two horns of the dilemma point to different, unrelated processes. Briefly: formulae help an oral performer; they are helped by a written background. There is a limit to how formulaic a totally oral work can be (without any help from writing, there is a limit on the emergence of a rigid, repetitive system). There is a limit on how developed a non-formulaic work can be, in a context which is to some extent oral (without formulae, it is difficult to create a developed work). I will therefore say this. Without writing, the formulaic system of Greek mathematics could not have been conserved and transmitted (though, as explained above, nothing in the actual *function* of these formulae owes anything to the specific written aspects of the text, such as layout). This in itself is speculative, yet the substantive claim is almost tautologic. Without writing, there would be no Greek mathematical continuity at all – as will be explained in chapter 7 below. We can therefore safely say that writing made the emergence of mathematical formulae possible. And yet, the specific shape of formulae represents pre-written assumptions.

4.2 *Formulae and cognition*

As explained above, formulae can help an oral performer. We know what Homer gained from his formulae. What was Euclid's gain, then? What was the gain for his readers? There are at least five ways in which formulae are directly helpful to proceeding deductively.

(a) Formulae strengthen the tendency of the lexicon as a whole (as noted in the preceding chapter) to be concise and thus manageable.
(b) The structures of formulae parallel the logical properties of the objects or procedures to which those formulae refer. In particular, the internal structure of an argumentation formula mirrors its logical content.
(c) Similarly, on a larger scale, the hierarchic, internal and external structures of formulae make the logical relations between their contents more transparent. Objects are repeated as object formulae, within clearly marked predicate and construction formulae.
(d) Going beyond the individual proposition, the tool of formulae serves as a means for transferring results from one proposition to another.
(e) Finally, the overall formulaic structure of the proposition serves as the basis for generality.

Point (a) is a continuation of chapter 3. Points (b)–(d) will be explained in detail and discussed in the next chapter; point (e) will be explained in detail and discussed in chapter 6.

How do such deductive contributions of the formulae explain their emergence? The question is not simple – nor is it simple in the Homeric case. It is not as if someone set Homer the task: 'improvise epic poems orally', and Homer came up with the idea 'I shall devise a system of formulae'. The process of emergence is much more complex: useful tools, easily transmittable tools, tend to accumulate – and to shape, in their accumulation, the task itself. The evolution of cognitive tasks and tools is intricate and reciprocal. As I have stressed at the beginning of this chapter, 'formulae' are ubiquitous. Wherever there are artefacts, there are repeated patterns. And this is not surprising: artefacts are repeatedly made, and the repetition of action implies some repetition of result. The Homeric singer sings day in day out, in each song frequently returning to the same themes.[52] Such repetition in subject matter naturally leads to the emergence of repeated patterns – in Homeric song as in any other product. Those patterns are conserved and transmitted which are especially useful for the practitioner and the audience. It is, then, possible to analyse artefacts according to their characteristic patterns, and these characteristic patterns will teach us about the requirements of the practitioner and the audience.

I will concentrate on the way in which logical relations are mirrored by linguistic structures. It is in this that mathematical formulae are not unlike Homeric formulae. Both kinds of formulae harness a natural linguistic skill for the specific task in hand, and thus make the task much easier. The natural skill, in the Homeric case, is prosodic awareness. In the mathematical case, it is awareness of form. Moreover, in both cases, the economy of the system makes it even easier to access the individual formulae.

In an expression such as 'as the *AB* is to the *CD*, so the *EF* is to the *GH*, therefore, alternately, as the *AB* is to the *EF*, so the *CD* is to the *GH*', it is essential that the expression is not perceived linearly. It is perceived structurally, as a 'tree'. The expression represents a proved result: this is why it is valid. But it is immediately convincing (which is what deduction requires) not because it was proved earlier, but because it is perceived as a structured whole. This has two aspects: first, it has the typical hierarchic structure of formulae; second, it is

[52] Hainsworth (1993) 12.

well known to its user, and is thus perceived directly, as a whole (as is true for formulae in general).

Greek mathematicians were not oral performers – they were not performers at all (see chapter 7 below). But then, imagine a Greek mathematician during the moment of creation. He has got a diagram in front of him, no doubt. But what else? What can he rely upon, what is available in his arsenal? There are no written symbols there, no shortcuts for the representation of mathematical relations, nothing besides language itself. He may jot down his thoughts, but if so, this is precisely what he does: write down, in full, Greek sentences.

The lettered diagram is the metonym of mathematics not only because it is so central, in itself, but also because it is the only tangible tool. There is no tangible mediation between the mathematician and his diagram – except, of course, his language. The immediacy of the mathematical creation is thus not unlike that of the Homeric performance. Both have nothing but words to play with, and therefore both must shape their words into precise, task-specific tools.

Return now to the Greek mathematician: we see him phrasing to himself – silently, aloud, or even in writing – Greek sentences. Most probably he does not write much – after all, there is nothing specifically written about his use of language. For four chapters, we have looked for the Greek mathematician. Now we have finally found him: thinking aloud (chapters 2–4), in a few formulae (chapter 4) made up of a small set of words (chapter 3), staring at a diagram (chapter 1), lettering it (chapter 2). This is the material reality of Greek mathematics. We now move on to see how deduction is shaped out of such material.

CHAPTER 5

The shaping of necessity

PLAN OF THE CHAPTER

My argument is simple. Some statements and arguments are seen as directly necessary – they are the building-blocks, the 'atoms' of necessity. These then combine in necessity-preserving ways to yield the necessity of Greek mathematics.

There are two types of atoms, and two types of combinations of these atoms. The two kinds of atoms are: first, assertions which are taken to be necessarily true in themselves. An assertion such as '*A* is either greater than *B*, or it is not' is taken as necessarily true in itself. I call such assertions the 'starting-points' for arguments. Note that this class is much larger than the 'axioms' used in Greek mathematics; in fact, it contains any assertions which are unargued for in the text. Section 1 discusses starting-points.

The other kind of atoms is 'arguments', but it must be understood immediately that I use the term 'arguments' in a technical sense. I will call an 'argument', in the singular, only what is an unanalysable argument. To explain, take the hypothetical derivation:

(1) $a = b$, (2) $b = c$, (3) so $a = c$, (4) $c > d$, (5) so $a > d$

This is not a *single* argument. There are two arguments here: one consisting of steps (1)–(3), the other consisting of steps (3)–(5). Section 2 discusses arguments. A method I often use in order to visualise the logical structure of arguments is to have them depicted as if in a 'tree',[1] e.g. the argument above will be represented as fig. 5.1:

[1] Not to be confused, of course, with the 'phrase-structure' trees of the preceding chapter.

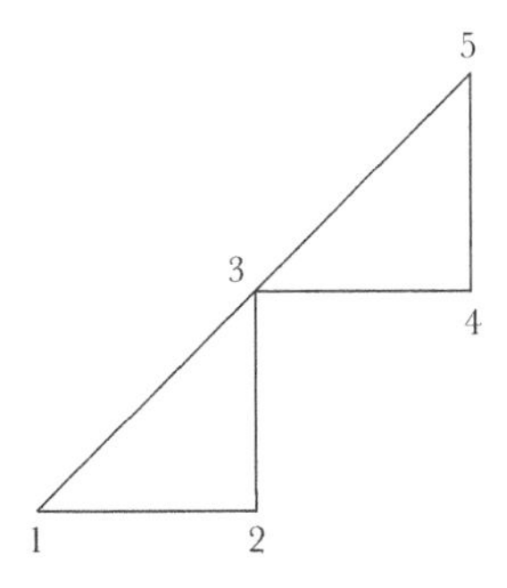

Figure 5.1.

In this 'tree', it can be seen how steps (1) and (2) combine to produce (3), which in turn combines with (4) to produce (5). In this way the concepts of 'starting-point' and 'argument' get a concrete sense. The starting-points of this derivation are (1), (2) and (4) – all the points 'with nothing below'. The arguments are the two triangles, (1)–(3) and (3)–(5).

By looking at such trees, then, it is possible to visualise the structure of combinations of starting-points and arguments within a single proof. This kind of combination is discussed in section 3.

The other kind of combination is the way in which the results of proofs become available for use in other, later proofs: how a 'tool-box' of known results is created and accessed. This is discussed in section 4. Section 5 is a summary.

1 STARTING-POINTS

The first thing to note about starting-points is that there are so many of them. I have surveyed 31 propositions.[2] In 18 of the 31, between 35% to 45% of the assertions are starting-points. Of the 5 which have less, almost the lowest is Archimedes' *SC* I.30, with 25%, still a considerable percentage; and this proposition has 8 assertions. It is safe to say that the typical proposition has more than a third of its assertions consisting of starting-points, but less than half (true of 23 of the 31).

[2] I will often refer to the same survey, so I give the list: Aristotle, *Meteor.* 373a5–18; Hippocrates of Chios' third quadrature (Becker's text); Autolycus' *Ort.* I.13; II.15, 18; Euclid's *Elements* I.5, II.5, III.5, IV.5, *Data* 61; Aristarchus' *On Sizes and Distances* I, 12; Archimedes' *SC* I.10, 20, 30, 40, *SL* 9, *Meth.* I, *Arenarius* 226–32, Apollonius' *Conics* I.9, 13, 15, 26, 37, 41, 46, II.2, III.17; (quoted in) Anonymous, *In Theatet.* cols. xxix.42–xxxi.28; Diocles' *On Burning Mirrors* 2 (Toomer's text); Ptolemy's *Harmonics* III.2, pp. 90–1.

For each assertion which is unargued for in the proposition, there are usually no more than two assertions (but no less than one) which are argued for.

Where are the starting-points? Everywhere in the proof. The only general rule is that the first assertion in a proposition tends to be a starting-point, while the last assertion tends not to be. Otherwise, the position of starting-points is flexible. It is true that, often, the frequency of starting-points is reduced as the proposition progresses. Take Apollonius' *Conics* 1.9: it starts with 2 starting-points, followed by 2 argued assertions; then a single starting-point followed by a single argued assertion; then another single starting-point followed by 3 argued assertions; again, one starting-point followed by three argued assertions; then a starting-point followed by 6 argued assertions; and, finally, a starting-point followed by 4 argued assertions. This is a relatively simple structure,[3] and even here the reduction in frequency is not monotonic.

The point of this is, first, that Greek mathematical proofs are the result of genuine cross-fertilisation. What I mean by the metaphor is the following: one way of doing mathematics, in principle, would be to take an assertion, develop some of its results, and then to combine these results until something interesting emerges. So one would get a few starting-points at the beginning, and then a continuous stretch of argued assertions. As a matter of fact, it is difficult to produce interesting things in this way: the cross-breeding of relatives tends to be barren. Derivatives of a single assertion must carry similar informative contents, whose intersection could be neither surprising nor revealing[4] (we shall return to this point when discussing the tool-box).

Second, we begin to see something about the global structure of proofs (to which I will return in section 3): proofs do not reuse over and over again materials, which have been presented earlier, once and for all. Instead, whatever is required by the proof is brought in at the moment when it is required. The introduction of new material for deductive manipulation goes on through the length of the proof.

[3] Try *Conics* 1.41 (P for starting-point, n for non-starting-point):

PPP n P n P n P n PP nnnnn PP nn P n P nnn PP n P nnn

[4] The lowest percentage of starting-points in my survey is *Elements* 1.5, the dreaded *pons asinorum* (for this name see Heath 1926, vol. 1, 415–16) which is in fact a mule, mechanically reprocessing over and over again the same equalities until the final result is ground out of the machine. Here, however, Euclid may say in his defence that the mulish quality of the proposition is due to the lack of materials, necessary in what purports to be the fifth link in a first chain. Life must precede sex.

Deduction, in fact, is more than just deducing. To do deduction, one must be adept at noticing relevant facts, no less than combining known facts. The eye for the obviously true is no less important than the eye for the obvious result and, as is shown by the intertwining of starting-points and argued assertions, the two eyes act together.[5]

1.2 The necessity of starting-points

The main distinction to be drawn is between relative and absolute starting-points. Think for a moment of Greek mathematics in its entirety as a huge, single proof (a completely ahistorical exercise, meant to clarify a point of logic), i.e. if proposition X relies on proposition Y, then *include* proposition Y inside proposition X. It is clear that some assertions which appear as starting-points now (in the context of a 'normal' proof as it occurs in, say, Archimedes) would immediately become results of other assertions (which are now contained in other propositions, say, in Euclid). Such assertions are only relative starting-points, that is, they are starting-points relative to the proof in which they appear. Relative to Greek mathematics taken as a global system, they are not starting-points but argued assertions. Other starting-points would remain starting-points even in this hypothetical case, and they are therefore absolute starting-points.

Relative starting-points occur in the following ways:

(a) Explicit reference. For instance, the first assertion of Apollonius' *Conics* 1.46[6] is:

> διὰ τὰ δεδειγμένα ἐν τῷ τεσσαρακοστῷ δευτέρῳ θεωρήματι
>
> 'Through the things proved in the forty-second theorem'.

Such references are very rare (and therefore there is no point in discussing in detail the – real – possibility that they are interpolations; but see section 4 below).

(b) The tool-box. The notion of the tool-box was raised above, and it can be better understood now, as being distinct from explicit

[5] One typical Greek mathematical method is to assume the desired result, and to work backwards until a starting-point is reached such as can be satisfied by the mathematician. This is known as 'the method of analysis'. I will not discuss it here, but I will point out that, in order to understand this method, it must be seen within the terms of this chapter – as a method of obtaining starting-points (this is in agreement with the argument of Hintikka and Remes (1974), that analysis obtains necessary auxiliary constructions).

[6] 140.24–5.

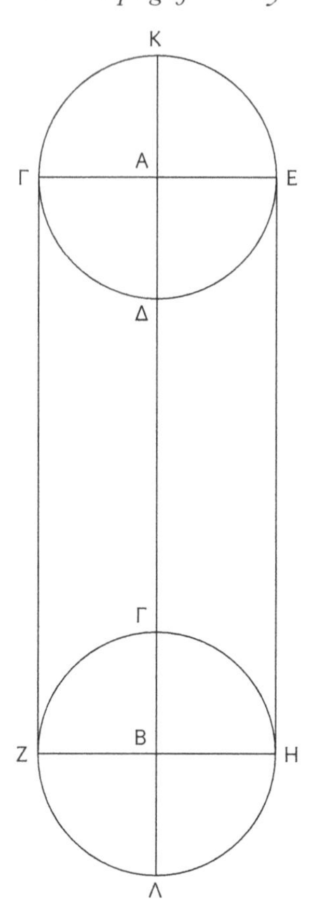

Figure 5.2. Aristarchus I.

references. The essence of the tool-box is that it is taken for granted. Consider, for instance, the following, from Aristarchus' first proposition (fig. 5.2):[7]

ἴση δὲ ἡ μὲν ΑΔ τῇ ΑΓ

'and ΑΔ is equal to ΑΓ'.

Nothing in the proposition so far supports the claim; still, it is simply put forward, without a hint that any justification is required. In fact, both ΑΓ and ΑΔ are radii of the same circle, so the result is indeed obvious. It can be seen to result from Euclid's definition

[7] Heath 358.2.

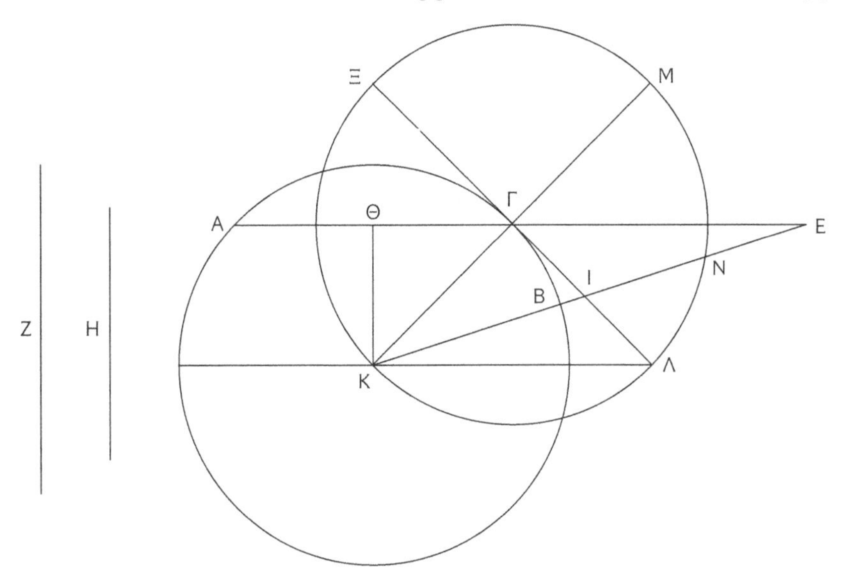

Figure 5.3. Archimedes' *SL* 9.

1.15,[8] but it can hardly be assumed that the ancient author, or the ancient audience, consulted their Euclid at this point. On the other hand, it is hard to imagine a geometer, of whatever level, who has not internalised the truth that all radii of the same circle are equal to each other. Unlike explicit references, starting-points taken from the tool-box are a common feature of Greek mathematics.

(c) More or less complex implicit arguments. These can be characterised as assertions which are used as if they were part of the tool-box, but probably were not in the tool-box. This is the one most frustrating feature for the reader of Greek (or any other) mathematics who is not as intelligent as the mathematicians themselves. Archimedes was more intelligent, which makes him more frustrating. Here is the first assertion made in Archimedes' *SL* 9 (fig. 5.3):[9]

> ἐσσεῖται δὴ μείζων [ὁ λόγος ὃν ἔχει ἁ Ζ ποτὶ τὰν Η] καὶ τοῦ, ὃν ἔχει ἁ ΚΓ ποτὶ τὰν ΓΛ
>
> '[The ratio of Z to H] is also greater than that which ΚΓ has to ΓΛ'.

[8] 'A circle is a plane figure comprehended by a single line, [so that] all the straight [lines which are] drawn from a single point, [which is one] of those inside the figure, towards it [i.e. the comprehending line] are equal to each other.'

[9] Heiberg 28.2.

The 'also' refers to an earlier comparison, where it was hypothesised that (algebraically paraphrasing) Z:H > ΓΘ:ΘΚ. The Archimedean reader sees immediately that the angle ΚΘΓ must be assumed to be right, the angle ΚΓΛ must be right through the properties of the tangent, and the angles ΘΓΚ, ΓΚΛ must be equal, since ΑΓ and ΚΛ were assumed to be parallel (remember as well, of course, the properties of parallel lines), so that the triangles ΘΓΚ, ΓΚΛ are similar, and in such a way that (algebraically paraphrasing) ΓΘ:ΘΚ::ΚΓ:ΓΛ, and the assertion is seen to obtain.

Is this a starting-point, an 'immediately obvious assertion'? The answer is partly negative, in the sense that one could reasonably ask what the grounds of the assertion are. The answer is also positive, in the sense that Archimedes did not ask. Was it a starting-point for his mind which we, with our puny brains, cannot quite grasp? Or may he have been impatient? Or just showing off? I offer a hypothesis in a footnote,[10] but it is impossible to get any certitude on such questions.

These are the three kinds of relative starting-points. There are also three kinds of absolute starting-points.

(d) Hypothesis. The most typical hypotheses in a Greek mathematical proof are those which are laid down in the construction. This is part of the story concerning the last example from Archimedes' *SL*. The intricate argument relied, ultimately, on the hypothesis that a certain ratio was greater than another:[11]

> δεδόσθω . . . καὶ λόγος, ὃν ἔχει ἁ Ζ ποτὶ τὰν Η, μείζων τοῦ, ὃν ἔχει ἁ ΓΘ ποτὶ τὰν ΘΚ
>
> 'Let a ratio, which Z has to H, be given, greater than that which ΓΘ has to ΘΚ'.

[10] The assertion becomes less mysterious once it is recognised that Archimedes went through the same territory in earlier propositions. (Most importantly, in the very preceding proposition, *SL* 8, 24.24–5, where the same assertion is made not as a starting-point but as an argued result of the relevant lines' being parallel. This is still a very deficient description of the necessary assumptions, but it provides at least the direction for the derivation.) The text demands a deep understanding of the mathematical issues, but this can be sustained if the text is read seriously. Archimedes is not just playing games with us. So this leads to the problem of what I call 'local tool-boxes' – assertions which become self-evident locally, rather than in the context of Greek mathematics as a whole. I shall return to this in section 4.

[11] Archimedes, *SL* 9 26.29–8.1.

This clause is governed by an imperative. It is not an assertion that such-and-such is the case. Instead, it is a demand that such-and-such will be the case.[12] Later in the proof this demand is picked up again, and an assertion is made, based on this demand. This assertion then is true *ex hypothesi.*

In this way, hypothesis is the most common starting-point.

(e) Next comes the diagram. Most Greek proofs include several starting-points which are simply the unpacking of visual information. Take, e.g. Euclid's *Elements* I.5 (fig. 5.4):[13]

> ἐπεὶ οὖν ἴση ἐστὶν ἡ μὲν ΑΖ τῇ ΑΗ ἡ δὲ ΑΒ τῇ ΑΓ, δύο δὴ αἱ ΖΑ, ΑΓ δυσὶ ταῖς ΗΑ, ΑΒ ἴσαι εἰσὶν ἑκατέρα ἑκατέρᾳ· καὶ γωνίαν κοινὴν περιέχουσι τὴν ὑπὸ ΖΑΗ
>
> 'So since ΑΖ is equal to ΑΗ, ΑΒ to ΑΓ, the two ΖΑ, ΑΓ are equal to ΗΑ, ΑΒ, each to each; and they contain a common angle, that <contained> by ΖΑΗ'.

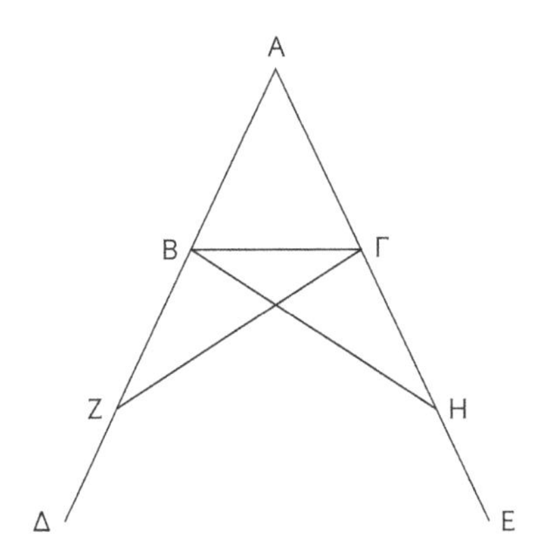

Figure 5.4. Euclid's *Elements* I.5.

Look at the last assertion, that 'they contain a common angle'. Is it compelling without the diagram? With some effort, it can be understood that the 'they' refers to the couples of lines mentioned immediately before. (The diagram is crucial even for understanding the reference of this demonstrative pronoun, but for the sake of the argument let us imagine that this can be done without the

[12] It is possible to look at the question 'what makes it permissible to lay down this piece of construction?'. To simplify the analysis, I have ignored such questions, i.e. I have concentrated on theorems instead of problems (further motivation for this simplification, in section 4 below).

[13] 20.15–18.

diagram.) Now, this is equivalent to saying that the angles ZAΓ, HAB are both identical with the angle ZAH. AΓ and AH are the same line, and AB and AZ are the same line (this again can be supported by some passages earlier in the text, but only with the greatest difficulty – while of course this is immediately obvious in the diagram). In this way, the claim can be propositionally deduced, and perhaps some New Maths crank may still inflict such proofs upon innocent children. But clearly the Greeks did not, and the assertion was immediately supported by the diagram and nothing else.

(f) Finally, assertions may be intrinsically obvious. It might be thought that Greek mathematics knows only a few of these, namely Euclid's axiomatic apparatus, but actually there are many intrinsically obvious assertions which the Greeks do not formulate as axiomatic. Take Euclid's *Elements* III.5, the clinching of the argument:[14]

> . . . ἡ EZ ἄρα τῇ EH ἐστὶν ἴση ἡ ἐλάσσων τῇ μείζονι· ὅπερ ἐστὶν ἀδύνατον

> '. . . so EZ is equal to EH, the smaller to the larger; which is impossible'.

The 'which is impossible' clause is a starting-point here, tantamount to saying that the smaller cannot be equal to the larger. I do not know of anything in Greek mathematics which legitimates this. This is not meant as a criticism – I see, together with Euclid, that the assumption is correct as far as he is concerned.[15] But it means that the assumption is indeed a direct intuition.

Incidentally, how do we know, in this case, that EZ is smaller than EH? The answer is that we see this (fig. 5.5). This should remind us that the distinction made between types of starting-points cuts across starting-points, not between them. Assertions may be obvious through a variety of considerations.

To recapitulate, then, the sources of necessity identified above were:

(a) Explicit references
(b) The tool-box

[14] 176.15–16.
[15] It ceases to be true as soon as set-theoretical discussions of infinity, so central to modern mathematics, are started. The differences in cognitive styles between Greek and twentieth-century mathematics owe much to such real differences in mathematical content.

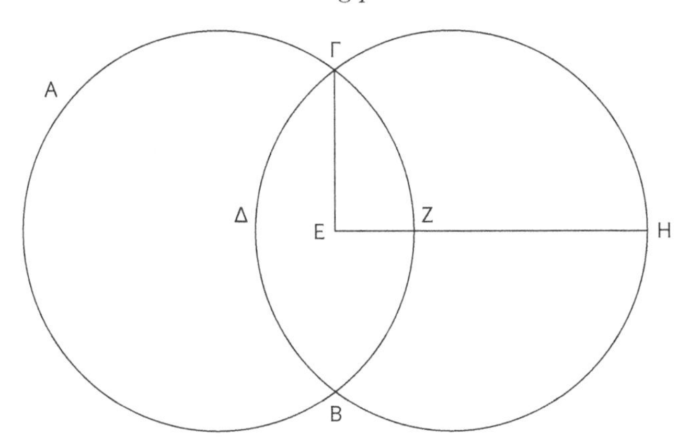

Figure 5.5. Euclid's *Elements* III.5.

(c) Implicit arguments
(d) Hypotheses
(e) Diagram
(f) Intuitions

It is clear that the last two are primary. The first three are relative. Hypotheses also do not add information beyond the fact that they were willed by the mathematician; and to the extent that they carry the information that the hypothesis involved was *legitimate*, other sources of necessity are required (mainly the tool-box, but also the diagram).

The diagram and intuitions, are (as far as starting-points are concerned) the only information-producing mechanisms. These two merit, therefore, a closer look.

1.2.1 The diagram and starting-points

What is the quantitative role of starting-points based on the diagram? Here the statistical limitations of my survey become serious. The obvious thing would be to count all the starting-points in the 31 proofs I surveyed, and to come up with percentages. I have even attempted to do this, but this exercise is almost meaningless. It should be stressed that the typology above is useful for large-scale analysis, not for detailed surveys, mainly because so many starting-points are combinations of different sources of necessity. Furthermore, a detailed survey shows that there is much variability, reflecting different subject matter. Different sources of necessity operate, according to the specific logical situations. Thus, the statistical value of such a survey is very limited. I can disclose my results,[16] but the reader is asked to forget them.

I will record some qualitative facts, instead. First, proofs without diagram starting-points occur. Archimedes' *SC* 1.30 is one such, and so is the much longer and more interesting Apollonius' *Conics* 1.41. But clearly most proofs have some diagram starting-points. Some proofs have several: e.g. Archimedes' *SC* 1.10 has six diagram starting-points.

An important consideration is that many starting-points have a 'diagrammatic' aspect, even if they are not just derived from the diagram. For instance, I said that Archimedes' *SC* 1.30 includes no diagram starting-points. It begins with the following assertion:[17]

> ἐν τῷ ΕΖΗΘ κύκλῳ πολύγωνον ἰσόπλευρον ἐγγέγραπται καὶ ἀρτιογώνιον
>
> 'In the circle ΕΖΗΘ an equilateral and even-sided polygon has been inscribed'.

This is a standard hypothesis starting-point (notice the typical perfect tense), but ΕΖΗΘ, interestingly, is not yet mentioned in the proposition. It refers back (perhaps) to an earlier proposition and (certainly) to the diagram; without them, the assertion is meaningless.

In short: a starting-point may be a diagram starting-point in a strong and in a weak sense. The weak sense is that the assertion would fail to compel had it not been for the diagram, even though the logical grounds for the assertion need not be related to the diagram. The strong sense is that the content of the assertion is contained, non-verbally, in the diagram, and the written assertion is an unpacking of this information. In the weak sense diagram starting-points are ubiquitous, but less startling. In the strong sense they are less common, say one or two per proposition on average (ranging from zero to a few).

Sometimes it is not clear whether the starting-point is a diagram starting-point in a strong or in a weak sense. This is when the starting-point is also an implicit argument. In such cases, you cannot easily say what the explicit argument would look like. I am thinking especially of symmetry arguments. Take Aristarchus' proposition 12[18] (fig. 5.6):

[16] A good third of the starting-points are hypothesis starting-points; another quarter are tool-box starting-points, and a sixth are diagram starting-points. This leaves about a quarter, divided between intuitions (somewhat more), implicit arguments (somewhat less) and explicit references (almost negligible).

[17] 110.25–6. [18] Heath (1913) 390.15.

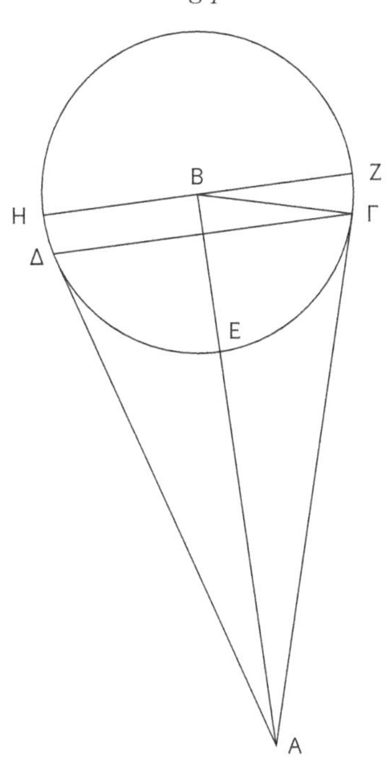

Figure 5.6. Aristarchus 12.

ἔστι τῆς ΓΕ β ἡ ΔΕΓ

'ΔΕΓ is twice ΓΕ'.

This is simply stated in the proposition, and the implicit argument is not difficult to reconstruct: say, angle ΔΒΕ would be equal to angle ΕΒΓ (due to another simple argument showing that the relevant triangles are congruent), so the relevant arcs must be equal. The argument is very easy to recreate. But *why* is it so easy to recreate? Because it asserts a manifest symmetry, so that one hardly *needs* an argument.

It is true that Greek mathematicians are no blind followers of appearances when symmetries are concerned. Euclid's *Elements* 1.5, for instance, sets out to prove exactly such a symmetry feature (that angles at the bases of isosceles triangles are equal). However, once such territories are conquered, Greek mathematicians become more relaxed, and allow themselves to be especially brief where the eye may profitably lead the mind. This does not mean that the diagram adds information, but it does mean that the diagram saves 'logical space', as it were, and thus makes deduction at least easier.[19]

In some cases it is possible to identify postulates which the Greeks 'left out' and which the diagram substituted. This is especially true concerning betweenness assumptions, already mentioned in chapter 1 above. For instance, in the case mentioned above from Euclid's *Elements* III.5, the diagram (fig. 5.5) was responsible for the obviousness of EH > EZ: E, Z and H were all on a straight line, and Z was between the other two. With some care, this could be *shown*. Instead, it was merely stated. It is more difficult to put one's finger on missing postulates in the following case. Quite often, diagram starting-points express a decomposition of the objects in the diagram. I shall take as an example Archimedes' *SC* I.20[20] (fig. 5.7):

> ἡ ἐπιφάνεια τοῦ ΑΒΓ κώνου σύγκειται ἔκ τε τῆς τοῦ ΕΒΖ καὶ τῆς μεταξὺ τῶν ΕΖ, ΑΓ[21]
>
> 'The surface of the cone ΑΒΓ is composed of that of ΕΒΖ and of that between ΕΖ, ΑΓ'.

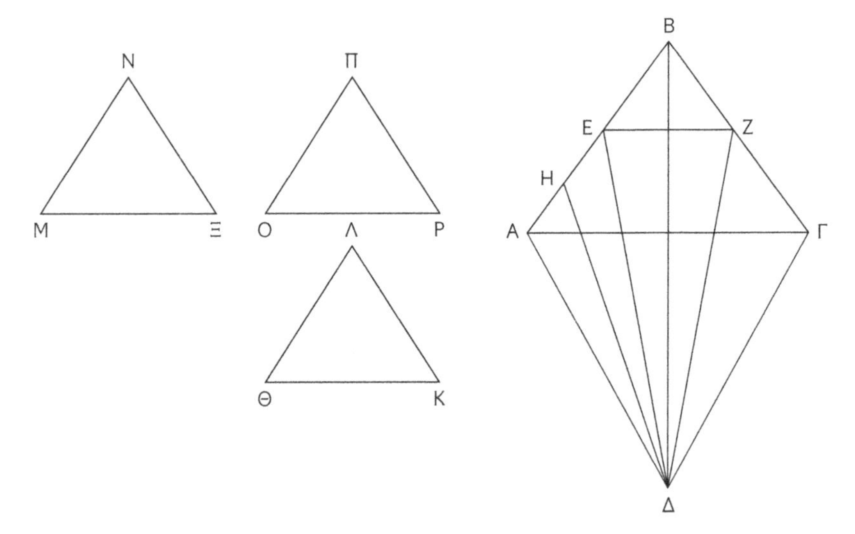

Figure 5.7. Archimedes' *SC* I.20.

[19] The third quadrature of Hippocrates of Chios relies in a strong sense upon symmetry assumptions (Becker 1936b: 418.39–45). Is this lax logic on the part of a mathematical pioneer? Or a simplification introduced by an ancient editor?

[20] 84.22–3.

[21] This is preceded by an ὁμοίως, referring back to proposition 19 (82.9). But this is no real complication for there, as well, no grounds for the assertion were made explicit.

Clearly, there is some general assumption involved (the whole as sum of its parts?) which could be spelled out. But what postulate could support the identification of these specific areas as parts of that specific whole? This is a particular, not a general claim, and could thus be supported only by some particular fact. None is forthcoming in the text, and the only source of necessity is the diagram. Such decompositions are very common, especially in formulae such as the κοινὸν προσκείσθω ('let . . . be added in common'), referred to above as 'equals added to equals'.[22] As is well known, Greek mathematics does not speak about plane figures being equal 'in area'. They are just equal (or unequal) *simpliciter*. The same is true for lines and lengths. What is being compared is not a certain function of the objects, but the objects, directly. When it is considered that the elementary form of comparison of sizes is superposition,[23] and that the relation between whole and parts is directly perceived rather than verbalised, the Greek practice becomes clear. The metric relations of Greek mathematics are not conceptual, but concrete. As such, they can easily be supported by an appeal to concrete evidence, namely the diagram.

Why is the diagram reliable? First, because references to it are references to a construction, which, by definition, is under our control. Had one encountered an anonymous diagram, it would have been impossible to reason about it.[24] The diagram which one constructed oneself, however, is also known to oneself, because it is verbalised. Note the combination: the visual presence allows a synoptic view, an easy access to the contents; the verbalisation limits the contents. The text alone is too difficult to follow; the diagram alone is wild and unpredictable. The unit composed of the two is the subject of Greek mathematics.

What in the diagram is referred to? As already noted above in chapter 1, the diagram is not directly relied upon for metric facts. My conclusion in chapter 1 was that the diagram was seen mainly through the relations between the lettered objects it contained. Those are exactly the betweenness and composition relations which we have seen above.

Diagrams were also used to help intuition in order to make arguments simpler. Hilbert's geometry is no less deductive than Euclid's, but it is much more cumbersome, and so its cognitive style is different. It is deductive through a different process, more 'strategic' in nature.

[22] Of course, decomposition and recomposition of objects are central to some specific theories, such as book II of the *Elements* and, interestingly, already Hippocrates of Chios' quadrature of lunules.

[23] As in Euclid's *Elements* I.4, 8.

[24] As noted by Plato – while making a rather different point – in *Rep.* 529d–e.

When we have surveyed all the elements of necessity in Greek mathematics, we may be able to spell out this notion of 'strategic'.

1.2.2 'Intuition' starting-points

I now move to the last group of sources of necessity. This was characterised above simply as 'intuitions', which would tend to make us think of it as a diffuse group made of unconnected sets of mental contents. But it is possible to see a certain affinity between the diagram-independent intuitions which are at work in Greek mathematics.

First, such intuitions cover arithmetic. Apollonius assumes in *Conics* 1.33, for instance, that – excuse the anachronism – $(2a)^2 = 4a^2$.[25] As noted already in chapter 1 above, arithmetic is common in Greek mathematics, which often (especially in astronomy) calculates things. Simple arithmetical facts are not proved, but seen and memorised. In terms of the practices of Greek mathematics, that $5+7$ equals 12 was a piece of true judgement, not of knowledge.[26]

That such judgements could be made with this sort of transparency reflects, perhaps, a 'tool-box' of a very rudimentary sort, like our multiplication-table;[27] or it may reflect the inherent simplicity of judgements such as $7+5=12$ which are similar, after all, to the geometrical decompositions noted above. They are the discrete equivalent of the continuous spatial intuitions of the diagram.

The structure of such abstract decompositions is partially captured by Euclid's common notions. I do not suppose that anyone really *relied* upon Euclid's common notions (although they are invoked, rarely, in Euclid's *Elements*),[28] but even these do not define the concepts of equivalence used in Greek mathematics, no matter how expansively we take the manuscript tradition. Take, for instance, Archimedes' *SC* 1.10:[29]

> τὰ ΑΗΕ, ΗΕΖ, ΓΕΖ τρίγωνα μετὰ τοῦ Θ ἐστιν τὰ ΑΕΔ, ΔΕΓ τρίγωνα
>
> 'The triangles ΑΗΕ, ΗΕΖ, ΓΕΖ together with Θ are the triangles ΑΕΔ, ΔΕΓ'.

[25] 100.18–20. In fact, the assumption may be purely geometric (that the square on a line twice the length is four times larger), but, even so, it does not appear that a *visualisation* is at work here.

[26] Cf. Plato's *Theaetetus* 195eff.

[27] The existence of such a – widely shared – tool-box is beyond doubt; see Fowler 1987 (270–9), 1995, for the evidence.

[28] See, e.g. *Elements* 1.1 12.10–11. This is related to the larger question of explicit references, to which I shall return later.

[29] 38.3–4.

This relies upon the provision made much earlier in the proposition:[30]

> ᾧ δὴ μείζονά ἐστιν τὰ ΑΕΔ, ΔΓΕ τρίγωνα τῶν ΑΕΗ, ΗΕΖ, ΖΕΓ τριγώνων, ἔστω τὸ Θ χωρίον
>
> 'That by which the triangles ΑΕΔ, ΔΓΕ are greater than the triangles ΑΕΗ, ΗΕΖ, ΖΕΓ, let the area Θ be'.

I am not denying that the first quotation is a hypothesis starting-point, but it must be realised that it is also an intuition. That $a + b = c$ is equivalent to $a = c - b$ may seem trivial, but it is still a necessary element in Archimedes' argument. What is more, this truth is not covered by Euclid's common notions.

Basic assumptions about relations of decomposition and equalities/inequalities permeate the use of proportion theory. Consider Apollonius' *Conics* I.41:[31]

> ὡς ἡ ΔΓ πρὸς ΓΘ, τὸ ἀπὸ τῆς ΔΓ πρὸς τὸ ὑπὸ τῶν ΔΓΘ
>
> 'As ΔΓ is to ΓΘ, the <square> on ΔΓ is to the <rectangle contained> by ΔΓΘ'.

No diagram, I believe, is required in order to see the force of the assertion. Of course, the assertion may come from the tool-box,[32] but its obviousness is related to the intuition of composition and decomposition behind it. You start with the lines, and you compose them with the same thing – so nothing was changed and the equivalence is retained.

Similar intuitions are at work in the composition-of-ratios structure: 'the ratio of *AB* to *CD* is composed of the ratio of *AB* to *EF* and of the ratio of *FE* to *CD*'. 'Composition' is indeed the right term. We saw in the preceding chapter the verbal underpinning of this operation, as a formula. The text is not laboriously read. The constituents are directly read off, and all one needs to do is to ascertain that they compose the formula according to the slots it prepares.

Just as the lettered diagram is the concrete substratum which supports diagram starting-points, so, in many cases, formulaic language is the concrete substratum which supports intuition starting-points.

[30] 36.8–10. [31] 124.15–16.

[32] The truth of this may be extracted from *Elements* VI.1, but the extraction demands some thought. Some ancient version of *Elements* may have contained the result referred to here directly, but I am in general against this line of interpretation (I shall return to this in section 4 below).

The most important intuition, perhaps, is yet another relative of 'decomposition' intuitions, only here the whole of logical space is decomposed. I refer to a starting-point such as in Archimedes' *SC* I.10:[33]

τὸ δὴ Θ χωρίον ἤτοι ἔλαττόν ἐστιν τῶν ΑΗΒΚ, ΒΖΓΛ ἀποτμημάτων ἢ οὐκ ἔλαττόν

'The area Θ is either smaller than the segments ΑΗΒΚ, ΒΖΓΛ, or it is not smaller'.

This is what I call a grid argument. A certain grid is laid over the logical space, and everything is said to fall under it. The grid is exhaustive, hence the necessity it conveys. After all the options have been surveyed, no alternative should be left open.

The quotation from Archimedes is a case where the grid divides logical space by a certain relation holding, or failing to hold. Here, then, the assertion is supported by the logical intuition behind the *tertium non datur.* In other cases, the intuition may be less logical and more spatial, as in Euclid's *Elements* IV.5 (fig. 5.8):[34]

συμπεσοῦνται δὴ ἤτοι ἐντὸς τοῦ ΑΒΓ τριγώνου ἢ ἐπὶ τῆς ΒΓ εὐθείας ἢ ἐκτὸς τῆς ΒΓ

'They will fall either inside the triangle ΑΒΓ or on the line ΒΓ or outside ΒΓ'.

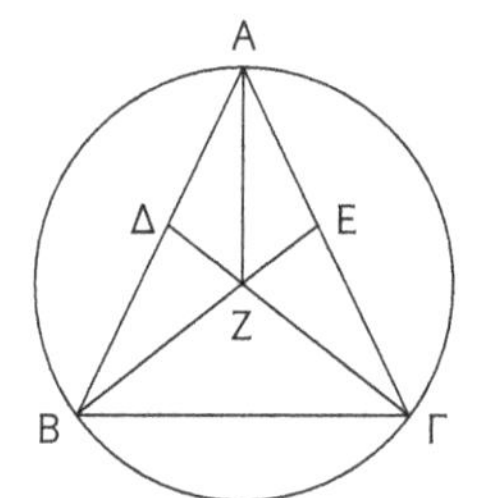

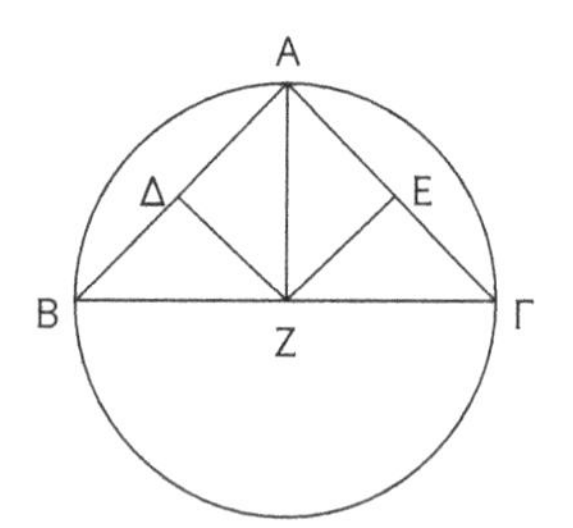

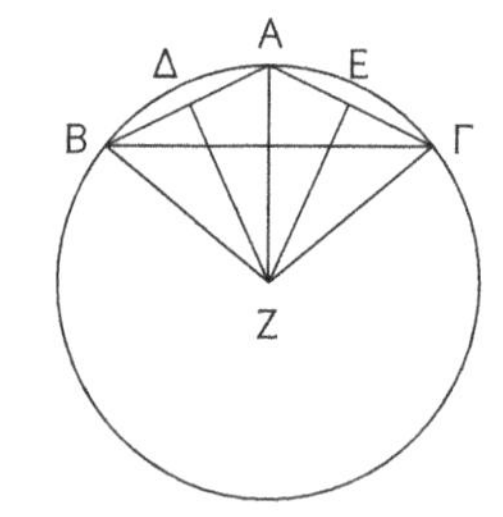

Figure 5.8. Euclid's *Elements* IV.5.

Here the diagram is required in order to grasp the necessity of the argument, but the logical structure is similar to the case above from Archimedes.

It should be noted that grid arguments are essential not only for argumentation by exhaustion of the kind we have just seen, but to any

[33] 36.10–12. [34] 280.22–4.

reductio argument. The refutation of the hypothesis of the *reductio* leads to the demonstration of the negation of the *hypothesis* through a grid-argument (often implicit).[35]

To survey quickly, then, the ground we have covered so far: we have seen that the diagram yields directly one set of starting-points, and is indirectly responsible for many other starting-points. Another important set of starting-points – intuitions – is sometimes mediated through the formulaic use of language. (And we will see in section 4 below that starting-points arising from the tool-box also sometimes rely on formulae.) On the whole, however, I would say that the linguistic cognitive tools employed by Greek mathematics are important mostly for arguments, for the co-ordination of arguments into clusters, and for the management of the tool-box, rather than for starting-points, where the diagram is most important.

2 ARGUMENTS

The typology of sources of necessity in starting-points can serve as a preliminary approximation. Of the six sources mentioned there, two are irrelevant: 'implicit arguments'[36] and, of course, 'hypotheses'. So we are left with four types of sources of necessity:

(a) References
(b) Tool-box
(c) Diagram
(d) Intuitions

The 'tool-box' and 'intuitions' are much richer in the context of arguments than they are in the context of starting-points.[37]

[35] The grid has a special interest, in that it can be connected to wider anthropological discussions of the role of tabulation in thinking; see Lloyd (1966); Goody (1977), chapter 4. Because of their role in the *reductio*, grid arguments are also relevant to the question of the relation between Greek mathematics and Greek philosophy, especially following Szabo (1969). Progress can be made only if we put grid arguments in the context of other decomposition arguments.

[36] It is pointless to distinguish 'completely explicit' and 'partly implicit' arguments, in the absence of a meta-mathematical theory of completeness. In Greek mathematical arguments, there are more or less immediately compelling arguments, and all of them are in a sense 'complete' – in the sense that they sufficed to convince someone.

[37] The statistical limitations of such surveys have been explained already. Granted this qualification, I will say that intuition arguments and tool-box arguments are responsible, each, for almost half the total arguments (often, in conjunction with the diagram). Of the remaining 10 per cent or so, almost all rely upon the diagram, with a very small minority being reference arguments.

2.1 *Reference*

Reference arguments are those where some explicit justification for the argument is made. That is: besides stating the premisses and the result, a reference argument asserts, in some form, 'when the premisses, then the results'.

This is rare, even in a minimal form. The following is an example (fig. 5.9):[38]

ἐπεὶ οὖν ὡς ὅλον ἐστὶ τὸ ἀπὸ ΑΕ πρὸς ὅλον τὸ ΑΖ, οὕτως ἀφαιρεθὲν τὸ ὑπὸ ΑΔΒ πρὸς ἀφαιρεθὲν τὸ ΔΗ, καὶ λοιπόν ἐστι πρὸς λοιπόν, ὡς ὅλον πρὸς ὅλον. ἀπὸ δὲ τοῦ ἀπὸ ΕΑ ἐὰν ἀφαιρεθῇ τὸ ὑπὸ ΒΔΑ, λοιπόν ἐστι τὸ ἀπὸ ΔΕ· ὡς ἄρα τὸ ἀπὸ ΔΕ πρὸς τὴν ὑπεροχὴν ἣν ὑπερέχει τὸ ΑΖ τοῦ ΔΗ, οὕτως τὸ ἀπὸ ΑΕ πρὸς τὸ ΑΖ

'Now since, as is the whole square on AE to the whole area AZ, so is the subtracted rectangle <contained> by AΔB to the subtracted ΔH, and the remaining is to remaining as whole to whole. But if the rectangle under BΔA is subtracted from the square on EA, the remaining is the square on ΔE; therefore as the square on AE is to the difference, by which AZ is more than ΔH, so is the square on AE to AZ'.

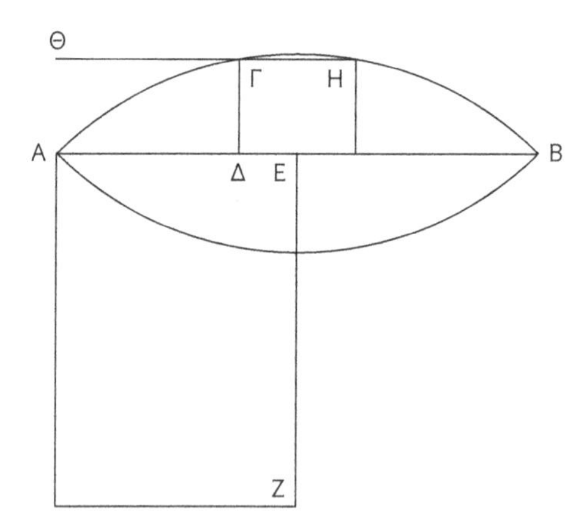

Figure 5.9. Apollonius's *Conics* 1.41 (Ellipse Case).

I have underlined the sentence in which a general principle is cited: an argumentation formula. Significantly, it is just cited, without giving

[38] Apollonius' *Conics* 1.41. 126.26–128.3.

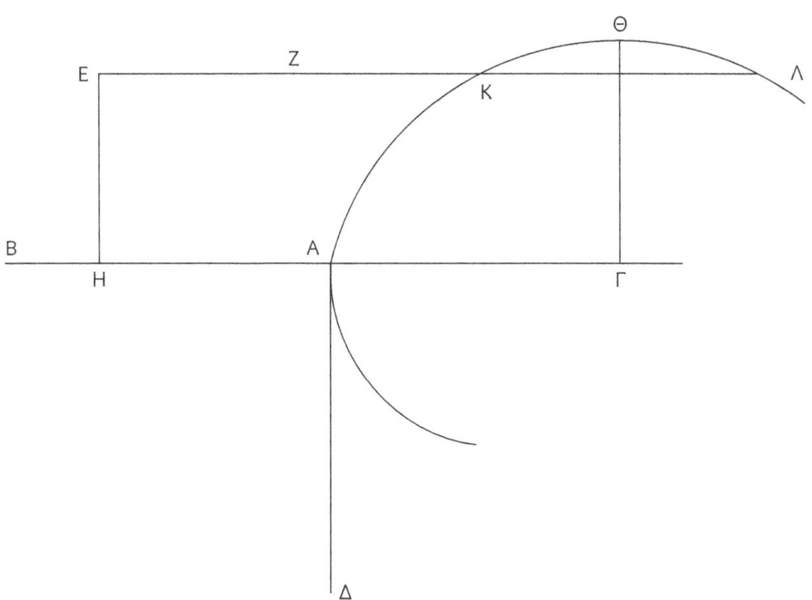

Figure 5.10. Apollonius' *Conics* I.26 (Parabola Case).

the reference.[39] In fact, the general principle is enigmatically alluded to in our example even by the words 'whole', 'remaining'. *This* is the more typical way of allusion in Greek mathematics. The underlined sentence is therefore an over-allusion. Even when *over*-alluding, the Greek mathematician still does not *refer* to his source by book and proposition numbers. This will be important in section 4 below.

The rarity of such meta-statements in Greek proofs is significant in another way. What Greek mathematics eschews is not just conditionals (assertions of the form 'but when *P*, then *Q*'), but also general statements ('but in general, when *P*, then *Q*'). Proofs move from one particular statement to the next, always remaining close to the particular objects discussed (and visualised in the diagram).

2.2 *Diagram*

The diagram is less important for arguments than it is for starting-points. Many arguments are mediated visually (so many tool-box arguments and intuition arguments are, partly, also diagram arguments), and sometimes arguments seem to rely upon diagrams directly, as in the following (fig. 5.10):[40]

[39] Euclid's *Elements* V.19. [40] Apollonius' *Conics* I.26, 82.8–9.

> ἡ EZ ἄρα ἐκβαλλομένη τέμνει τὴν ΘΓ· ὥστε καὶ τῇ τομῇ συμπεσεῖται
>
> 'The line EZ, produced, cuts the line ΘΓ; so it will meet the section as well'.

ΘΓ is a line inside the section, and EZ is a line outside it, so it does seem obvious that EZ cannot cut ΘΓ unless it cuts the section as well. This is a betweenness argument which could never rely, in Greek mathematics, upon anything except the diagram.

Similarly, diagrammatic 'composition' may be relevant for an argument rather than for a starting-point, as in the following (fig. 5.11):[41]

> ἐπεὶ ὅλη ἡ AZ ὅλῃ τῇ AH ἐστιν ἴση, ὧν ἡ AB τῇ AΓ ἐστιν ἴση, λοιπὴ ἄρα ἡ BZ λοιπῇ τῇ ΓH ἐστιν ἴση
>
> 'Now since the whole AZ is equal to the whole AH, of which [both] AB is equal to AΓ, therefore the remaining BZ is equal to the remaining ΓH'.

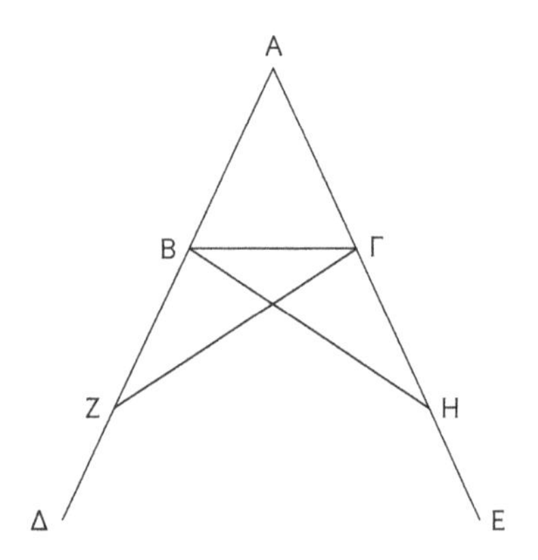

Figure 5.11. Euclid's *Elements* I.5.

2.3 *Tool-box*

The fact that the tool-box is important for arguments shows that the tool-box contains conditionals (assertions of the form $P \rightarrow Q$). This should not surprise us, as the tool-box is made mostly of theorems, which are usually of the form $P \rightarrow Q$.[42]

The tool-box will be discussed in section 4. Here I will note just that it is a source of necessity. Our evidence is that Greek mathematicians

[41] Euclid's *Elements* I.5, 20.23–5. [42] I will discuss this issue in detail in chapter 6.

are unworried about it. Tool-box arguments are presented just like any other arguments. Consider Apollonius' *Conics* i.37:[43]

(1) ὡς ἡ ΕΖ πρὸς ΖΒ, ἡ ΖΒ πρὸς ΒΔ. (2) ἴσον ἄρα εστὶ τὸ ὑπὸ ΕΖΔ τῷ ἀπὸ ΖΒ

'(1) As ΕΖ to ΖΒ, so is ΖΒ to ΒΔ. (2) The rectangle under ΕΖΔ is, therefore, equal to the square on ΖΒ'.

The result assumed here is relatively simple, but it is compelling only by virtue of its being proved elsewhere.[44] However, this is not marked here (by a 'for it is proved', say). Remember that arguments are generally clearly charted by logical connectors, by 'therefore' and 'since' relations. The unmentioned premiss, *Elements* vi.17, is therefore marked by its absence. Apollonius effectively asserts that the grounds for the claim in (2) are the claim in (1) and nothing else. *Elements* vi.17 is a background for the derivation, not a part of it. *Elements* vi.17 functions here in the same way as Greek grammar does. It is a necessary piece of background, but it is not even noticed. I shall discuss and qualify this in section 4.

2.4 *Intuition*

This is not a homogeneous group. Some intuition arguments, for instance, rely upon arithmetic. Most belong, however, like their counterparts in starting-points, to a well-defined family.

Consider, for instance, the following:[45]

ἐπεὶ ἔστιν, ὡς τὸ ἀπὸ ΜΨ πρὸς τὸ ἀπὸ ΨΙ, τὸ ὑπὸ ΑΠΒ πρὸς τὸ ὑπὸ ΔΠΕ, ἀλλ' ὡς τὸ ὑπὸ ΑΠΒ πρὸς τὸ ὑπὸ ΔΠΕ, τὸ ἀπὸ ΛΤ πρὸς τὸ ἀπὸ ΤΙ, καὶ ὡς ἄρα τὸ ἀπὸ ΜΨ πρὸς τὸ ἀπὸ ΨΙ, τὸ ἀπὸ ΛΤ πρὸς τὸ ἀπὸ ΤΙ

'Since, as the <square> on ΜΨ to the <square> on ΨΙ, so is the <rectangle contained> by ΑΠΒ to the <rectangle contained> by ΔΠΕ, but as the <rectangle contained> by ΑΠΒ to the <rectangle contained> by ΔΠΕ, so is the <square> on ΛΤ to the <square> on ΤΙ, therefore also as the <square> on ΜΨ to the <square> on ΨΙ, so is the <square> on ΛΤ to the <square> on ΤΙ'.

[43] 110.18–19. The numbers (1) and (2) are mine. [44] *Elements* vi.17.
[45] Apollonius' *Conics* iv.46, 72.4–8.

The structure of the argument is ($a{:}b{::}c{:}d$ and $c{:}d{::}e{:}f$) $\rightarrow$ $a{:}b{::}e{:}f$. This is the transitivity of proportionality. By understanding what proportionality is, we see that the relation between $a{:}b$ and $c{:}d$ which is implied by $a{:}b{::}c{:}d$ is such that, in other proportionality contexts, $a{:}b$ and $c{:}d$ may be substituted *salva veritate* (i.e. with truth-values remaining the same). For instance, we see that, if $a{:}b{::}c{:}d$, then $c{:}d{::}a{:}b$. Changing the order is immaterial.

This intuition may be backed by the underlying symbolism. In '$a{:}b{::}c{:}d$', for instance, both the distinction into two parts and the symmetry between the two parts are clearly preserved. The Greek formula – formula 55 of the preceding chapter – has the following structure:

$P_{55}[O_{54},O_{54}]$ =
as $[O_{54}]$ so $[O_{54}]$ =
as [[*a*] to [*b*]] so [[*c*] to [*d*]]

This is the best possible oral approximation to $a{:}b{::}c{:}d$.[46]

The most important thing for both systems of symbolism, the typographic and the oral, is that both support the *salva veritate* substitution intuition by supplying slots in which elements may be substituted: the two flanks of the '::' in the typographic symbol, or the two clauses in the oral symbol.[47]

The concept of substitution *salva veritate* is the key to most proofs. The structure is usually the following: a certain property is asserted for *A*; a substitution *salva veritate* between *A* and *B* is established; the property is transferred to *B*; the proof, or a major result needed for the proof, is thereby settled. Take Euclid's *Elements* IX.6. This proves that if A^2 is a cube, so is A. The proof starts off by laying down B as A^2, and Γ as (A*B). Then the argument moves on to the seemingly quite irrelevant 1:A::A:B. This has nothing obvious to do with the proof, but it allows a certain interchangeability. Indeed, the proof then continues to arrive at 1:A::B:Γ. Here comes the crucial argument:[48]

> ἔστιν ἄρα ὡς ἡ μονὰς πρὸς τὸν A, οὕτως ὁ B πρὸς τὸν Γ ['1:A::B:Γ'; the conclusion of the second line of reasoning]. ἀλλ' ὡς ἡ μονὰς πρὸς τὸν A, οὕτως ὁ A πρὸς τὸν B ['1:A::A:B'; the first

[46] '$a{:}b{::}c{:}d$' is, of course, a matter of typography, and in oral contexts a modern reader may well supply 'as *a* is to *b* . . .' as the reading of that typographical symbol.

[47] Going one level lower, one may substitute individual constituents of a ratio instead of a complete ratio.

[48] 348.20–3.

line of reasoning is recalled] καὶ ὡς ἄρα ὁ Α πρὸς τὸν Β, ὁ Β πρὸς τὸν Γ ['Α:Β::Β:Γ'; the *salva veritate* argument is effected and the proof can now unfold, having secured the central result].

One type of substitution is the *salva veritate*, where a single element is replaced within a single slot. A more complex type of substitution is where the substitution of *A* by *B* is compensated by the substitution of some other *C* by *D*. Most argumentation formulae in proportion theory belong to this type, e.g. formula 56 in chapter 4:

$S_{56}[P_{55}[O_{54},O_{54}],P_{55}[O_{54},O_{54}]] =$
$S_{56}[P_{55}[$[*a*] to [*b*], [*c*] to [*d*]], $P_{55}[$[*a*] to [*c*], [*b*] to [*d*]]] =
$S_{56}[$[as [*a*] to [*b*], so [*c*] to [*d*]], [as [*a*] to [*c*], so [*b*] to [*d*]]] =
[as [*a*] to [*b*], so [*c*] to [*d*]] therefore, *enallax*, [as [*a*] to [*c*], so [*b*] to [*d*]]

All that has happened is that *b* and *c* have been interchanged. Yet how much is involved! One must identify at a glance the correct slots (hence the all-importance of the awareness of form, harnessed by the formulaic structure). And the result is vast: for instance, the arithmetical equivalent, *Elements* VII.13, is directly used by Euclid in his arithmetical books 16 times. And then, *enallax* is only one of a set of similar formulae (discussed in chapter 4, subsection 3.2). No other single set is as important.[49]

Is there an ancient Greek proof that $(a{:}b{::}c{:}d \;\&\; c{:}d{::}e{:}f) \rightarrow (a{:}b{::}e{:}f)$? There is: Euclid's *Elements* V.11. But this is the exception. Book V of the *Elements* is a rare treatise in its logical completeness, and even there, crucial theorems are left implicit. For instance, Euclid never proves that $(a{:}b{::}c{:}d) \leftrightarrow (c{:}d{::}a{:}b)$, and, therefore, it is not a proved result of $(a{:}b{::}c{:}d \;\&\; e{:}f{::}c{:}d)$ that $(a{:}b{::}e{:}f)$.

Excluding the (partial) descriptions of the logical structures of equality (in the *Common Notions*) and proportionality (in book V), no effort to capture such logical structures was made by Greek mathematicians. However, it is chiefly this logic of relations which is responsible for the original component (as opposed to the tool-box) in the arguments made in Greek mathematics.[50]

[49] I shall discuss the relative importance of various parts of the tool-box in section 4 below. While this is the most important set of argumentation formulae, it is not the most important set of results. The most important results are those which do not rely on formulae alone (i.e. are not book V), but those which mediate between diagram and formulae (i.e. book VI).

[50] Other relations whose logic is only implicit in Greek mathematics include: inequality, proportional inequality, similarity, congruity, addition and subtraction.

Just as starting-points based on the intuition of decomposition of objects into their components are directly compelling by virtue of their visual appeal, so arguments based on substitution are directly compelling by virtue of their linguistic appeal, arising from the formulaic use of language. There is a parallelism between the two triads:

Starting-points – Decomposition – Diagram
Arguments – Substitution – Formula

I will now widen the concept of substitution, parallel to the way I have widened the concept of decomposition in the preceding section. Consider the following:[51]

παρὰ τὴν δοθεῖσαν ἄρα εὐθεῖαν τὴν ΑΒ τῷ δοθέντι εὐθυγράμμῳ τῷ Γ ἴσον παραλληλόγραμμον παραβέβληται τὸ ΣΤ ἔλλειπον εἴδει παραλληλογράμμῳ τῷ ΠΒ ὁμοίῳ ὄντι τῷ Δ

'Therefore the parallelogram ΣΤ, equal to the given rectilinear Γ, deficient by the parallelogram ΠΒ (this parallelogram being similar to Δ), has been applied on the given straight line AB'.

This is the conclusion of an argument, perhaps the most crucial in its proposition. But it is nothing but a renaming, a reidentification of the objects. The objects are reidentified as falling under specific descriptions. This is a very typical ending. The point is often (as in the case quoted here) to equate the assertion arrived at with that which the proof set out to prove. Or reidentifications may be important in other ways, e.g. the main argument in Archimedes' *SC* is a reidentification of the solid created by rotating a polygon with a series of cones and truncated cones (fig. 5.12).[52] This allows Archimedes to carry over properties shown for polygons to cones. The possibility of seeing the same thing as equivalent with some other allows the author to move back and forth between the two equivalent representations, yielding deductively fertile combinations.

What makes such substitutions workable? The structure is: (1) *A* occurs once in the *substitution warrant*, the assertion stating its inter-

[51] Euclid's *Elements* VI.28, 166.11–14.
[52] This is *SC* I.23, the first proposition of this kind; there are several in this book. Another case where reidentifications are the key to argumentation is the identity, used in Archimedes' *Method*: area ↔ set of lines.

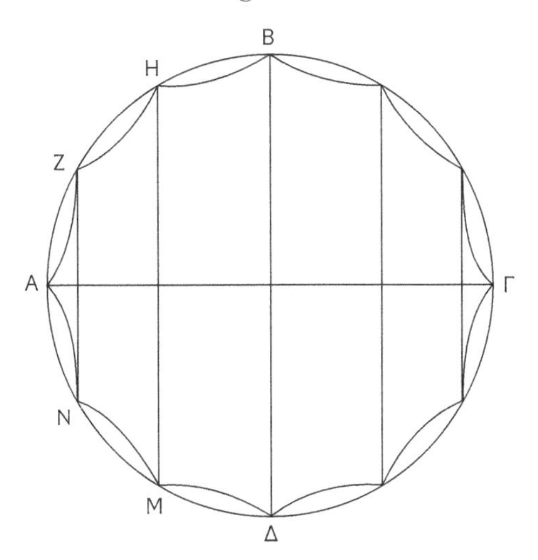

Figure 5.12. Archimedes' *SC* 1.23.

changeability with *B*; (2) *B* occurs in another context; (3) *A* is inserted in that context, replacing *B*.[53] This requires two things: *A* must be labelled consistently, so that its identity is secure; and the context in which *B* occurred and, later, *A* occurred, must be identifiable, i.e. the context in which *B* occurred must be conceptualised as an open function, in which the slot used by *A*/*B* is left empty. The existence of verbal formulae is the basis of both these conditions.

The constant reshuffling of objects in substitutions may be securely followed, since it is no more than the refitting of well-known verbal elements into well-known verbal structures. It is a game of decomposition and recomposition of phrases, similar, indeed, to the game of decomposition and recomposition of visual objects which we have seen in the preceding section.

I now move to a detailed and more technical example (readers without Greek may prefer to skip it). I take the first three arguments of Apollonius' *Conics* 1.50 (fig. 5.13):

(1) καὶ ἐπεὶ ἴση εστὶν ἡ ΕΓ τῇ ΓΚ, (2) ὡς δὲ ἡ ΕΓ πρὸς ΚΓ, ἡ ΕΣ πρὸς ΣΘ, (3) ἴση ἄρα καὶ ἡ ΕΣ τῇ ΣΘ. (4) καὶ ἐπεί ἐστιν, ὡς ἡ ΖΕ πρὸς ΕΗ, ἡ ΘΕ πρὸς τὴν διπλασίαν τῆς ΕΔ, (5) καί ἐστι τῆς ΕΘ

[53] The precise sequence may vary, of course.

ἡμίσεια ἡ ΕΣ, (6) ἔστιν ἄρα, ὡς ἡ ΖΕ πρὸς ΕΗ, ἡ ΣΕ πρὸς ΕΔ. (7) ὡς δὲ ἡ ΖΕ πρὸς ΕΗ, ἡ ΛΜ πρὸς ΜΠ (8) ὡς ἄρα ἡ ΛΜ πρὸς ΜΠ, ἡ ΣΕ πρὸς ΕΔ

'(1) And since ΕΓ is equal to ΓΚ, (2) but as ΕΓ to ΚΓ, ΕΣ to ΣΘ, (3) therefore ΕΣ is equal to ΣΘ, (4) and since, as ΖΕ to ΕΗ, ΘΕ to twice ΕΔ, (5) and ΕΘ is half ΕΣ, (6) therefore, as ΖΕ to ΕΗ, ΣΕ to ΕΔ. (7) But as ΖΕ to ΕΗ, ΛΜ to ΜΠ (8) therefore as ΛΜ to ΜΠ, ΣΕ to ΕΔ'.

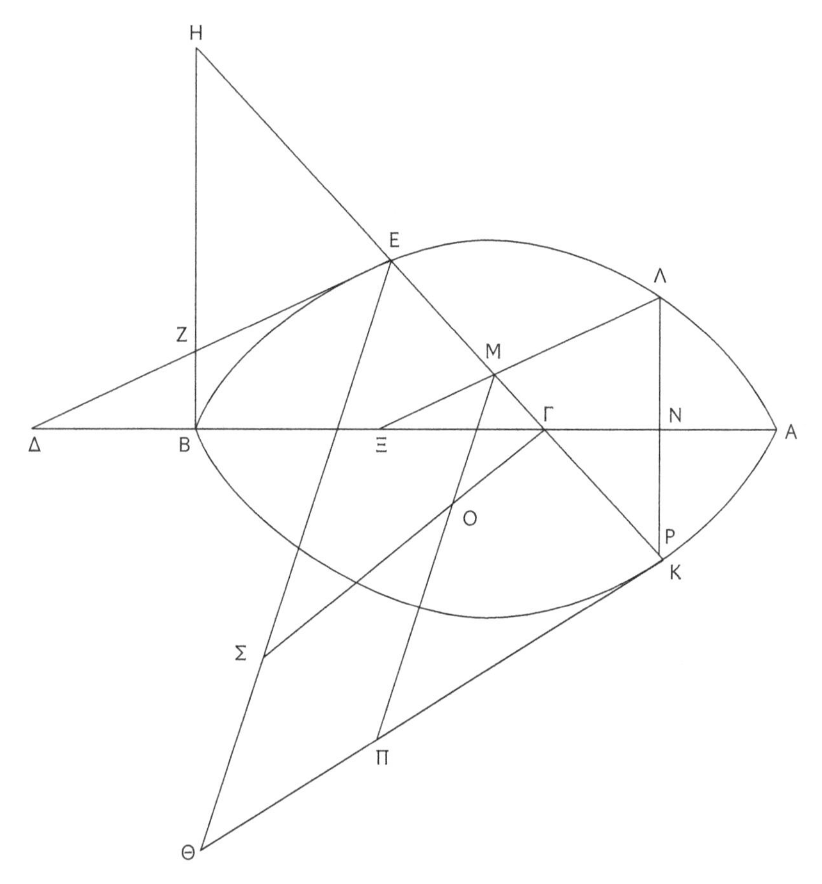

Figure 5.13. Apollonius' *Conics* 1.50 (Ellipse Case).

The development of the argument can be visualised as in the following 'tree' (fig. 5.14a):

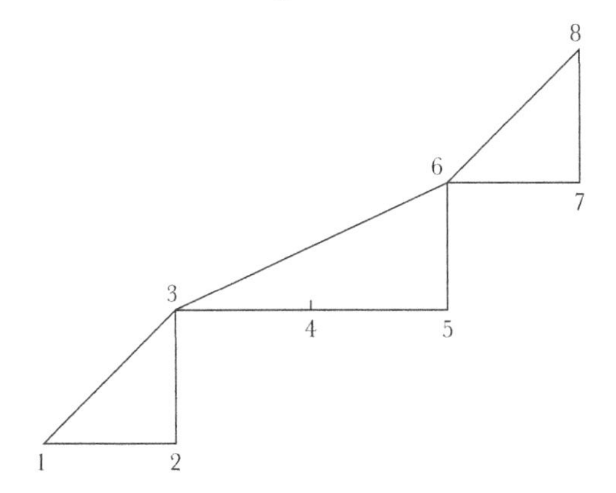

Figure 5.14a.

The formulae are (numbers are those of the preceding chapter):[54]

2 – Line 55 – Proportion
54 – Ratio 78 – Equality

Fig. 5.14b is the phrase-structure trees of the assertions:

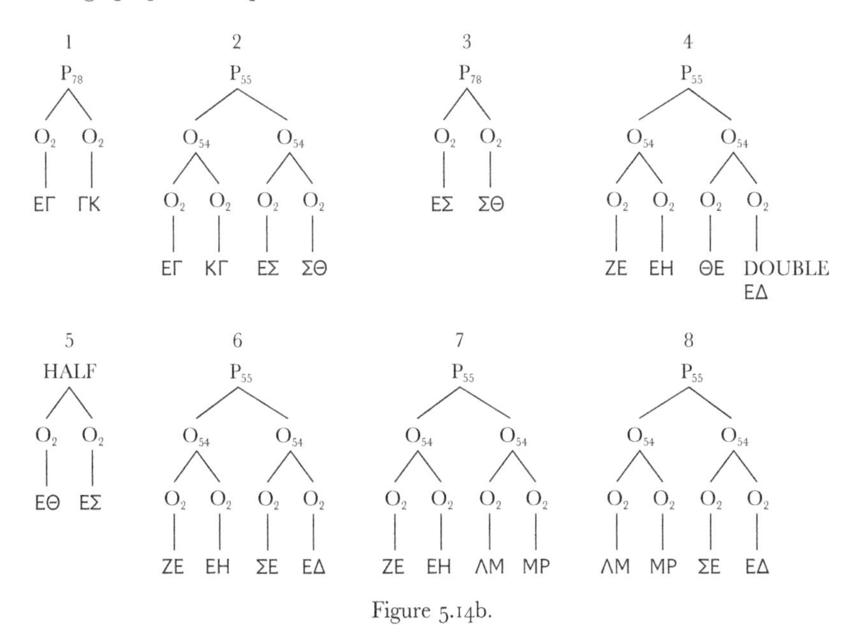

Figure 5.14b.

[54] 'Twice' and 'half' I do not view as formulaic. As explained in chapter 4, numbers tend to be non-formulaic. The reciprocal relation between 'twice' and 'half' is directly perceived: an arithmetical intuition.

All the formulae govern, ultimately, 'line' formulae, in turn governing pairs of letters. It is solely by substituting structures of such couples of letters that the text proceeds.

The substitutions are:

First argument (assertions 1–3)
The phrase-structure tree of assertion 1 is one of the constituents of the tree of 2; 2 asserts an equivalence between its two topmost consituents. Hence 1 may be inserted into 2. This is 3.
Second argument (assertions 4–6)
On an unformulaic basis, we understand that 'twice' is the reciprocal of 'half'. Hence moving from 'twice' to the thing itself, or by moving from the thing itself to its 'half', the equivalence is retained: 4 directly contains one member as a 'twice' (twice ΕΔ), and its correlate member, ΕΣ, is identified, in 5, as that whose half is ΕΘ. Hence the two may simultaneously transform with the equivalence kept. All this pertains to only one wing of 4, and, this wing changed, 6 results. Not so many formulae – but notice how much awareness of form is required here!
Third argument (assertions 6–8)
The simplest possible case: 7 asserts an equivalence between two wings, one of which occurs in 6. The substitution results in 8.

Everywhere, substitution and awareness of form is the crucial element – no doubt, much facilitated by the accessibility of the referents via the diagram (notice the substitutability, so well known from chapter 2, of, e.g. ΓΚ and ΚΓ).

Before the section on substitution is concluded, an even wider generalisation is required. What is it about Greek mathematics which makes it so amenable to the operation of substitution? Of course, formulae are the material reality which make substitution work – this requires no further argument. But why substitutions to begin with?

The answer must confront the subject matter. The crucial thing is that Greek mathematics relies so much upon relations of equivalence, such as identity, equality, proportionality. These formulae operate in a double role: once as a substratum for manipulation, once again as a licence for manipulation. *a*:*b*::*c*:*d* is both a set of objects, in which '*a*:*b*', for instance, is ready for substitution by other, equivalent ratios, and it is also a statement about objects, asserting the substitutability of '*a*:*b*' and '*c*:*d*'. Equivalence relations are both the raw material and the machines in the factory of Greek proofs.

On a wider view still, another small class, of relations which are transitive without being equivalence relations, governs much of what remains of Greek proofs:[55] mainly, the relations 'greater/smaller'. The combinations of transitive relations and equivalence relations yield the set of legitimate substitutions which is the core of Greek mathematical argumentation.

I have counted the assertions in the first ten propositions of *Elements* III and the first five propositions of *Elements* VI.[56] Of a total of 276 assertions, 199 assert relations, while 77 assert single-place predicates. Of the 199 relations, 152 are equivalences, 45 are transitive, and only two are neither.[57] The majority of *assertions* is that of equivalences. I have already noted in the previous chapter the enormous repetitiveness of the relation 'equality' in Greek mathematics. We now see the logical significance of this centrality.

So the logicists were right. It is the logic of relations which sets mathematics going.[58] Mathematical relations fall under specific logical relations, and this is why the deductive machine (as far as formulae are concerned) is capable of dealing with them. So what shall we say then? Perhaps, that to the – significant – extent that deduction works with formulae, we can say that Greek mathematics is ultimately deductive, because it deals with transitive relations.[59] This answer is partly valid. The empirical world is recalcitrant, it does not yield to logic, and this is because it behaves by degrees, by fine shades, by multiple dimensions. Shading into each other, the chains of the relations operating in the real world break down after a number of steps: the quantity of liquid, transferred again and again from vessel to vessel, will finally reduce; the preferability of A to B and of B to C does not always entail that of A to C. Mathematical objects are different.

Or are they just assumed to be different? Are they *constructed* as different? We are historians – we do not have to answer such questions. All we have to note is that there is a decision here, to focus on

[55] A transitive relation is such that pRq (signifying 'p stands in the relation R to q'), with qRs, entails pRs. A subset of the class of transitive relations is equivalence relations, which are also symmetric (pRq entails qRp) and reflexive (pRp for every p).

[56] The combination is meant to yield a more or less 'representative' sample, III being free of proportion, while VI is all about proportion.

[57] Twice, the relation 'twice' in VI.1.

[58] Russell (1903) 23: 'A careful analysis of mathematical reasoning shows (as we shall find in the course of the present work) that types of relations are the true subject-matter discussed.' By his 'analysis' Russell meant a remaking of mathematics, not a historical appreciation. I now offer a historical vindication of Russell's claim.

[59] I focus on transitivity, a more difficult concept than symmetry or reflexivity.

relations in so far as they are transitive. Whether they really exist independently of the decision is a question left for the philosopher to answer; the historian registers the decision. At some stage, some Greeks – impelled by the bid for incontrovertibility, described in Lloyd (1990) – decided to focus on relations in so far as they are transitive, to demand that in discussions of relations of area and the like, the make-believe of ideal transitivity should be entertained. Here is finally the make-believe, the abstraction truly required by Greek mathematics. Whether the sphere is made of bronze or not is just immaterial. The important requirement – the point at which mathematics takes off from the real world – is that if the sphere is equal in volume to some other object, say 2/3 the circumscribed cylinder, and this cylinder in turn is equal to some other object *X*, then the sphere will be equal to *X*. This is true of 'equal' only in an ideal sense, a sense divorced from real-life applications and measurements. And this is the *qua* operation, the make-believe at the heart of Greek mathematics.

Finally, a paradox. It is just because there is an inherent make-believe in the diagram that the make-believe of transitivity is naturally entertained. 'This is equal to that, and this to that, so this to that' – 'Oh really? Have you measured them?' – 'Come on, don't be a fool. There's nothing to measure here – it's only a diagram.'

'Nothing to measure here': I have invented this retort, but it is there in the original – in the behaviour of the diagram. It is precisely this metric aspect, these relations of measurement, that the diagram does not set out to represent. Such relations were represented by a system of formulae. Diagrams and formulae are thus functionally related in a single structure.

The diagram and (more generally than just formulae) the technical language are the two complementary tools, yielding atoms of necessity. We now move on to see how these atoms are combined in necessity-preserving ways.

3 THE STRUCTURE OF PROOFS

3.1 Size

Proofs are combinations of arguments (in the limited sense of 'argument' used above). What size of arguments? And how many?

As always, the statistical value of my survey is limited, but I will venture this on the size of arguments: roughly, about half (or more) of all arguments are two-assertion arguments, i.e. of the form $P \rightarrow Q$ (The assertion P yields the assertion Q). Of those remaining, the great majority are three-assertion arguments, i.e. of the form $(P \,\&\, Q) \rightarrow R$ (the assertions P and Q, together, yield R). A few arguments (less than 10%) are four-assertion arguments, i.e. of the form $(P \,\&\, Q \,\&\, R) \rightarrow S$. Larger arguments exist, but are a rarity.

We see therefore that arguments are short – and this is one way in which they are easy to follow. Another way is the explicitness of arguments, i.e. the use of logical connectors, which I have already discussed in chapter 3 above. The system of logical connectors is small and relatively rigid. So a proof consists of short, clearly marked arguments. How many of these?

There are 343 arguments in the 31 propositions I have surveyed, somewhat more than 10 arguments per proof. However, I have tended to choose longer proofs, so the true average is in fact below 10. This can be understood better when we analyse proofs according to their main types. These are best seen by means of 'trees'. As was explained in the introduction to this chapter, it is possible to draw a 'tree' – a diagram depicting the logical progress of a proof. Look at the following trees, then. First, Archimedes' *SC* I.30 (fig. 5.15):[60]

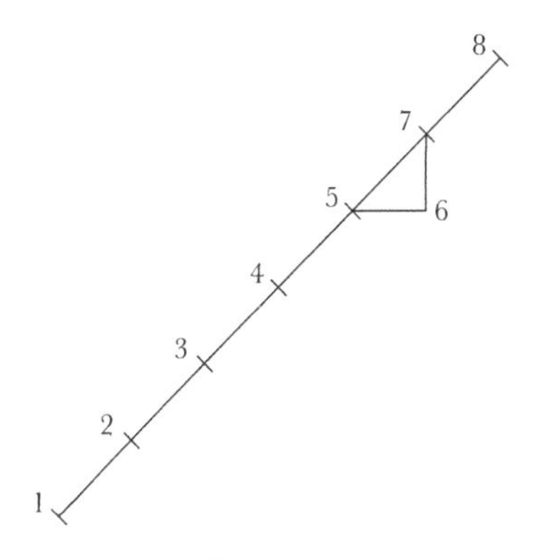

Figure 5.15.

This is a good example of a simple proof. It consists of a single, direct argument. There are no asides, no breaks, no internal structure.

Compare this to Apollonius' *Conics* I.41 (fig. 5.16):

[60] I follow Heiberg's judgement on what is 'authentic'. Since this is just an example, not much hinges on this.

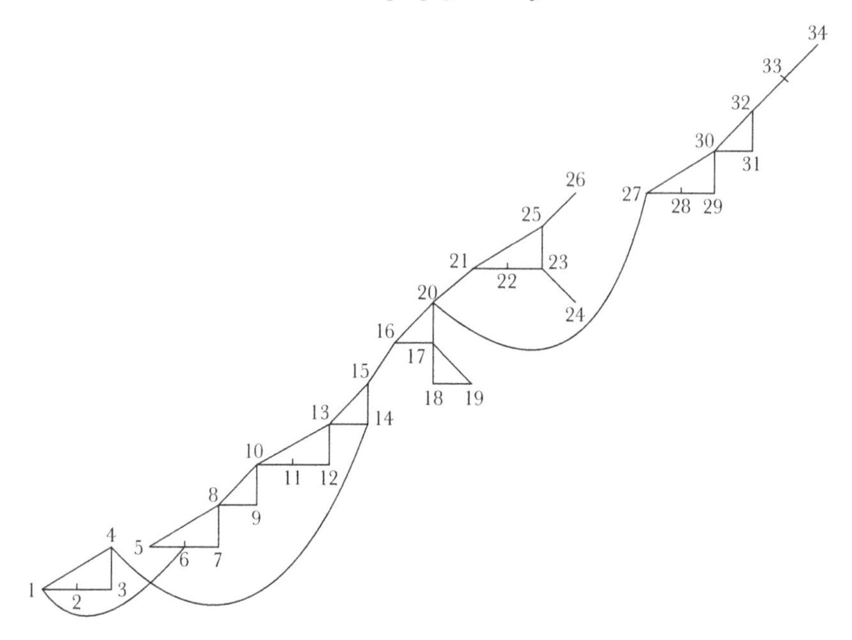

Figure 5.16.

This is a good example of a more complex type of proof. Here some internal structure is noticeable; asides and breaks occur; the proof is less direct.

SC 1.30 has 6 arguments; *Conics* 1.41 has 16 arguments. The line dividing the two types is fuzzy, but is probably at around 10. With 4 or fewer arguments, there is hardly any proof. With 10 or more, structure becomes necessary.

Logical size, measured by assertions and arguments, represents some cognitive reality. Absolute size, measured by words (or, better, syllables?), is roughly equivalent to the logical size, and has its own significance: for instance, the time necessary for reading the proof. The proof of Archimedes' *SC* 1.30 has 17 Heiberg lines; the proof of Apollonius' *Conics* 1.41 has 54 Heiberg lines. It is interesting that the ratio between absolute sizes is somewhat larger than that between logical sizes:[61] the assertions of more complex proofs tend to be more complex, individually.

[61] The logical ratio is 16/6 = approx. 2.7. The absolute ratio is 54/17 = approx. 3.2. The difference is about 20%, which seems significant.

The 10-or-fewer type will be called 'micro-proofs', the 10-or-more type seen here will be called 'meso-proofs'. There is a third type, 'mega-proofs'. By these I refer to propositions such as Apollonius' *Conics* v.51.[62] I did not count the number of assertions there, but it is a three-digit number: a very special proposition within a very sophisticated treatise. The mega-size is the result of the way in which many different results are crammed into a single proposition and then proved separately.[63] Each of these is a normal meso-proof.

It is difficult to survey the logical sizes of proofs, since much work is required before the logical structure of any given proof can be discerned. The *absolute* size of proofs is much easier to tabulate, and is related to their logical size. I have surveyed the length of proofs in the extant Greek works of Apollonius: 119 proofs are 13 to 25 lines long – which is almost always a micro-proof; 31 proofs are even shorter than that; 76 are longer, but 25 of these are 30 lines long or less (i.e. some of them are still micro-proofs). This is a relatively complex treatise, and we see that more than two-thirds of the proofs are micro-proofs (and none is more than a meso-proof). So Greek mathematics is very much composed of micro-proofs.

And yet, meso-proofs prove stronger and more interesting results than micro-proofs. My problem in this chapter is the ways in which proofs are necessity-preserving. The answer is obvious for micro-proofs, which are direct, simple combinations of their components (whose necessity was explained in the preceding sections). Meso-proofs are the locus of the problem; this is why most of the examples I have chosen to survey in detail are meso-proofs. But it is helpful to note that the standard fare in Greek mathematics was the micro-proof. Greeks were used to necessity-preservation of the micro-proof kind. We may formulate our problem as that of describing the ways in which meso-proofs converge towards micro-proofs.

3.2 Structure of meso-proofs

Euclid's *Elements* II.5 can be taken as an ideal type. I will repeat the literal translation of its proof (part of my 'specimen of Greek

[62] This is preserved only in the Arabic (Toomer (1990) 145.14–157.20), but the translation probably did not affect such global structures.

[63] Six different proofs on Toomer's count: the Arab translators supply the phrase 'proof of that' whenever they see a new proof, and Toomer adds one in 155.8.

mathematics', following the Introduction), and then give its 'tree'. I shall then discuss the important aspects of this tree – what makes it 'ideal'.

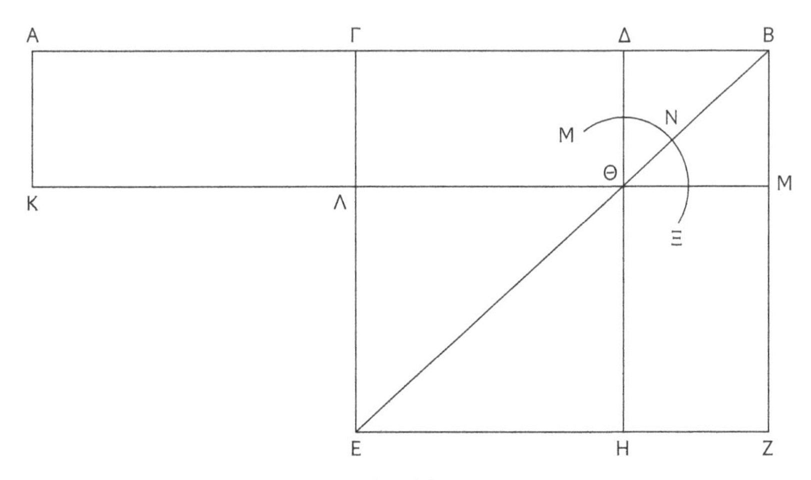

Figure 5.17. Euclid's *Elements* II.5.

(1) . . . since the complement ΓΘ is equal to the complement ΘZ,
(2) let the <square> ΔM be added in common;
(3) therefore the whole ΓM is equal to the whole ΔZ.
(4) But the <area> ΓM is equal to the <area> AΛ,
(5) since the <line> AΓ, too, is equal to the <line> ΓB;
(6) therefore the <area> AΛ, too, is equal to the <area> ΔZ.
(7) Let the <area> ΓΘ be added in common;
(8) therefore the whole AΘ is equal to the gnomon MNΞ.
(9) But the <area> AΘ is the <rectangle contained> by the <lines> AΔ, ΔB;
(10) for the <line> ΔΘ is equal to the <line> ΔB;
(11) therefore the gnomon MNΞ, too, is equal to the <rectangle contained> by the <lines> AΔ, ΔB.
(12) Let the <area> ΛH be added in common
(13) (which is equal to the <square> on the <line> ΓΔ);
(14) therefore the gnomon MNΞ and the <area> ΛH are equal to the rectangle contained by the <lines> AΔ, ΔB and the square on the <line> ΓΔ;
(15) but the gnomon MNΞ and the <area> ΛH, <as a> whole, is the square ΓEZB,
(16) which is <the square> on the <line> ΓB;
(17) therefore the rectangle contained by the <lines> AΔ, ΔB, with the square on the <line> ΓΔ, is equal to the square on the <line> ΓB.

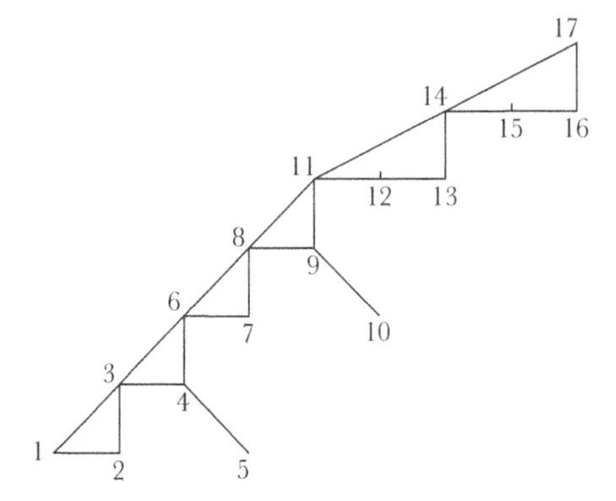

Figure 5.18. The tree of Euclid's *Elements* II.5.

3.2.1 Few backward-looking arguments

Two arguments – in assertions 5 and 10 – look backwards, justifying what comes before, not what follows next. These are the minority of arguments, and they are small, logically (they both consist of a single assertion as a premiss) and absolutely (in number of words). This is typical: the arguments of Greek mathematics look forwards, not backwards.

Of the 31 proofs I surveyed, 12 contain no backward-looking arguments, and in 8 others only 10% or less of the assertions are contained only in backward-looking arguments. Propositions which contain significantly more than this are those we will expect to be atypical. These include: Diocles' *On Burning Mirrors* 2, with 22% – which we have in the Arabic of some thousand years' distance from the original; Hippocrates of Chios' third quadrature, with 20%[64] – no less distant from the original; Aristotle's *Meteorology*, with 56% – a non-mathematical context; and, finally, Archimedes' *Method* 1, with 21%, and the *Arenarius* proposition, with 19%, as well as Ptolemy in the *Harmonics* – all originating from a context in which the mathematical text is embedded within a discursive setting.

That Greek mathematicians tend to avoid backward-looking arguments is noticeable in the logical size of such arguments, as well. As in the 'ideal type' in *Elements* II.5, such arguments tend to be small-sized, i.e. to occur as the premisses of arguments using only a single premiss. There are altogether 50 assertions which occur in such arguments in my survey. Ten of these belong to arguments with two premisses,

[64] This may be statistically meaningless, however: the size of this proof is 10 assertions, so that 1 less would result in an admissible 10% of backward-looking justifications.

40 belong to arguments with a single premiss: i.e. there are eight times more backward-looking arguments with a single premiss than there are with two premisses.[65] What is extremely rare is double backward-looking arguments, i.e. *P*, supported by a later *Q*, which is supported by a later *R*. There are two such cases in my survey: one in the non-mathematical passage in the *Meteorology*,[66] another in the *Arenarius*, where Archimedes tries very hard to get a point through to a non-mathematical reader.[67] I can mention one other case in a mathematical context, Apollonius' *Conics* I.33 100.18–22;[68] presumably there are a few others, but no more than a few. Other than these, such sustained backward-looking arguments occur frequently in one and only one context – scholia. A scholiast is the opposite of a mathematician: he aims at perfection. Having explained one thing, he then discovers he needs to explain the explanation, and this goes on until the scholiast finally tires of the game.[69]

3.2.2 The absence of recycling

No assertion in *Elements* II.5 is stated more than once. Assertions are made where they are required, and then they are, as it were, completely forgotten. There is no recycling of the same assertion.

There are a few cases of recycling in Greek mathematics in general. They are relatively few in number – no more than one or two on average per proposition in general. They are also a relatively short distance apart – usually 10 or fewer assertions between use and recycling.[70] Interestingly, when repetitions occur at longer distances, they tend to become explicit, using the indicator *edeichthē*, 'it was proved' (formula 45 of the preceding chapter).[71] Long-range recycling was problematic, at least in the sense that it had to be explicitly referred to in order to carry conviction.

Again, propositions which are atypical in other ways are atypical also in having long-range implicit recycling, this time the main offender being Archimedes' *Method* I.

[65] In normal arguments, about the same number of arguments have one as have two premisses.

[66] *Meteor.* 373a14–15.

[67] *Arenarius* 230.1–7.

[68] Interestingly, this passage is also – as noted in the preceding section – peculiar in another way, namely in using the genitive absolute, which is employed to compress the argument.

[69] See, e.g. *SC* I.34 126.26–128.2, which is almost certainly a scholion.

[70] There are 46 cases of recycling in my survey, 35 of which are over 10 assertions or fewer apart. The average distance apart of recycling is about 8 assertions, but it should be borne in mind that my proofs are relatively long and complicated, i.e. with more, and longer, recyclings than the real average.

[71] Apollonius' *Conics* I.13 52.12: 14 assertions apart; ibid. 41, 126.6: 10 assertions apart. Euclid's *Elements* I.5: throughout, distances ranging from 6 to 12 assertions apart.

3.2.3 No hiatuses

Hiatuses are where the argument gets to a certain point, and then suddenly switches to another line of reasoning, momentarily setting aside that point. This does not happen in *Elements* II.5 and is generally rare.

The absence of hiatuses is less easy to quantify than earlier features, because hiatuses may sometimes be demanded by the subject matter. Consider Euclid's *Elements* III.5: the main argument of the proof is where the assertions 'EΓ = EH' and 'EΓ = EZ' are combined. Now it is impossible to prove both by the same path of reasoning. That is: a proof that EΓ = EZ cannot use EΓ = EH, and vice versa. On the other hand, both must be proved. Thus a hiatus is necessary. The argument must move once towards one of the premisses, then begin anew, working towards the other premiss. Only then the two can be combined.

So hiatuses imply recycling as well. In general, if assertion *n* does not lead to assertion $n + 1$, it must lead somewhere else. So assertion *n* would be recycled later on. Avoidance of hiatuses is therefore related to avoidance of recycling.[72] The different parameters are interrelated by inner architectonic principles, which helps to explain how proofs converge around a single model.

When hiatuses are inevitable, where are they made? Consider the proof of Apollonius' *Conics* I.41. There, a hiatus occurs after the fourth assertion, which is later recycled as assertion 14.[73] This means, in effect, that assertions 1–4 contain one logical development, assertions 5–13 contain another, and both are combined in 13–15. It is logically indifferent whether we go first through 1–4 or through 5–13. The difference between the two is merely of size. In fact, 1–4 takes 5 Heiberg lines, 5–13 takes 17 Heiberg lines. Clearly, changing the order would have made the proof as a whole less linear. As it stands, it has a long continuous stretch from 5 to the end, and the unconnected 1–4 is no more than a brief preliminary, which does not stand in the way of the main linear development.

The tendency is therefore to make hiatuses as early as possible in the proof, so as to get them out of the way. This is clearly the case in other propositions; but, again, such an assertion is difficult to quantify.

[72] Recycling may occur without hiatuses, when the same assertion is developed and then reused in a different context. This occurs, of course, most notably in propositions such as Euclid's *Elements* I.5; but, as noted in section 3 above concerning such propositions, a recycling of this sort implies a certain logical barrenness, which the Greeks tended to avoid for logical, as well as for aesthetic, reasons.

[73] Assertion 4: 124.18. Assertion 14: 126.5–6.

Before making a concluding remark on the relative absence of hiatuses in Greek proofs, I would like to make a hiatus in my argument and to raise two new issues. The first is what I call the cadenza effect; the second is the use of *toutestin.*

3.2.4 The cadenza effect

Quite often, proofs end with a series of truly simple and brief derivations, containing few if any starting-points. Arguments which may have been taken for granted in earlier parts of the proofs are here spelled out explicitly. The effect of such an ending is that the conclusion of the proof is quickly and easily read, and is also more impressive. In Apollonius, where the general shape of meso-proofs is relatively simple, this effect may be seen from the use of arguments based on a single premiss. Such arguments tend to be both easy and strongly compelling – often, they are reidentifications.[74] In the much more complex Archimedes, the relatively smoother cadenza is reflected more by the use of a continuous series of arguments based on two premisses towards the end.[75]

The definition of cadenzas is difficult, since many proofs are smooth and simple from beginning to end: e.g. Diocles' *Burning Mirrors* I, or indeed Euclid's *Elements* II.5. The impression, however, is that when the mathematical realities force a more complex structure, there is a tendency to make the ending, at least, more akin to a micro-proof.

The term 'cadenza' evokes something which is more elaborate and impressive, rather than more simple and direct. However, I think the term is fitting. A simple, direct ending is also a directly compelling ending, an ending which is more successful in the job of persuasion. It is obvious that the point where it is most important to persuade your audience is just near the end, near the *goal* of the persuasive act. The structure of cadenzas, therefore, offers a clue to the nature of persuasion at work here.

3.2.5 Toutestin

Toutestin – Greek for 'i.e.' – is a fine instrument, used by Greek mathematicians to minimise the presence of elements deviating from the

[74] *Conics* I.9, 17–26: 4 (2-sized) arguments.
Conics I.13, 16–23: 3 arguments.
Conics I.15, 39–41: 2 arguments; 48–54: 4 arguments.
Conics I.26, 6–12: 3 arguments; 16–18: 2 arguments.
Conics I.37, 8–17: 7 arguments.
Conics I.41, 32–4: 2 arguments.

[75] *SC* I.10, 10–18: a series of 4 two-to-one arguments.
Meth. I, 26–34: a series of 3 two-to-one arguments (with a complex post, however). Compare the *Arenarius'* proof, 27–37: 3 one-to-one and 3 two-to-one arguments.

smooth, linear development of the proof. Consider the following argument from Apollonius' *Conics* I.13:[76]

> (10) ὡς μὲν ἡ AK πρὸς KB, οὕτως ἡ EH πρὸς HB, (11) τουτέστιν ἡ EM πρὸς MΠ

> '(10) As AK to KB, so EH to HB, (11) i.e. EM to MΠ'.

(10) asserts that AK:KB::EH:HB; the combination of (10) and (11) virtually asserts that AK:KB::EM:MΠ. This is mediated via the implicit argument EH:HB::EM:MΠ, which is what (11) comes closest to asserting. An implicit argument based on two premisses is reduced to an argument based on a single premiss. The length of the conclusion, (11), is no more than five words; and, perhaps most importantly, a logical connector such as *ara* is avoided, and thus the logical structure is not mirrored by a verbal structure. Why do this? Because 10–11 (as well as their parallel, 12–13) are nested within a larger argument: 8, 9, 11 and 13 are the premisses from which 14 is concluded;[77] 10–11 and 12–13 are intruders into this structure (complex enough as it is). The *toutestin* makes it possible to minimise their presence.

3.2.6 Hiatuses and logical structure

Finally, another way in which hiatuses are avoided has to do with the very choice of the logical components of proofs, rather than with their arrangement. Greeks could have chosen a completely different architecture.

To see this, it is worth noticing, first, that some proofs do include hiatuses as a matter of course. This happens with what I have called 'grid arguments'.

Thus, for instance, whenever the double method of exhaustion is used, a hiatus must occur between the first part (say, 'assuming it greater') and the second part (say, 'assuming it smaller'). Or grids may be spatial: Autolycus often proves for the case where the sun is in the stretch *A*, then for the case of stretch *B*, etc., until the whole ecliptic is covered.[78] In arguments of this sort, the proof can be seen to consist of two different stages. One is the preparation of material: proof for the various cases. Within this stage, there are several hiatuses. The second stage is the integration of the material, yielding the required result. The first part consists of several independent pieces of information; the second part synthesises them. It is reasonable to think of the several

[76] 50.26–8. The numbers are those in my numbering of the assertions for the survey.

[77] Ibid., 50.22–52.2.

[78] See, e.g. *Ort.* II.15, with two hiatuses.

pieces in the starting-point part as so many premisses, the second part as a conclusion. The entire proposition, therefore, if conceptualised in such a way, becomes a typical argument. It is a fractal: its overall structure is identical with the structure of its elements. To pursue for a moment the metaphor: fractals are characterised by their ambiguous dimensionality. If a river is like a one-dimensional line, its fractal delta is more like a two-dimensional area. Levels of description become blurred. And this may happen with arguments, too. In a Greek mathematical treatise, there are three very clearly demarcated levels of description: individual assertion, the mathematical proposition, and the treatise as a whole. The borders between the levels of description are clear. In a truly complex structure of arguments, such as, say, Plato's *Republic*, it is impossible to tell apart levels of description in the same way. Individual sentences may play a key role in the main argument of the treatise as a whole; the point where one unit ends and another starts is always difficult to detect. So this is another possible way of doing argumentation: with fuzzy borders between the units and between the possible levels of description.

Excluding the case of grid arguments, Greek proofs are not fractal. They could easily be transformed into such. Look at the proof of Archimedes' *SL* 9. I supply its 'tree':

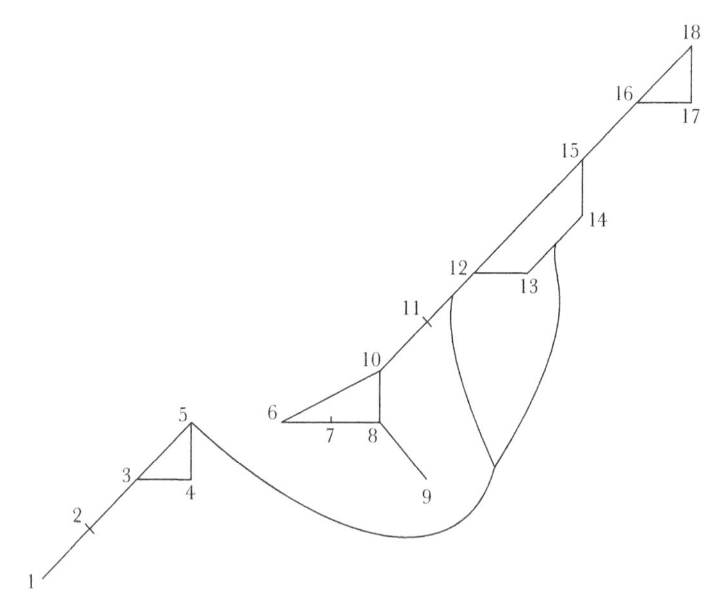

Figure 5.19. Tree of *SL* 9.

Here is a clear linear structure, though with a hiatus at 5/6: 8–9 is backward-looking, and 13–14 is nested within a larger argument. These two are therefore detachable: 17, a starting-point, refers back to a hypothesis.

Imagine therefore the following structure:

First part – preliminaries:

* 1–5 (as they are in the proof);
* 9, therefore 8;
* 13, therefore 14.

Second part – synthesis:

> Since 6 and 7, so (by 8 – see above) 10, therefore 11, therefore 12 and (by 14 – see above) 15, so 16 and (by 17 – a hypothesis) 18.

The difference between the structure of the actual proposition and that of the counterfactual version above is that 8–9 and 13–14 are brought by Archimedes into the proposition as it develops, rather than being prepared in advance, as in the example above. When the structure of the proof must contain such an advance preparation – that is in grid arguments – Greeks allow such structures. Otherwise, structures are linear (and not fractal). In this section, we have seen various ways in which this is the case.

But of course there is one class of preparations which does come before the argument. These preparations are then even recycled; indeed, they are often unconnected. A counter-example, then? But note what preparation this is: the preparation of the construction. Now finally some explanation begins to emerge. For why is the linear argument unimpaired by *this* preparation, by *this* recycling? Clearly, because the reader and the author, when referring to the construction, do not refer to an earlier *text*. They refer to the *diagram* – as shown above again and again.[79] So the text repeatedly assumes only what

[79] The overwhelming rule is that starting-points based on the construction are tacitly introduced even where there is a considerable distance between the construction which validates them and the assertion where they are picked up. Take Apollonius' assertion in *Conics* 1.41 (fig. 5.16) 126.11:

> ἰσογώνια γάρ ἐστι [τὰ ΗΓΔ, ΑΕΖ]
>
> 'For they [ΗΓΔ, ΑΕΖ] are equiangular',

which relies upon 124.5–6:

> καὶ ἀπὸ τῶν ΕΑ, ΓΔ ἰσογώνια εἴδη ἀναγεγράφθω τὰ ΑΖ, ΔΗ
>
> 'And upon ΕΑ, ΓΔ, let there be constructed equiangular figures, <namely> ΑΖ, ΔΗ'.

The distance is considerable, yet no explicit reference to the construction is made in the proposition; the construction is simply assumed. This should be compared with the tendency to signal with the formula 'for it was proved . . .' *assertions* which are recycled.

is directly available: either because it was just asserted in the text, or because it is present in the diagram. Again, we see the complementary nature of text and diagram.

3.3 Why are proofs the way they are?

The section above started from a 'typical' proposition. This section looks more closely at an 'untypical' proposition. By understanding why the proofs of such untypical propositions are strange, I hope to explain, as well, what makes proofs in general so linear.

The two 'ideal untypical', as it were, propositions, are Archimedes' *Method* 1 and Aristotle's *Meteorology* passage: the one a masterpiece of deduction, the other hardly worthy of being called mathematics. And yet, I believe, the second may explain the first.

The proof of Aristotle's *Meteorology* passage is replete with deviant structures, as can be seen by its 'tree':

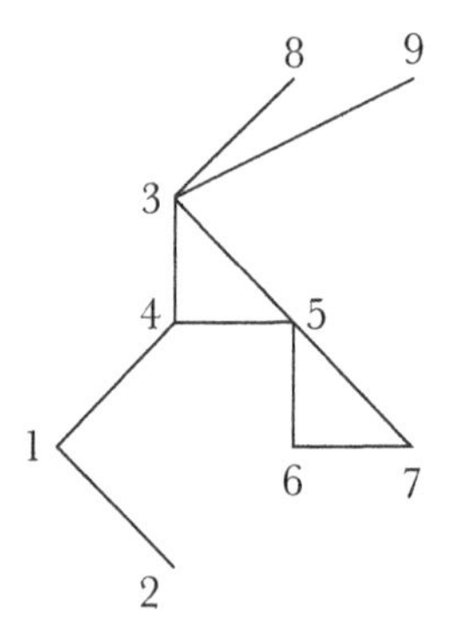

Figure 5.20. Tree of Aristotle's proof.

Instead of a linear arrangement, the proof is organised around a central triangle – the *post-factum* justification of 3 in 4, 5. Three offshoots spring out of this triangle: a justification of 4 is 1–2, a *post-factum* justification of 5 is 6–7, and two results rely on 3, namely 8 and 9. The proof is recognisably fractal: if 1–2, 5–7 and 8–9 are all perceived as a single assertion, the resulting proof is that of fig. 5.21, which is uncannily similar to the elements of fig. 5.20.

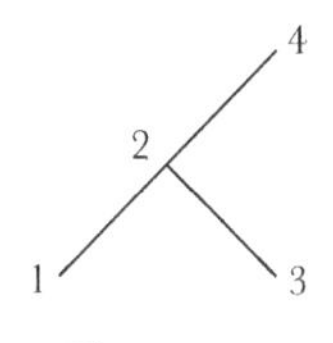

Figure 5.21.

If this is a fractal, the strangeness of the whole should be reflected by the strangeness of the constituents. So let us look closely at the passage 1–2, together with some of the background[80] (fig. 5.22):

ἴσαι δὲ αὗταί τε αἱ ΑΓ ΑΖ ΑΔ ἀλλήλαις, καὶ αἱ πρὸς τὸ Β ἀλλήλαις, οἷον αἱ ΓΒ ΖΒ ΔΒ. καὶ ἐπεζεύχθω ἡ ΑΕΒ, (1) ὥστε τὰ τρίγωνα ἴσα· (2) καὶ γὰρ ἐπ' ἴσης τῆς ΑΕΒ

'And [let] ΑΓ ΑΖ ΑΔ [be] equal to each other, and those towards Β [equal] to each other, as e.g. ΓΒ ΖΒ ΔΒ; and let ΑΕΒ be joined, (1) so that the triangles are equal; (2) for they are also on an equal line, namely ΑΕΒ'.

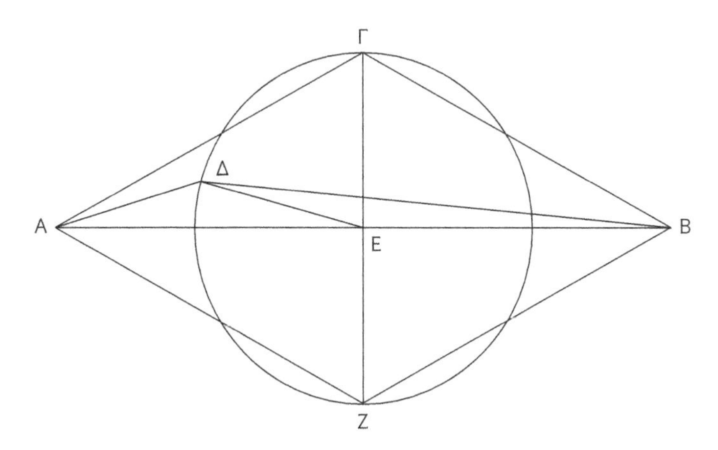

Figure 5.22. Aristotle's *Meteorology* 373a.

Why a *post-factum* justification? Consider the alternatives. Aristotle could have made the construction of ΑΕΒ earlier, and here he could have said, instead of 'and let ΑΕΒ be drawn', something like 'and they stand on ΑΕΒ, a common line', moving on to '*ergo* they are equal'. It seems that Aristotle had little patience for the construction, and

80 Aristotle, *Meteor.* 373a8–11.

that he passed as soon as he could to the actual derivation; when the construction was required, it was brought into the proof. Indeed, the overall structure of the proposition repeats this small structure, with another belated construction at 373a11–14.

In short: Aristotle does not compartmentalise, does not distinguish firmly between construction and proof, hence hiatuses, non-linearity, chaos. To develop the proof in an orderly fashion, it must be clearly separated from the construction. To co-operate, diagram and text must first be set apart.

Indeed, why should Aristotle compartmentalise his proof? After all, one thing that is clear about this proof is that it is not compartmentalised from a more general, non-mathematical discussion. The proof starts immediately from a discussion of the rainbow, and ends, just as immediately, with an identification of the letters in the proof with the topics under discussion – eye, sun, cloud. The proof is embedded within a larger discursive context, and borrows the discursiveness of this context.[81]

Archimedes' *Method* 1 is not a fractal:

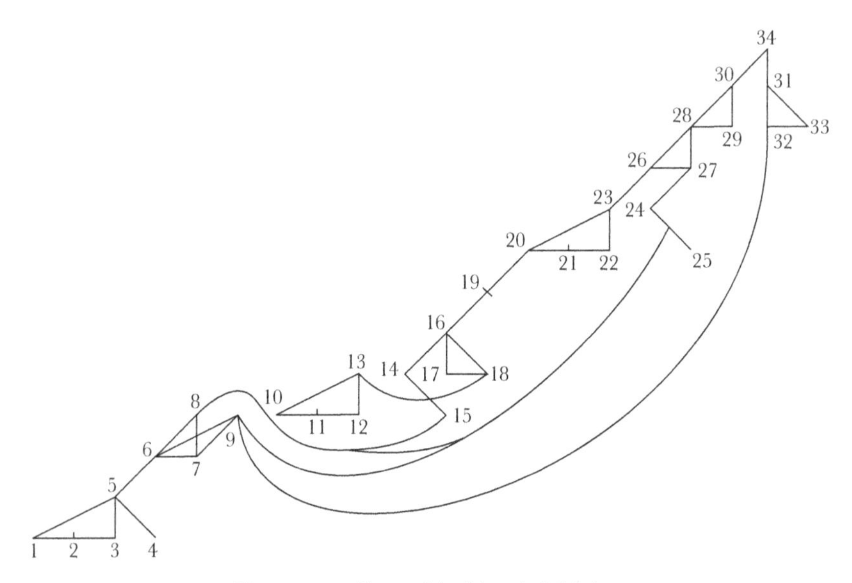

Figure 5.23. Tree of Archimedes' *Meth.* 1.

[81] Does it derive from a lecture? It is interesting that the text is especially diagram-dependent; not a single letter is fully defined. A lecture, referring to a present diagram on a *pinax*, seems plausible. Ordinary proofs are soliloquies following a diagram; Aristotle's proof is a dialogic response to a diagram.

The proof consists of three parts: 1–9, 10–13 and 14–34, of which 1–9 is an ordinary linear micro-proof; 10–13 is an ordinary argument;[82] 14–34 is a brilliant meso-proof, peculiar only in its frequency of backward-looking arguments. Of these, 25, like 5, is an explicit reference (a kind of reference which tends to be backward-looking in general). The remaining are 15, 17–18, and 32–3.

Let us look closely at the structure 13–18. Here are six assertions, centred around a single derivation: 14–16. This is, to paraphrase into English, that since N is the centre of gravity of MΞ, therefore MΞ and TH are balanced. I suppose the reader is surprised by such a derivation, but it actually works – thanks to a belated construction in 436.12–14, and the parenthetical *post-factum* justifications of 15 and 17–18. Now 16 is the key to the proposition (together with the 'falsehoods' 21, 22). Here is the prize, the surprising fact discovered by Archimedes' ingenuity; and I have no doubt that Archimedes was very happy to keep it as surprising as possible. The proof is a showpiece, meant to impress, and clarity of structure is not the only way to impress. An element of surprise may be just as important.[83]

Is there anything special about this proposition which may explain why Archimedes is so keen to impress here in this particular way? *Method* 1 is interestingly located in terms of first-order/second-order distinctions. It ends remarkably, 438.16–21[84] (in a paraphrase): 'now this is not quite a proof, but it supplies a ground for thinking the result should be true'. This can only be read as a continuation of the long methodological passage of the introduction, 428.18–430.26, so the axiomatic material and proposition 1 are sandwiched by these methodological comments. The introduction makes it clear that methodology is the real issue of this treatise – after all, results such as proposition 1 are proved elsewhere – so the sandwiching is significant. Proposition 1 is an example of a way of proving, not just a proof. It is an element in a second-order discourse. Furthermore, the addressee here is not just a token individual. The person Eratosthenes is present before Archimedes'

[82] I would not lay too much stress on the fact that 1–9 is logically larger than 10–13, though earlier than this. In absolute size, they are almost identical (51 words in 1–9 – a few of which may be interpolations – 42 words in 10–13), and the result of 10–13 is by far the stronger, hence its natural later position.

[83] 32–33 were not discussed in the above; I think they play a similar role. Archimedes tries to get to his result as quickly as he can, so as not to lose the momentum of surprise. To get 32–33 out of the way, they are relegated to a *post-factum* position.

[84] Heiberg, inexplicably, puts this as the beginning of what *he* calls proposition 2 (it seems there is no basis for this in the only manuscript of this work).

mind. The treatise, especially in the first proposition, is a dialogic, second-order discussion. It is more compartmentalised than Aristotle's proof is, but it is not compartmentalised in the way in which most Greek mathematics is.

I suggest therefore that one part of the answer to 'why are Greek mathematical proofs the way they are?' is that proofs are compartmentalised from broader discussions, so that their structure is wholly autonomous. When doing mathematics, one does nothing else. Instead of the multidimensional structure of interests and implications of natural discourse, Greek mathematics abstracts mathematical relationships. This is perhaps obvious for a science, but then Greek mathematics had no earlier science to imitate in this respect.

We saw that the ideal behind the structure of Greek mathematical proofs is a direct, uninterrupted act of persuasion. Paradoxically, this ideal can be more perfectly achieved only when the mathematical discourse is abstracted from any context, when it is no longer part of the real-life acts of persuasion, but, rather, an artificial exercise in a compartmentalised domain. The ideal of real-life persuasion survives and reaches perfection when abstracted from its origins. Is this perhaps parallel to the development of the Socratic discourse, from its real-life origins, through its gradual Platonic idealisation, to its Aristotelian abstract, general reformulation?

3.4 *How is necessity sustained by proofs?*

As usual, it is useful to consider a possible alternative. One way of preserving necessity is, let us say, strategic. In this strategic mode, we have some general grounds for believing in the necessity-preserving properties of the proof. In a computer-aided proof, for instance, it may be impossible to follow the steps of the argument in detail. However, it is possible to understand the mechanism which generates the steps, and to see that it is such that it will be necessity-preserving.[85] This is of course an extreme case. More generally, any proof which is too complex to allow a synoptic view must gain credence by such metamathematical, strategic considerations. These considerations may be as simple as 'surely this one knows what s/he is doing', or as complex

[85] Computer-aided proofs are 'strategic' in more ways than one. They may be strategic because of their final presentation, or because of their method of discovery. There, they involve the systematic search for a proof, which must be governed by some 'strategy' – a technical term. The essence of such proofs is that a general procedure, a strategy for finding proofs, replaces the individual acts of human intuition. See, e.g. Wos et al. (1991), chap. 9, and 547ff. for the technical term 'strategy'.

as in the computer-aided proof, but in some form such considerations must be present.

Arguments which rely upon an understanding of their overall strategy rather than upon an understanding of specific assertions may occur in more concrete forms: e.g. mathematical induction is an argument of this form. Another 'strategic' form of argumentation is that from isomorphism: you show the equivalence between two sets of objects A and B, prove the proposition P for A, and *ipso facto* consider it proven for B as well. Had the Greek method of analysis/synthesis relied on analysis alone (leaving the synthesis to the reader) it could also have been an indirect, 'strategic' way of proceeding in an argument.

This, then, is one option: preservation of necessity by indirect, meta-mathematical, strategic considerations. Another option is the preservation of necessity by immediate inspection. My claim is embarrassingly obvious. The necessity-preserving properties of Greek mathematical proofs are all reflected by their proofs, and no meta-mathematical considerations are required. As a rule, the necessity of assertions is either self-evident (as in starting-points) or dependent on nothing beyond the immediate background. Rarely, an immediate *post-factum* justification is required, made briefly so as not to yield a hiatus; sometimes a recycling is made, but this is done only once or twice in a given proof and such recyclings are made at relatively short distances. Finally, the structure of derivation is fully explicit. Immediate inspection is possible; this, and no meta-mathematical consideration, is the key to necessity.

As mentioned above, the Greeks did not use the method of analysis/synthesis as an indirect, 'strategic' form of proceeding in an argument. They did not have mathematical induction,[86] and they certainly did not argue through isomorphism.[87] The one 'strategic' weapon in their arsenal was the *reductio* and, more generally, grid arguments. Otherwise, meta-mathematical considerations are excluded.

It should be noticed that one sort of meta-mathematical consideration is ruled out by these properties of proofs. I refer to the 'surely this one knows what s/he is doing' principle. Greek mathematical proofs offer nowhere to hide. Everything is inspectable. It requires an Archimedes to think up an Archimedean result, but anyone equipped with sufficient intellectual stamina – and with an acquaintance with

[86] I shall return to this in chapter 6, subsection 2.5 below.

[87] Not only are the Greeks careful not to import from one species to another, but even importing from genus to species is problematic: book VII of the *Elements* re-proves much of book V, as is well known (genus: magnitudes; species: numbers); less famously, Serenus re-proves much of Apollonius (genus: ellipses; species: cylindrical sections).

the tool-box – can check an Archimedean proof. One of the most impressive features of Greek mathematics is its being practically mistake-free. An inspectable product in a society keen on criticism would tend to be well tested.

But, as suggested right now, there is yet another element to this inspectability – the direct accessibility of the tool-box. What is this tool-box? And how is it accessed?

4 THE TOOL-BOX[88]

4.1 Definition

Every starting-point or argument whose truth is not obvious from the diagram or from some other intuitive basis must reflect a more specialised knowledge. When there is no explicit reference, this more specialised knowledge is assumed to be known to the audience. Sometimes this is easily secured: the relevant piece of knowledge is known to the audience because it was recently proved, in the same treatise. The absence of explicit reference shows then how fresh the result is in the mind of the audience (or at least the author). In other cases, the result invoked is not proved in the same treatise. Such results are the tool-box.

Some qualifications must be made. First, we need not look further than the immediately invoked result. For instance, when invoking *Elements* I.5 (isosceles triangles have equal base angles), one must assume, logically, the validity of I.4 (the congruity of triangles with two equal sides containing an equal angle), which is invoked at the proof of I.5 and is necessary for that proof. Cognitively, however, this need not be invoked. It is possible that I.4 served as a stepping-stone, used locally by I.5 and forgotten thereafter.

Second, not everything which can be reconstructed as 'the result assumed' stands for something specific in the author's or the audience's mind. *Elements* I. Def. 15 ensures that all the radii in a circle are equal to each other, and this is often used by Greek mathematicians.[89] But it would be an exaggeration to say that the mathematicians *refer* to

[88] I adopt the term used by K. Saito, who reopened the study of the tool-box in 1985, and has since brought it to ever higher levels of rigour. No other part of this book is so deeply indebted to another's work, as this section is to Saito's. I am extremely grateful for his willingness to let me consult his database of the tool-box of Pappus VII, at the moment of writing unpublished (and only a part of Saito's tool-box project, at the moment of writing, still at its first stages). Of course, the responsibility for the claims made in this chapter lies with me alone. The best way of consulting Saito's tool-box project is at: HTTP://HEART.CIAS.OSAKAFU-U.AC.JP/~KSAITO/

[89] I gave this above, in subsection 1.2, as a simple example of the 'tool-box'. I now refine my concept.

Elements I. Def. 15 when assuming the equality of radii. The equality is so fundamental as to be directly intuitive. It does not take up any specialised cognitive storage space. It is unlike, say, Pythagoras' theorem, which you must *learn* in order to know, and must henceforth *remember*. No one needs to *remember* that the radii in the circle are equal. Of course, the question as to what needs to be remembered and what does not is difficult and to some extent subjective.

'The tool-box', with a stress on 'the', is a concept standing for a reality. There simply was such a unique set in Greek antiquity. But while stressing the existence of 'the tool-box', it is clear that different persons must have internalised it to varying degrees, or that learners must have known it less than initiates. Differences and qualifications existed, but the basic unity is the more significant fact.

4.2 *Preliminary quantitative description*

Some highly sophisticated works take for granted propositions which are not proved in the *Elements*: for instance, Archimedes may take his own mechanical results for granted. Results in conics, sphaerics and, later, spherical trigonometry are required in some advanced works. However, I will argue that the bulk of the tool-box is made of the *Elements*. This is not a textual thesis about the shape of *Elements* in antiquity, but a logical thesis about the assumptions required by Greek mathematical works.[90]

This thesis has two parts. One is that Greek mathematicians extensively assume Euclid's *Elements*. The other is that they do not assume extensively anything else. I shall begin with the first.

I have surveyed the propositions from the *Elements* used in two medium-level works (those surveyed in chapter 1): Apollonius' *Conics* book I, and Euclid's *Elements* book XIII.[91] Archimedes' *Spiral Lines*, a very advanced work, was surveyed as well.[92]

The data have been organised in tables of the form of table 5.1.

[90] I often use the phrase 'Euclid's *Elements*' (or just '*Elements*') as a shorthand for 'the mathematical contents which we associate with Euclid's *Elements*'. The exceptions to this will be obvious to the reader.

[91] The reader of the *Elements* is in a special position, becoming an initiate as he is reading the text. The last book of the *Elements* therefore occupies an even more special position, as the place where the reader finally emerges as a complete initiate. I believe this is the role of book XIII: not to add new material to the tool-box, but to be a sort of a 'test yourself' supplement: have you become an initiate? Have you 'got' the tool-box?

[92] My source was, almost always, Heiberg's judgement. This is far from perfect (see, e.g. Mueller (1981), nn. 30, 33 to chapter 2, 115). It can be seen as a lower limit on the number of references, and in this way it is very useful.

Table 5.1. *Euclid's* Elements *book XIII*:[93] *uses of* Elements *book I*

Elements book I	1	2	3	4	5	6	7	8	9	10	11	12	13	14	15	16	17	18	19	total
4							2	1		1	1		1						1	**7**
5								1	2	2										**5**
6							2	1												**3**
8							2										1			**3**
26										1										**1**
29																1				**1**
31																1				**1**
32								2	3	2	1								1	**9**
33																2				**2**
43	1			1	1															**3**
47												1		3	1		4	2		**11**

Elements book XIII

[93] The text contains an unnumbered proposition, following proposition 18, to which I refer as '19'.

Elements I.8, for instance, is invoked twice in XIII.7, and once in XIII.17. There is a difference between the total number of uses of the tool-box (46 in this case) and the number of propositions invoked (11 in this case). This is because the same proposition may be invoked more than once, either in the same proposition or in different propositions. And there is an important difference here. I.29 is invoked only once, and therefore it may be a mere fluke – it may not be deeply internalised at all. But I.8 is invoked twice, and in two propositions that are distant from each other (so it is not as if the author had a passing obsession with I.8). It must therefore represent something about the author's knowledge and his expectations from the audience.

The total numbers of uses of the tool-box are 195 in Apollonius (a work of 60 propositions), 126 in Euclid (a work of 19 propositions), and only 36 in Archimedes (a work of 28 propositions). These are all minima. Clearly Euclid relies more on the tool-box (this is what book XIII is for), Archimedes much less (the *SL* is an independent work, very much propelled by its own logic). We see the averages: a few uses of the tool-box per proposition. This is in line with the results accumulated so far on the nature of starting-points and arguments.

The number of *propositions* invoked is smaller, of course. Apollonius invokes 55 propositions from the *Elements*, Euclid invokes 43, Archimedes 17. This should be compared with the number of geometrically related propositions in the *Elements*, excluding books IV and X (exotic developments) and, of course, XIII itself: this is 215. We see that a large proportion of these are covered (note, however, that 8 of the propositions used in *Elements* XIII are from books IV and X. This is a special feature of book XIII, with its aim to cover all the materials of the *Elements*).

Picking 55, 35 and 17 results at random from among 215, we should expect the following degree of overlap: about 7–8 propositions common to Apollonius and Euclid, about 4 propositions common to Apollonius and Archimedes, about 2–3 propositions common to Euclid and Archimedes. In fact 19 propositions are common to Apollonius and Euclid, 10 are common to Apollonius and Archimedes, and 4 are common to Euclid and Archimedes. The three works have nothing common in their subject matter. The degree of overlap shows that some propositions are routinely more prominent than others, and their general identity is clear. First, results from book VI (referred to 75 times by Apollonius, and 47 times by Euclid). Book VI contains results in which plane geometry (mainly developed in books I and III) and proportion-theory (mainly developed in book V) are combined. Book VI is the book of geometrical proportions. This is the mainstay of Greek mathematics, the place where the visuality of geometry and the diagram meets the verbality of proportion and the formula. Trailing book VI are books I, III and V themselves. Between them, then, these four books account for what the Greek mathematician simply *had* to know.

Did he know all of them? Did he know the results of the remaining books? The answer seems positive. The degree of overlap is above chance levels, but not overwhelmingly so. Clearly, a slightly different subject matter would entail a slightly different tool-box. Diocles' *Burning Mirrors*, for instance, relies heavily upon Euclid's *Elements* III.7.[94] This is a proposition which is never invoked by the works I have surveyed. Obviously, the only reason is that in Diocles' work, but not in Euclid's or Apollonius', III.7 is useful. Most propositions in my survey are referred to more than once (33 of the 55 propositions referred to by Apollonius, for instance), and even when a proposition is referred to only once, this is done in the usual taken-for-granted way, without any comment being made. So the impression is that whatever is being invoked from the *Elements* is part of the tool-box, is supposed to be known, and that, if we enlarge our survey, we shall get a much larger proportion of the *Elements* referred to. My working assumption therefore is that the entire geometrical material of the *Elements* (excluding books IV, X and XIII) was part of the tool-box.

Saito's complete survey of Pappus VII may be consulted at this point. This is a far more thorough survey than those made by Heiberg, and it covers a very long work. Of the 215 basic Euclidean geometric propositions, 89 are invoked. This is already a considerable proportion. True, Pappus is something of a name-dropper: it is clear that he is showing off his mastery of the tool-box. But it is also clear that this mastery is real, and was at least an ideal projected by his writing.

The size of tool-boxes other than Euclid's *Elements* is more difficult to estimate, of course. Autolycus' *Moving Sphere*, probably intended as the tool-box for sphaerics for its date, is remarkably small – 12 propositions. The trigonometric 'tool-box' of Ptolemy is summarised in 7 propositions.[95] Pappus sometimes uses Euclid's *Data*, which can be seen as a tool-box for the analysis/synthesis operation: altogether, he uses (in book VII) 17 propositions from the *Data*, far fewer than the number of propositions he uses from the *Elements*. And some of these *Data* propositions are so simple that they probably need not be

[94] See, e.g. Toomer (1976) 72 (p. 56). [95] All in the *Almagest* I.10.

'remembered' (e.g. *Data* 3: that if two magnitudes are given, so is their sum!)

The size of pre-Apollonian *Conics* could only be guessed at, of course, but something may be made of the fact that Apollonius projected the first four books as his version of elementary *Conics*,[96] that his third book contained many new propositions,[97] and that book IV was new in most of its subjects.[98] As the first two books alone contain 113 propositions, this still leaves quite a lot, but clearly, in conics, not every result was part of the tool-box. There are good studies of Archimedes' conic background, in Heath (1896) and Dijksterhuis (1938). Those discussions go beyond the tool-box (what Archimedes invokes directly) and try to analyse the mathematical contents assumed, indirectly, by Archimedes. The interest, for Heath and Dijksterhuis, is to try to reconstruct as much as possible of the lost Euclidean *Conics*. Furthermore, Dijksterhuis also lists Archimedes' original contributions to *Conics*, obviously outside our scope here. The 54 dense pages of Dijksterhuis 55–108 thus turn out to contain much more than the list of *Conics* results required immediately by Archimedes. This contains no more than 8 propositions from (what is to us) Apollonius' book I, and a single proposition from all the remaining books:[99] remarkably little, given the depth and breadth of Archimedes' involvement in conics.

Apollonius himself was careful to point out the precise possible applications of books II and III,[100] which strengthens the impression made by the Archimedean data, namely, that book I alone of Apollonius' *Conics* was a 'tool-box' in any strong sense, something to which the geometer repeatedly returned. Books II and III have a more specialised interest, similar to, say, the role of book IV in Euclid's *Elements*.

To sum up the quantitative results, then: propositions often contain a few uses of the tool-box. A typical treatise has tens, or even hundreds,

[96] 2.22–4.1: τὰ πρῶτα τέσσαρα πέπτωκεν εἰς ἀγωγὴν στοιχειώδη.

[97] 4.12–13: ὧν τὰ πλεῖστα καὶ κάλλιστα ξένα.

[98] 4.19–20: καὶ ἄλλα . . . ὧν οὐδέτερον ὑπο τῶν πρὸ ἡμῶν γέγραπται.

[99] I give the earliest references in each work (often, the same proposition is referred to more than once in the course of a given work): *CS* 3 → *Conics* III.17, *CS* 3 → I.11, *CS* 7 → I.1, *CS* 9 → I.21, *CS* 15 → I.26, *CS* 15 → I.10, *QP* 1 → I.46, *QP* 2 → I.35, *QP* 3 → I.20. A further complication is that we have only the enunciations of several Archimedean propositions, whose proofs must have assumed some conic results (see, e.g. Dijksterhuis (1938) 80). The problem is that we cannot tell in such cases where the Euclidean apparatus ended and the Archimedean proof started. For our purposes, therefore, such reconstructed proofs are no help at all. But clearly 8 is only a minimum figure.

[100] Book II has results χρεῖαν παρεχόμενα πρὸς τοὺς διορισμούς (4.8); book III's results are χρήσιμα πρός τε τὰς συνθέσεις τῶν στερεῶν τόπων καὶ τοὺς διορισμούς (4.11–12).

of such uses. These typically involve a double-digit number of propositions from the *Elements*, and (sometimes) a handful of propositions from other, more specialised sources for the tool-box. The main, Euclidean tool-box clearly has a special role. A close-up is required.

4.3 *A close-up on the* Elements

Pappus uses 89 propositions from the *Elements* in book VII. What happened to the rest? Let us look even more closely at the first book. Pappus uses 30 propositions out of 48. He does not use: 1, 2, 6, 7, 9, 16, 17, 20, 22, 24, 25, 39, 40, 42, 43, 45, 46, 48. Does Pappus not 'know' 16 and 17? This is impossible. Here they are (in Heath's translation):

> 16. If one of the sides of any triangle is produced, the exterior angle is greater than each of the interior and opposite angles.
> 17. Two angles in any triangle taken in any way are less than two right angles.

Compare this to 32, used by Pappus:

> 32. If one of the sides of any triangle is produced, the exterior angle is equal to the interior and opposite angle, and the three interior angles of the triangle are equal to two right angles.

It is at once apparent that the contents of 16 and 17 are contained in those of 32.[101] Such relations of containment mean that the whole of the *Elements* is less than the sum of the parts. One is not really required to learn by heart 215 separate results: they are structured so as to occupy less than 215 units of content. One useful structural relation is that of containment. Another is that of clustering. Proposition 32 is a cluster: it clusters together two claims which are closely related. It is a reasonable guess that the Greek mathematician remembered the two facts simultaneously. A special case where this is clearest is that of converse theorems. Consider 5 and 6:

> 5. The angles at the base of isosceles triangles are equal to one another, and if the equal straight lines are produced further, the angles under the base will be equal to one another.
> 6. If two angles of a triangle are equal to one another, the sides which subtend the equal angles will also be equal to one another.

[101] This is a result of the well-known Euclidean decision, to postpone use of the parallels postulate to proposition 29.

Proposition 5 argues (with a small complication) that if *P* is true, so is *Q*; 6 argues that if *Q* is true, so is *P*. Euclid separates the two for his own ease: after he proves 5, he can use it for the proof of 6. Clearly the Greek mathematician remembered the two as a single unit – 'the equivalence of equal angles and equal sides'. So Pappus' failure to use 6 is again meaningless, as he did use 5 (similarly, 39 and 40 are the converse of 37 and 38, respectively, and 48 is the converse of 47).

This leaves us still with a number of propositions. Most of these, however, are not theorems. I have already suggested the possibility of assertions which are so obvious as to need no 'remembering'. Problems may be quite complex. But, in a sense, they do not require 'remembering'. This is because of the following. When a theorem asks us, in its construction, 'to draw a line perpendicular to a given one', it is an interesting question whether the required operation can be done by Euclidean means. *But this does not prejudice the truth of the proposition.* If the operation cannot be done with Euclid's postulates alone, then this means that the truth of the theorem can be shown only with means stronger than Euclid's postulates: perhaps, for instance, it is necessary to use a special machine to draw the required lines. But the *truth* of the theorem is left unaffected. On the other hand, if a theorem claims that 'figure *A* is equal to figure *B*', it is absolutely necessary to know that this has been proved: for otherwise the claim may be false, and with it the entire theorem. It is therefore necessary to know, on-line, the truth of the theorems used in the proposition. As for the problems used in the proposition, suspension of judgement is possible.

Furthermore, problems are much simpler than theorems. This is because they fall into two main kinds: the absolutely simple ones (such as raising a perpendicular) and the problematic ones – essentially, the three great problems (squaring a circle, trisecting an angle and duplicating a cube). The border between the two was very well known to any geometer. This border, in fact, was at the heart of Greek mathematics, largely structured around proposed solutions to the three main problems.[102]

It is therefore possible to ignore problems, which results in a remarkable simplification. We are now left with only five propositions from book I which are not used in Pappus VII: 7, 20, 24, 25, 43. Of these, I believe proposition 7[103] was never part of the tool-box. Euclid

[102] Knorr (1986).

[103] 'On the same straight line there cannot be constructed two other straight lines equal to the same two straight lines and at a different point, in the same direction, and having the same extremities as the original straight lines.'

inserts it there because it is required for proposition 8, not because it has any significance in itself. Propositions 24 and 25 are part of a single cluster. So we are left with three clusters:

20. Two sides of any triangle taken in any way are greater than the remaining side.
24 + 25. If two triangles have the two sides equal to two sides respectively but the angle contained by the equal straight lines greater than the angle, they will also have the base greater than the base. If two triangles have the two sides equal to two sides, respectively, but have the base greater than the base, they will also have the angle contained by the two equal straight lines greater than the angle.
43. The complements of the parallelogram around the diameter of any parallelogram are equal.

I am sure Pappus' tool-box contained 20 and 43, and only an accident prevented him from using them in book VII; 24–25 may possibly be a rare case: a result intended to form part of the tool-box, but failing to become such.

I have made an analysis of the clusters into which the theorems of book I fall. There are 17 such clusters:

Independent clusters: 15, 20.

All the following are properties of triangles:

Clusters of congruence theorems (each, I allow, individually remembered): 4, 8, 26.

Non-congruence relations: 21, 24 + 25.
Side/angle relation: 5 + 6 + 18 + 19.
Sum of angles: 16 + 17 + 32.

The following have to do with parallels:

Conditions: 27 + 28 + 29.
Relations of parallels: 30, 33 + 34.

The following have to do with areas:

Triangle/parallelograms relations: 35 + 36, 37 + 38 + 39 + 40, 41.
Other area clusters are: 43, 47 + 48.

The above analysis can be given as a figure (fig. 5.24).

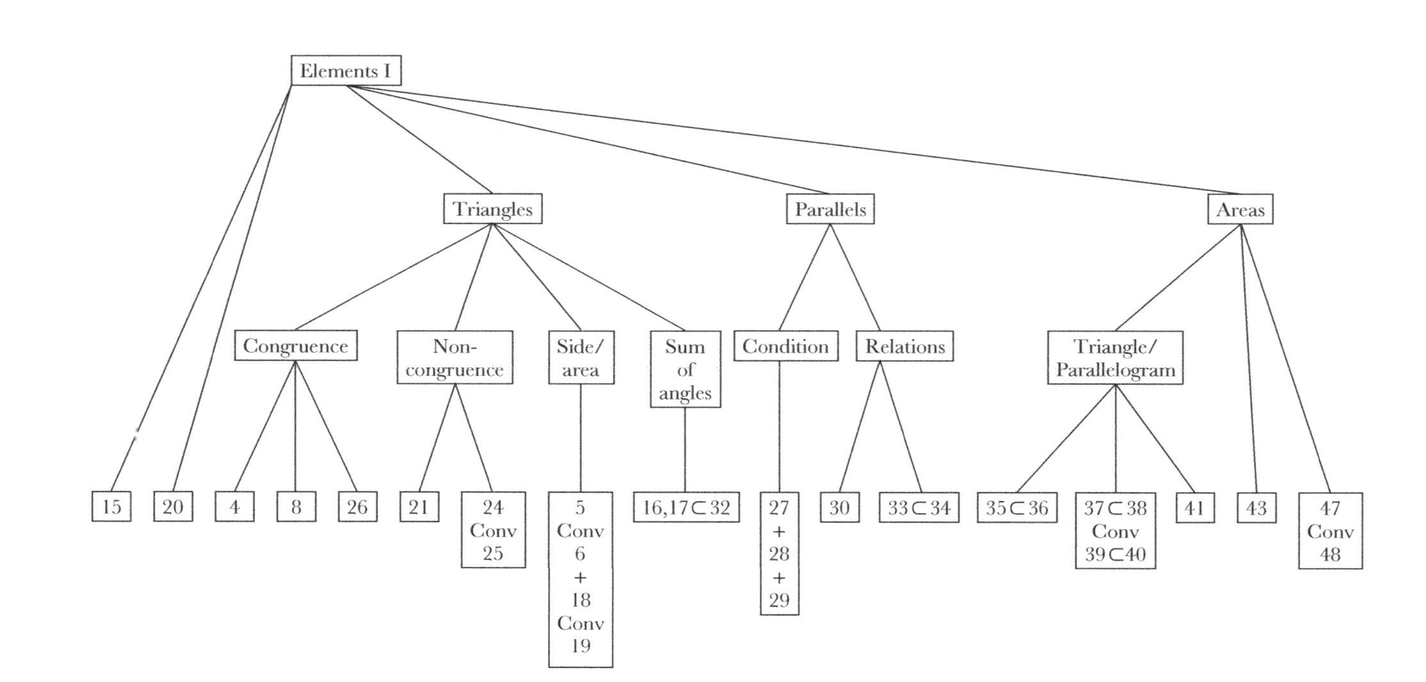

Figure 5.24. The structure of Euclid's *Elements* I as a tool-boox.

There are six hierarchic levels here. The highest is that of the book. This is an implicit kind of organisation: there is no need to believe that the mathematician explicitly remembers his tool-box according to books. Below that comes the level of general subject matter (e.g. 'results having to do with triangles'). This may be explicit in the mathematician's mind, but it does not have to be so. The next level is that of specific topics, such as 'congruence theorems'. Here there is no question that the mathematician has this level of organisation explicitly. Then we reach the level of individual 'clusters', the ultimate level of analysis as far as the mathematician's own perception is concerned. Below that is the level of Euclidean individual theorems (which there is no reason to think the mathematician remembers as such), and going even further, it is possible to analyse the Euclidean result into its atomic constituents (whenever the Euclidean formulation is a conjunction of two assertions or more). But here clearly the analysis is important for our purposes rather than for the ancient mathematician.

The claim that the ancient mathematician thought at the level of clusters (and higher organisations) can, perhaps surprisingly, be supported from ancient material. Thus, for instance, when Archimedes recalls a set of Euclidean results from book XII,[104] he gives a list of results arranged according to subject matter, not according to the order of the propositions in Euclid (i.e. this is the level of topics) and, most interestingly, he amalgamates two Euclidean propositions into a single sentence:

> Cones having an equal height have the same ratio as the bases; and <those> having equal bases have the same ratio as the heights.

This should be compared with *Elements* XII.11, 14:

> 11. Cones and cylinders which are under the same height are to one another as their bases.
> 14. Cones and cylinders which are on equal bases are to one another as their heights.

Archimedes need not have had our text of the *Elements*, but this is not the point. The point is precisely that our text of the *Elements* need not represent the way in which knowledge of geometrical contents was organised.

[104] Following *SC* I.12 (72.25–74.12).

I have not attempted a complete analysis of the *Elements*, but it is generally organised on the lines of *Elements* I. Thus the 215 Euclidean propositions boil down to, say, around 75 basic clusters, and about 50 specific topics (or about 20 topics). That this tool-box can be easily mastered is now clear.

But is this the only relevant tool-box? We are now in a position to ask whether elementary results outside Euclid were used in the way in which Euclid was. I will concentrate on an example discussed in Gardies (1991) and Saito (1994). The situation is the following: Euclid's *Elements* V.14 proves that:

V.14: if $a{:}b{::}c{:}d$ and $a = c$ then $b = d$.[105]

Very often in later books the proposition actually implicit in Euclid's reasoning is the following:

V.14a:[106] if $a{:}b{::}c{:}d$ and $a = b$ then $c = d$.

V.14a is nowhere proved in book V, though it can be easily proved on the basis of the definitions of this book.[107]

Most spectacularly, it sometimes happens that the starting-point for this piece of reasoning is of the form:

Starting-point as given: $a{:}b{::}c{:}d$ and $a = c$.

Instead of employing V.14 immediately, Euclid then transforms the ratio $a{:}b{::}c{:}d$, via V.16, to $a{:}c{::}b{:}d$, thus arriving at the transformed starting-point:

Starting-point as transformed: $a{:}c{::}b{:}d$ and $a = c$.

It is then that Euclid employs V.14a and obtains the result $b = d$ which, of course, could have been obtained directly through the original starting-point and V.14.[108]

V.14a may therefore be taken as an example of a result used by Greek mathematicians although not proved by Euclid. How can this

[105] Also $a > c \rightarrow b > d$, $a < c \rightarrow b < d$; however, the case of equality is the typical one.

[106] 'V.14a' is Saito's useful title.

[107] As pointed out by Gardies.

[108] The evidence is all contained in Gardies (1991). I will add that in the first two books of Apollonius' *Conics*, V.14a is unambiguously used four times (I.57, II.11, 14, 20), while two other occasions are ambiguous between V.14 and V.14a (II.37, 53).

be accounted for? Gardies suggests a major textual corruption, i.e., at least in the spectacular cases, some 'unskilful hand'[109] transformed good Euclidean propositions using V.14 into bad pseudo-Euclidean propositions using V.14a. Gardies' theory involves a textual event occurring later than Euclid himself. Saito's theory, on the other hand, involves a textual event prior to Euclid. His assumption is that, prior to Eudoxus (who supposedly is responsible for book V), the version used by Greek mathematicians was V.14a; book V did not cater for this, but added V.14 instead, for its own specific purposes. Saito is right in pointing out the textual extravagance of Gardies' theory. His theory, however, is made vulnerable by Apollonius' use of V.14a (see n. 108). Surely the problem need not be approached textually at all. Once the problems of text are forgotten, the issue becomes clear: V.14a is far more intuitively appealing than V.14, hence the tendency of Greek mathematicians to rely upon V.14a rather than upon V.14. Do they need to rely upon some other, Ur-Euclid, or simply a different version of Euclid, in order to do so? Of course they may, but they do not have to. An acquaintance with the contents of book V as they stand would suffice to make anyone feel happy with the claim of V.14a. V.14a is so directly deducible from them as to put no extra pressure on the mathematician's memory. That V.14a is not stated in Euclid's *Elements* would be a problem only if accessing the tool-box involved the physical reference to the text of Euclid – which, I will argue below, it does not.[110]

So it should be said once and for all: those useful editorial interventions, supplying the 'exact reference' which the Greek omitted – the '[V.14]' – are in a sense misleading. I will immediately note how Greeks in fact access their propositions, but they truly do not access propositions as units. Often, after all, the modern editor has to supply a *number* of propositions as a reference. Or the editor supplies a single proposition, but the Greek probably memorised that proposition as belonging to a larger cluster; or finally, the Greek simply *knew* this was true, and forgot about anything specific in Euclid. To see this, we must finally see how the tool-box is accessed.

[109] Gardies' expression, quoted from Simson.

[110] In Saito's survey of Pappus, *all* references which are not 'in' Euclid are such direct and simple extensions of Euclidean results. The only exception is 'Taisbak's theorem', which is a complex result of proportion theory as applied to circles. There is a tiny logical lacuna in Euclid: book VI combines proportion with plane geometry, but it concentrates on book I-style plane geometry, and says almost nothing on circles. I can even imagine we have lost something from the original text. More probably, the elementary proportion theory of circles was rediscovered by the main geometers independently.

4.4 *Accessing the tool-box*

Look, for instance, at the following (fig. 5.25):[III]

καὶ ἐπεί ἐστιν, ὡς ἡ ΚΗ πρὸς ΗΒ, οὕτως ἡ ΓΑ πρὸς ΑΒ . . .

'And since it is, as ΚΗ to ΗΒ, so ΓΑ to ΑΒ . . .'.

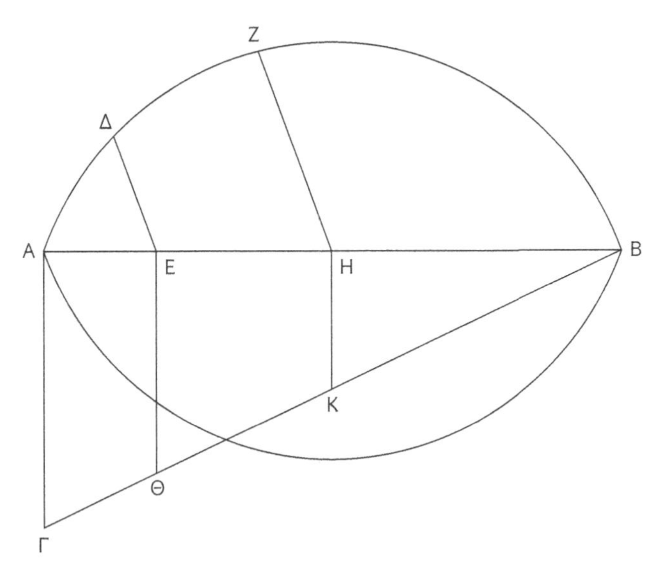

Figure 5.25. Apollonius' *Conics* I.21 (Ellipse Case).

This clause – an antecedent for a piece of reasoning – is presented by Apollonius as a truth in need of no support. The diagram probably helps us to remember what was implied in the set-up of the situation, namely that the triangles ΒΗΚ, ΒΑΓ are similar, which, through Euclid's *Elements* VI.4, makes the claim above true. Heiberg – very usefully – writes '[Eucl. VI. 4]' in his Latin translation, but this reflects nothing in the original. Such a reference offers no clue at all to the question *what* it is in VI.4 which is being referred to by Apollonius. We simply have no access to his mind on this question. On the other hand, the very fact that he refers to, or rather assumes, VI.4 in this fashion shows something interesting about *how* the tool-box is accessed.

For the moment, however, I concentrate on the relatively uncommon occasions where the way in which results are used offers

[III] *Conics* I.21, 74.17–18.

some hints concerning the precise contents which are being used: a redundant word or two refer, obliquely, to the original formulation. It is difficult to give exact estimates, since the definition of such 'redundancies' is very fuzzy. Still working with Apollonius' *Conics* I and *Elements* XIII, I found in each work around 30–40 cases where such hints are given. But the reference is often so brief that it may be attributed to chance. Take *Elements* XIII.14 298.22, ἐν ἡμικυκλίῳ γάρ, 'for <it is> in a semicircle'. The word γάρ, 'for', is a connector, so the reference is two words long in the Greek. These two words do echo III.31 very well, but this need not reflect any internalisation of the verbal formulation of III.31.

Often, however, the verbal echoes which do occur are unmistakable. I will now note a case where a long quotation occurs (but most often what we get is a short formula which, while being short, is also clearly formulaic):[112]

> αἱ δὲ τὰς ἴσας τε καὶ παραλλήλους ἐπιζευγνύουσαι ἐπὶ τὰ αὐτὰ μέρη εὐθεῖαι ἴσαι τε καὶ παράλληλοί εἰσιν
>
> 'The lines joining the [*sic*] equal and parallel <lines>, in the same direction, are equal and parallel'.

This reads like a quotation, and the similarity to I.33 is obvious:[113]

> αἱ τὰς ἴσας τε καὶ παραλλήλους ἐπὶ τὰ αὐτὰ μέρη ἐπιζευγνύουσαι εὐθεῖαι καὶ αὐταὶ ἴσαι τε καὶ παράλληλοί εἰσιν
>
> 'The lines joining the equal and parallel <lines>, in the same direction, are themselves equal and parallel'.

The similarity is not an identity. In the original, ἐπιζευγνύουσαι comes after ἐπὶ τὰ αὐτὰ μέρη; in the quotation the order is reversed.[114] The quotation leaves out the words καὶ αὐταὶ (the δὲ of the quotation is of course insignificant). This is enough to make clear that Euclid is not even attempting to quote verbatim – after all, the length of the original sentence is just 19 words, so an omission of two and a permutation of another is a significant change. But the similarity is clear.[115]

To sum up, then, the position so far: mostly, reference is a perfectly tacit taking-for-granted affair. When a quotation is made, there is no attempt to make it verbatim. The psychological implication would be

[112] *Elements* XIII.16, 306.9–11. [113] 78.20–2. [114] This cannot be mirrored by English syntax.

[115] For other 'badly' quoted Euclidean results, see, e.g. ibid., 320.1–5; 338.1–2. There is no need to invoke textual corruptions here. Misquotation is a well-attested ancient practice, for which see, e.g. Whittaker (1990) xviiff.

that the Greek geometer simply knew his Euclid very well, as a set of logical facts in need of no reference. There is nothing impossible about this – modern readers develop a similar grasp of the Euclidean system as they read Greek mathematics, and, as I have shown above, the numbers involved are such as to allow this: certainly less than 100 clusters would cover the entire geometrical tool-box.

I am offering here a hypothesis according to which there is an essential difference between the way in which earlier results are used *inside* a given proof, and *outside* it. I have argued in the preceding section that the internal structure of proofs tends to be linear, i.e. results tend to be used immediately upon formulation rather than being put aside for later use. Inside proofs, results relate to each other like the steps of a ladder. The relevant information consists of the recent results only, which implies a constant re-editing on the mental 'page' of the relevant information available directly to the working mathematician.

On the higher level, however, of systems of propositions, I claim that the metaphor should be quite different. It should no longer be a ladder; rather, it should be that of an open plane. At any given stage of the discussion, the set of accessible prior results is all equally accessible, and no changes are made in its content as one goes on.

It is true, of course, that 'ladders' occur in large-scale mathematics. I will mention in n. 121 below Saito's analysis of *Conics* I.36–51, which is a series of lemmata leading to more central results. In general, I am not speaking here about the organisation of particular treatises, but rather about the use of the Euclidean tool-box. And yet, it is useful to see that even within particular treatises the ladder is often not the correct metaphor. In Apollonius' *Conics* I, most propositions do not rely upon immediately preceding propositions.[116] The main organising principle in the book is thematic, and 36–51 is one of several themes developed in the book.[117] Euclid's *Elements* book II is a famous example of a work which could easily be given a ladder structure, but which was not given that shape by Euclid.[118] Mueller was especially interested

[116] There are 60 propositions in this book, 59 of which might have referred to immediately preceding propositions. 2, 4, 5, 7, 8, 22, 30, 37, 38, 44, 48, 51, 53, 55, 57, 58, 60 actually do so – 17 out of 59, i.e. rather less than a third.

[117] The rest are: 1–10: cones and circles on them; 11–16: the basic conic tool-box; 17–35: elementary 'topological' discussion of tangents and intersections (with the intrusion of 20–21, a major addition to the conic tool-box); and 52–60, the construction of the sections.

[118] Heron did this, to judge by the *Codex Leidenensis* (Besthorn and Heiberg (1900) 8, 14, 30, 42, 62–4); but it is significant that Heron's treatment of the material remained uncanonical, and was probably never intended as more than a *jeu d'esprit*. Moreover, Heron's point was not the ladder structure, but simply the use or avoidance of lines in diagram, as noted in chapter 1, n. 61. There was no premium, in Greek mathematics, on ladder structure *per se*.

in the ladder structure of treatises, and a central part of Mueller (1981) is occupied by a description of the deductive organising principle of Euclid's *Elements* book I. But Mueller himself admits that thematic rather than deductive structure is the rule in Euclid.[119]

The communication situation this implies is intriguing. The taking-for-granted manner also takes for granted that the readers are the initiated. This is not to say that the readers had all to be equally initiated. Apollonius (like any other of the Greek mathematicians whose works are still extant) certainly knew the tool-box by heart, but some of his audience may have known it less well, and Apollonius seems to have made few concessions to them. Some readers, at some points, must have shared the perplexities which a commentator such as Eutocius sought to clarify. But it took many years for Apollonius to gain a commentary, and the impression is that the audience was expected to accept Apollonius' assumptions as valid without questioning them. The very fact that an argument was made, without any intuitive or diagrammatic support for that argument, must have signalled for the audience that that argument was sanctioned by the *Elements*. Once this is the expectation, the need to refer explicitly to the *Elements* declines, which would in turn support the same tendency: the regular circle in which local conventions are struck without explicit codification. Again, the communication situation implications are interesting: the author–audience relation in Greek mathematics was much more relaxed than in other areas of Greek thought. Paradoxically, even as he is engaged in proving his assertions in the most incontrovertible way, the Greek mathematician is relying upon uncommon trust from his audience. The Greek mathematician speaks to his friends, not to his rivals.

In this connection, we should repeat the fact mentioned often above, that one sort of reference – the only one which could really be checked by the audience – is almost completely missing from Greek mathematics: the explicit reference. In the books I have surveyed, there is *one* such reference, in *Elements* XIII.17, referring to XI.38.[120] Significantly, this is an *internal* reference, from Euclid to himself.

I have insisted in 3.1 above that my 'tool-box' does not cover internal references to results which were proved in the same treatise. Such

[119] Mueller (1981) 31. Of course, one should distinguish between treatises which survey a field, and which tend to be 'planes', and problem-solving treatises which prove a specific, original result, such as most of the Archimedean corpus, which approximate more closely a 'ladder'.

[120] 322.19–20: τοῦτο γὰρ δέδεικται ἐν τῷ παρατελεύτῳ θεωρήματι τοῦ ἑνδεκάτου βιβλίου. Note also *Arenarius* 232.3–10 – which may, however, be a scholion.

results, I claimed, could be memorised. Alternatively, they could be referred to textually, by rolling the papyrus backwards a bit. One would expect, perhaps, that references to results proved in the same treatise would need no explicit reference, unlike, perhaps, references to earlier, *Elements*-style material. In fact, most explicit references in Greek mathematics are to results proved *inside* the same work.[121]

A few external references do occur in Greek mathematical works, such as the well-known mention of Euclid by 'Archimedes'.[122] These are few and far between, and while some may be authentic, such references are the staple of Byzantine scholia, and most are probably textual accretions. Rarely, with fleeting verbal echoes, the audience was given some clue as to why the tool-box-assertion was correct, but they could very seldom check those clues unless they already knew them very well to begin with.

This, then, is the hypothesis: repeated exposure to the very fact that an assumption is taken for granted would lead the reader, if not to trust the assumption, at least to know it well. And thus Greek mathematics does not require second-order tools for its propagation: it is self-replicating. The texts teach their own background.

We may perhaps reach a similar result via an independent route. It should be remembered that, after all, a considerable minority of the references do include some hints concerning the reasons why the tool-box-assertion is true. Such cases, again, show that references were not offered in order to placate imaginary critics. In the following,[123]

ὡς . . . τὸ ἀπὸ τῆς ΖΗ πρὸς τὸ ὑπὸ ΒΗΑ, οὕτως τὸ ἀπὸ ΔΕ πρὸς τὸ ὑπὸ ΒΕΑ· ἐναλλάξ, ὡς τὸ ἀπὸ ΖΗ πρὸς τὸ ἀπὸ ΔΕ, οὕτως τὸ ὑπὸ ΒΗΑ πρὸς τὸ ὑπὸ ΒΕΑ

'As . . . the square on ΖΗ to the rectangle <contained> by ΒΗΑ, so that on ΔΕ to that <contained> by ΒΕΑ – *enallax*, as that on ΖΗ

[121] The explicit references in Apollonius' *Conics* I – a work very rich, relatively speaking, in explicit references – are: 132.19–20: 43 referring to 41; 136.7: 44 to 43; 138.19–20: 45 referring to 41; 140.24–5: 46 to 42; 142.19: 47 to 43; 162.24: 52 to 11; 164.25: 53 to 49; 170.26–7: 54 to 12; 174.21: 55 to 50; 180.14–15: 56 to 13. There is no doubt that this reflects, among other things, a passing whim on the part of the author (or his editor?). But *what* is being referred to explicitly, now that the decision to refer explicitly has been made? (For, of course, even in Heiberg 130–180, most propositions are referred to implicitly.) In fact, almost all these explicit references are very short-range indeed, and none are to the *Conics* 'tool-box'. Saito (1985) 37–43, a very valuable analysis of the structure of I.36–51, shows how some of the results are lemmata, important merely as stepping-stones for later results. Finally, the references to 11–13 are to Apollonius' own invented *definitions* of the sections, certainly not part of the tool-box as far as Apollonius and his readers were concerned.

[122] *SC* I.2, 12.3. [123] Apollonius' *Conics* I.21, 74.24–7.

to that on ΔE, so that <contained> by BHA to that <contained> by BEA',

the word *enallax* is what I would call a reference – it has the right sort of redundancy. But it would not make any sceptic see the light. Worse: from the point of view of the sceptic, such a 'reference' must seem arbitrarily dogmatic. This is not to say that the reference is useless, it's just saying that it is not the sceptic who is being envisaged here.

Enallax is a case of extremely brief allusion, but in fact about half the references in Apollonius' *Conics* fall into this specific category, of single-word redundancies relating to proportion theory (all related to formulae 56–60 presented in chapter 4 above). By attaching the word to an argument, the argument is characterised in a certain way. It is marked. And it is very easy to internalise the correlation between a certain set of arguments and the corresponding set of names, such as *enallax*. It is true that one does not need the formula in order to know that the argument is valid. But the set of formulae, taken as a whole, is useful. To master proportion theory, one needs a system by which its valid rules would be memorised. It is interesting to see that the system adopted by the Greeks is more 'oral' than, say, a written reference table might have been.

We have moved, then, from the 'why' question back to the 'how' question. It is clear that the tool-box was internalised (not just passively available in external texts) – this is the significance of its taking-for-granted manner. It is also clear that it was internalised 'orally' – this is the significance of verbal echoes which are not precise. As Aristotle said, the elements of geometry were like the multiplication table.[124] They were *not* like, say, a logarithm table.

The sense of 'orality' adopted here, I repeat, is limited. There is no question that the transmission of the mathematical text is done via the written mode. The initiate gains access to mathematics via written texts. Then, however, the contents are internalised in such a way that constant reference to written texts becomes superfluous. The results are accessible not through written symbols but (if anything) through aural associations. Often, probably only the mathematical gist is memorised (perhaps aided by visualisation based on the diagram).

The tool-box is limited – fewer than 100 items – and is structured in such a way as to make it easily internalised. This is the strength of

[124] *Topics* 163b23–8. Bear in mind, of course, that the multiplication table, for the Greeks, was not a written object, based on a specific layout, but rather (as it still very much is for us), a system of memorised formulae of valid multiplications (see Fowler (1987) 238–9).

Greek mathematics, what allows it to be deductive while capable of being firmly under the control of a single individual. That this has its drawbacks is of course true; I shall return to this immediately.

In brief, then, what is it about the process of Greek mathematics which makes its results accessible to the tool-box? In a limited sense, it is its use of formulae. In a wider sense, it is its on-line, oral nature. The general rule is that the tool-box may be created interactively, by exposure to the texts, assuming that the reader adopts an active relation to the texts, but a relation which is not critical or polemical in nature.

Perhaps a useful comparison may be the use of the writings of philosophers by their followers. A Hellenistic philosophical school was organised around a 'tool-box' – the canonical texts.[125] Clearly, this tool-box was transmitted in the written mode and, naturally, it was irreducible to a memorisable set.[126] However, even this larger tool-box is often accessed through aural associations, through verbal echoes of the original, rather than through explicit references.[127] The mathematical practices are not unlike those of wider Greek groups. The main difference, perhaps, is the relative absence of polemic concerns – the way in which Greek mathematical authors are not defensive. We begin to approach the question of the historical setting of Greek mathematics.

5 SUMMARY

5.1 The shaping of necessity

The primary sources of necessity are almost always either the diagram, perceived as a small discrete system (chapters 1–2), or another small and discrete system, namely that of formulae (chapter 4). Formulae, in turn, are simplified considerably because they are based on the small,

[125] This claim is based on Sedley (1989).

[126] Though Epicurean κύριαι δόξαι or Platonic *Definitions*, for instance, may go some way towards the fulfilment of such a need.

[127] Alcinous refers explicitly to particular works by Plato in two contexts only (chapter 6, logic; chapters 27–8, leading to the *summum bonum*), while verbal echoes of Platonic passages permeate the entire text. This reflects the general nature of the *Didascalicus*; more technical, specific works, such as Ps-Plutarch's *On Fate*, often refer explicitly to particular Platonic works. A commentator on Plato, say, would sometimes need to look closely at particular Platonic works, referring to them textually, in order to create his own interpretation of Plato. Euclid was not subject to such deconstructions, and was simply assumed, in the straightforward way in which Plato is assumed by Alcinous (J. Mansfeld has patiently tried to explain to me the issues relevant to this footnote, but I cannot guarantee that he would agree with it).

manageable lexicon (chapter 3). These sources of necessity are combined on an on-line principle. That is, what matters is always what is directly accessible. This is an idealised interpretation of oral persuasion (subsection 5.3).

This gives rise to a secondary source of necessity: results which are kept as part of the tool-box. Without this, no complicated results can be possible. This is yet another small and discrete system – and yet again, it is not mediated by the advantages of the written form. Acquaintance with it is taken for granted. In other words, an expertise is taken for granted (subsection 5.4).

We see therefore the role of tools: the verbalised diagram, the regimented language. We see the role of a wider cultural background: the ideal of oral persuasion, the possible role of expertise. Such combinations, of tools and of cultural expectations concerning the tools, are what constitute cognitive methods. I have given above a partial description of the cognitive method used by Greek mathematics. In the final chapter, I will try to explain this method in its wider historical context.

But instead of concluding this chapter by congratulating the Greeks on their brilliant methods – a congratulation they no doubt deserve – I think it is fitting to note the limitations of the cognitive method described above. Such methods may be understood not only by their achievements but also by their limits.

5.2 *On the limits of Greek mathematics*

The remarkable equilibrium of cognitive tools and cultural background achieved by Greek mathematics could equally be represented as stagnation. Knorr is scathing:[128] 'Such stasis is virtually incomprehensible in comparison with the developments of mathematics since the Renaissance. Each interval of fifty or one hundred years has since embraced fundamental changes in the concepts, methods and problems studied by mathematicians.' This is indeed disturbing. What hampered progress in Greek mathematics? One is perhaps inclined to say that the general pace of change was slower in the past than what it has been – in an ever-accelerating mode – since the Renaissance. This is true to an extent, but then, classical culture at the time at which Greek mathematics developed, or failed to develop, was far from static. Long

[128] Knorr (1981) 176.

and Sedley invite Aristotle's ghost on a trip around the Athens of 272 BC, and they note his shock as he finds there a totally transformed intellectual landscape.[129] The story of Hypatia is emblematic just because she was a mathematician, and thus a relic of the past in a world much changed.[130] But mathematics survived even Christianity, and in the totally dissimilar culture of Islam the very same mathematics went on. The image of mathematics as an invariant over history is false, but least so for Greek-style mathematics.

The stability of mathematics is not due to any stability in its background. Knorr imagines a coercive, maiming background, in a 'scholastic attitude toward mathematics . . . [which] must have been felt through the curriculum of higher education, by encouraging students toward this scholastic attitude and by limiting their exposure to heuristically useful methods'.[131] Quite apart from Knorr's interpretation of the *Method* (one of the least understood passages in Greek mathematics and his sole evidence!), there is the basic impossibility of this curriculum, apparently constant throughout antiquity and forced upon Greek youths. But I really do not think the Classical Greek Ministry of Education is to blame. It did not exist, and this is not a trivial point: Greek mathematics was backed by no institutions. Lacking any institutional backing, the stability of Greek mathematics must be located internally.

There are two points here. One is that, in the given communicative background, the cognitive style of Greek mathematics was self-perpetuating in various ways. Another point is the following. It may be said that Greek mathematics ceased to be productive not only in methods but in results as well. This may be considered an over-generalisation, but it is qualitatively true from Apollonius onwards. Hipparchus' trigonometry is subtle, but it is not deeper than Archimedes' studies in conicoidal volumes or Apollonius' more advanced studies in *Conics*. The development of Greek mathematics becomes horizontal rather than vertical: instead of recursively ever-richer results, similarly rich results are being added.

The lack of recursiveness is an important clue, as well as the fact that both Archimedes' and Apollonius' great achievements are related to *Conics*. Greek mathematics stops at a certain point, and the point is well defined. It is the third floor.

[129] Long and Sedley (1987) vol. 1, 2ff.

[130] Indeed, in the changing conditions of late antiquity, the traditional role of mathematics was exploited, by mathematicians and non-mathematicians alike; see O'Meara (1989), Cuomo (1994).

[131] Knorr (1981) 177.

The first floor of Greek mathematics is the general tool-box, Euclid's results. To master it, even superficially, is to become a passive mathematician, an initiate.

The second floor is made up of such works as the first four books of Apollonius' *Conics*; other, comparable works are, e.g. works on trigonometry. Such results are understood passively even by the passive mathematician, the one who knows no more than Euclid's *Elements*. They are mastered by the creative mathematician in the same way in which the first floor is mastered by the passive mathematician.

The third floor relies on the results of the second floor. This is where the professionals converse. They have little incentive and occasion to master this level in the way in which the second floor is mastered.[132] What is more important, the cognitive situation excludes such mastery, for oral storing and retrieval have their limitations. Already the second floor is invoked only through some very specific results.[133] When Archimedes entered the scene, the tool-box was already full. Mathematics would explode exponentially only when storing and retrieval became much more written, and when the construction of the tool-box was done methodically rather than through sheer exposure.

In a sense, my claim in this subsection is a mirror-image of Eisenstein's thesis on the role of printing in early modern science.[134] Already in early modern science, monographs lead on to monographs, in the recursive pattern which we now associate with the development of science. The Greek author, even when referring to extant monographs, cannot assume that the audience is acquainted with the same monographs, and thus he refers to them differently. The typical way of going one better than the preceding monograph is not to do something even more spectacular with the results of that monograph, but to cover again the materials of that monograph in a more thorough and scientific way. This is Apollonius' attitude to Euclid in the introduction

[132] Little incentive – for the professional discourse is obtainable simply by mastering the second floor. Competition ends here. Little occasion – for, after all, results such as the higher results of Archimedes are deliberately presented in startling ways, rather than in the textbook presentation of Euclid. Just because these are professional *tours de force*, they cannot be useful textbooks.

[133] See the census above, based on Dijksterhuis, of Archimedes' premisses taken from *Conics*.

[134] Eisenstein (1979), part 3 – a study of the liberating impact of the invention of printing on the development of science. Eisenstein herself notes the significance of the mirror-image (how the absence of printing halted scientific development) in the context of medieval science (498ff.). My suggestion here is that such limitations were characteristic of ancient science no less than of medieval science, and led to the construction of a cognitive method consistent with such limitations.

to *Conics* I; this is his attitude to Conon, in the introduction to *Conics* IV; and this is the attitude of Hypsicles to Apollonius, in the introduction to *Elements* XIV; and this is the background for the accumulation of solutions to the three problems. The limitation of results to the third floor at most, a corollary of the cognitive method, is itself a self-perpetuating feature. But the system is self-perpetuating mainly through the mechanism of self-regulating conventionality, explained in chapters 2–4. The historical background of this will be examined in chapter 7. Before that, we must tackle the issue of generality.

CHAPTER 6

The shaping of generality

INTRODUCTION AND PLAN OF THE CHAPTER

Greek proofs prove general results. Whatever its object, a Greek proof is a particular, an event occurring on a given papyrus or in a given oral communication. The generality of Greek mathematics should therefore be considered surprising. And this is made even more surprising following the argument of chapter 1 above, that Greek mathematical proofs are about specific objects in specific diagrams. The following chapter, then, tries to account for a surprising feature, a paradox.

The nature of this chapter must be different from that of the preceding one. I have explained necessity in terms of atomic necessity-producing elements, which are then combined in necessity-preserving ways. But there are no atoms of generality, there are no 'elements' in the proof which carry that proof's generality. Generality exists only on a more global plane. This global nature has important implications. That which exists only on the abstract level of structures cannot be present to the mind in the immediate way in which the necessity of starting-points (chapter 5, section 1), say, is present to the mind. The discussion of necessity focused on a purely cognitive level. It was psychological rather than logical. This chapter will have to be more logical.

This is not to say that the chapter should be judged by its success in reconstructing a lost logical theory, once fully developed by Greek mathematicians. The theory which explicates and validates a practice may be only partially understood by those who follow that practice. I shall therefore develop a theory according to these criteria:

(a) It should give a good reason for judging the results of particular proofs to be general (in other words, it should be philosophically attractive to some degree).

(b) It should be reflected in the practices of Greek mathematics (so that it is plausible to ascribe to Greek mathematicians a certain groping towards such a theory, even if they had not fully articulated it).
(c) Finally, it should be a theory of a rather general nature. This is because a very precise and sharply articulated theory could not be the sort of thing towards which people 'grope'.

A further clarification of our objective is required, and can be made by a certain comparison. The following chapter is to a great extent nothing more than a commentary on three pages written by Mueller.[1] There, when he is about to explain the source of generality in Greek mathematics, Mueller asserts: 'I do not believe the Greeks ever answered this question satisfactorily [i.e. the question of what validates a move from a particular proof to a general conclusion].' In this Mueller is probably right, i.e. the Greek philosophy we have is probably a good sample of the Greek philosophy there was, and a satisfactory theory of generality (in a fairly strong sense of 'satisfactory') is not found there. But Greek mathematicians asserted general conclusions on the basis of particular proofs, and at some level they must have felt that this was a valid move. To do justice to this, we must unearth their unsatisfactory theory – even supplying them with the articulation they may have lacked.

The issue is not what made Greek mathematics valid. The question is what made it felt to be valid, for felt to be valid it certainly was. So logic collapses back into cognition, in a sense.

My plan is to proceed, as usual, from the practice. This offers what I call 'hints' for a solution: certain practices which tackle the issue of generality in specific contexts, and which may – with caution – give us an idea of how generality was normally conceptualised. Section 1 discusses those hints. Section 2 starts with my solution for the problem of generality, relying upon a certain practice central to Greek mathematics. Later in the same section, further developments and implications of the theory are considered. Section 3 is a brief summary.[2]

[1] Mueller (1981) 12–14.

[2] For reasons of space, I do not include here my analysis of the ancient Greek views of the generality of mathematics. I have analysed the theories of Plato, Aristotle and Proclus, and argued that they are all compatible with my interpretation. I hope to publish this elsewhere.

1 HINTS FOR A SOLUTION

1.1 Explicit generalisation

Our investigation would have been much easier had Greek mathematicians used expressions such as 'and therefore this has been proved generally'. They do not. The obvious word, *katholou*, 'generally', occurs once in Archimedes,[3] where it seems to be no more than a local variation on the word *pan*, 'all': another second-order synonym. In Euclid the word occurs a few times: twice in the *Elements*, twice in the *Optics*.[4] The occurrences in the *Elements* are especially interesting. These are in the (spurious?) corollaries to VI.20.[5] In both cases, following a proof for all polygons (interpreted as n-gons, $n > 3$), it is mentioned that the same result is known for triangles. Hence, the result holds 'generally' for all figures (that is, both for 'polygons' in the limited sense, and for triangles), i.e., this sort of explicit 'generalisation' is what is known as 'perfect induction'.[6] It is not generalisation into an unsurveyable infinity. It is a generalisation following the actual survey of an actual totality.

The Greeks do not say, then, that results are proved 'in general'. They do say, however, on occasion, that the proof of a certain particular result yields another result, which sometimes may be seen as more general. I will take this practice as my starting-point. The key term for the following discussion is not *katholou*, 'generally', but *homoiōs*, 'similarly'. This occurs in the context of formula 41 of chapter 4, *homoiōs de deixomen/deichthēsetai hoti*, 'so similarly we will prove/it will be proved that . . .'. This is very common, especially in Euclid.[7] The formula introduces an assertion which is not proved explicitly. The absent proof is meant to derive from an extension of the foregoing proof.

Most often the situation has little to do, directly, with generalisation. Consider, for instance, Euclid's *Elements* I.15 (fig. 6.1). This is the proof for the equality of vertical angles. Naturally, vertical angles come in sets of pairs, so after the proof for the equality of ΓEA, BEΔ has been

[3] *SC* I.180.2.
[4] Following Heiberg's ascriptions – not much depends on the identity of the author.
[5] *Elements* VI.20, 130.14, 25.
[6] For the meaning attached to this term, see especially Aristotle, *APr.* 68b27–9.
[7] There are about 40 occurrences of this formula in the first four books of the *Elements*. Archimedes has about 20 to 30 occurrences of the formula in all. Generally, the formula is the type of pointer which the more experienced geometers would require less.

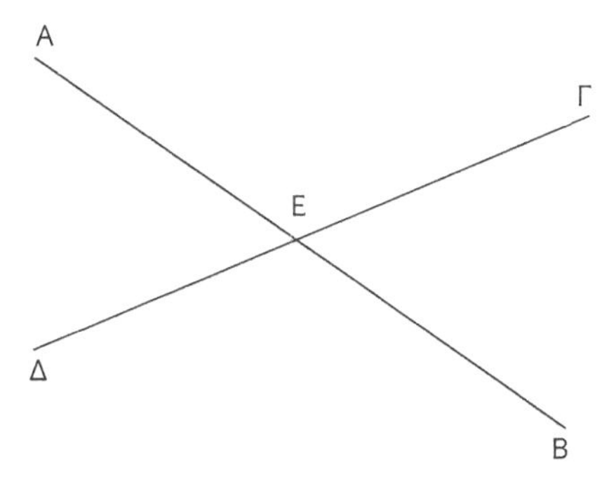

Figure 6.1. Euclid's *Elements* I.15.

effected, there remains another pair: ΓΕΒ, ΔΕΑ. The proof is not repeated for this pair however, and the *homoiōs* is used instead:[8]

> ὁμοίως δὴ δειχθήσεται, ὅτι καὶ αἱ ὑπὸ ΓΕΒ, ΔΕΑ ἴσαι εἰσίν
>
> 'So similarly it will be proved, that the angles <contained> by ΓΕΒ, ΔΕΑ are equal, too'.

While this is not a case of generalisation, it is well worth our attention. Euclid does not say that the proof for one pair immediately supplies the proof for the other pair as well (for which he could use something like the verb *sunapodedeiktai*, 'simultaneously proved').[9] The proof for the first pair is a proof for that pair and no other. What Euclid says is that the other pair *will* be proved similarly (the future has the force of something like a command-cum-permission, impersonal with the passive *deichthēsetai*, self-addressed with the active *deixomen*).

Especially in such a simple proof, it is difficult to imagine someone actually going through it, explicitly, with the other pair. Proclus, for instance, does not even bother to note the use of the *homoiōs* here, making only the following brief – and just – comment about Euclid's proof:[10] ἀλλὰ τὸ μὲν Εὐκλείδιον θεώρημα φανερόν, 'But Euclid's theorem is obvious'. What Euclid asks his audience to do is to notice the obviousness of the extendability of the proof to the other pair, no more than that. Here, indeed, the obviousness is so transparent that the *homoiōs* is hardly noticed.

[8] 40.21–2. [9] See Archimedes' *SC* 212.17 (which may be an interpolation).
[10] Proclus' *In Eucl.* 299.12. On the other hand, there's a very interesting scholion to Euclid's *Elements* III.20 (the scholion is at 267.13–14). The Euclidean text had ὁμοίως δὴ δείξομεν, on which the scholiast remarked σκόπει, μή σε παρέλθῃ τὸ νόημα – 'check this, to make sure you get the idea' – sound advice, especially in III.20, an example of a Euclidean ὁμοίως hiding a real difference between cases.

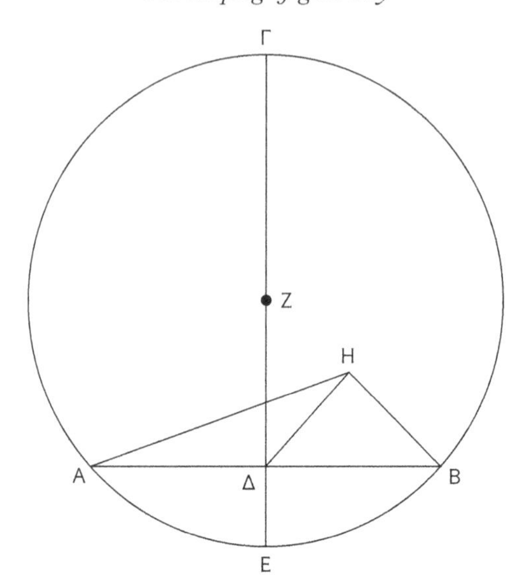

Figure 6.2. Euclid's *Elements* III.1.

The following point must be stressed: what is extendable here is not the result, but the proof.

It is possible to have results extended without an extendability of the proof. For example, in modern mathematics it may often happen that an equivalence is established between domains, so that results can be transferred from one to the other. One is hard put to it, however, to identify the *proofs* in one domain which mirror the proofs in the other. But this is not the type of extension which occurs with the *homoiōs* we have seen. The extension in Greek mathematics is exactly of what is explicitly said to be extended in the *deixomen/deichthēsetai* verbs: the *proof* is being extended and, derivatively from that proof, so is the result. The *homoiōs* demands one to glance, briefly, at the shadow of a proof. That shadow being removed, the result would no longer be valid. It is only because we can imagine ourselves going through the proof that we allow ourselves to assume its result *without* that proof.

An even more interesting situation arises with a certain variation. Consider Euclid's *Elements* III.1 (fig. 6.2). This is a problem, to find the centre of a circle. Euclid offers the construction, and then proves the impossibility of the centre being any other than the one found in the construction. This is done by picking a point H as the putative centre, assumed different from the one found, Z, and obtaining the absurd (angle ΑΔΗ) = (angle ΑΔΖ). Then Euclid concludes:[11]

ὁμοίως δὴ δείξομεν, ὅτι οὐδ' ἄλλο τι πλὴν τοῦ Ζ

'So similarly we will prove, that no other except Z'.

This is a very interesting logical move, given that Euclid practically promises to give an infinite number of proofs. This is in fact a genuine case of explicit generalisation from a particular proof to a universal result (not, of course, the universal result that all *circles* behave in a certain way, but the universal result that all the *points* in this *particular* circle behave in a certain way).

But while we may be impressed by the new logical ground which is being covered by this move, it should be seen that nothing in Euclid's practice points to *his* being very excited about this form of argument.[12] His calm is explicable. There is an infinite number of points in the same section of the circle in which H is, but there is no problem in imagining the proof for all those cases: namely, it must be *the very same proof – the very same sequence of words* – no matter which point you take. The three remaining sections of the circle require a slightly modified proof, which is, however, very similar (the *homoiōs* vouches for no more). The cases of the point H falling outside the circle, on its circumference, on the given line or on the diameter, are so obvious as to demand no proof.[13]

The moral of this is therefore interesting: a proof may be extended to a finite range of cases or to an infinite range, as long as it is clear that the variation involved in that range does not affect anything in the wording of the proof.[14]

[11] 168.7–8.

[12] There are three similar cases in book I of the *Elements* – for which see n. 14 below – and it is interesting to note, again, Proclus' silence concerning the logic of the situation.

[13] However, their proof *would* be substantially different from the one outlined in the text. I do not think that this shows anything significant concerning the sense of ὁμοίως. The Greek tendency is not to count 'trivial' cases as cases at all: for instance, a point-sized circle, whose circumference coincides with its centre, is not a circle for the Greeks. Correspondingly, there is a tendency to ignore such 'trivial' cases within proofs – and my hypothetical cases are 'trivial' in this sense. This tendency is a deductive blemish – it cannot be tolerated if complete generality is aimed at. Such blemishes show, once again, that no explicit codification, motivated by some meta-mathematical ideal, ever constrained Greek mathematicians.

[14] In the first four books of the *Elements*, this sort of infinite extension occurs five more times: I.14 38.23–4, I.39 92.23–94.1, I.40 94.23, III.18 216.11–12, III.19 218.9–10. For further cases, see the typical arithmetical case VIII.6, 290.5–6; also Archimedes, *CS* 13 314.18–19.

Yet we have not quite reached our goal. The universal generalisation in the case discussed above is, as explained, universal in a very limited sense. Greeks do not say things like 'and similarly it will be proved generally' when passing from particular proofs to general results. The *homoiōs* is reserved for a limited class of situations. This class, however, offers a hint of a solution. This is that generalisation may apply to the proof rather than directly to the result. What we have seen in this subsection is not generalisability, but, effectively, repeatability. Generalisability is then a derivative of repeatability. The proof may be *repeated* for the cases to which the *homoiōs* refers, and this is what constitutes the ground for asserting the generalisability. Repeatability of proof rather than generalisability of result: how far will this principle get us?

1.2 Quantifiers

Quantifiers are words such as 'all', 'some', 'the', occurring at the head of noun phrases. The quantifiers most important for generality are *tis*, as well as the adjective *tuchon* and its cognates (the last, met already in chapter 4 as formula 73).[15]

Accustomed as we are to the modern careful setting out of quantifiers, we may expect the relevant words to be systematically applied. As with other Greek mathematical practices, however, no exception-free, explicitly codified system can be deduced. Very often, Greek quantifiers are 'weaker' than what we may expect. To take a typical enunciation, that of *Elements* III.5:[16]

ἐὰν δύο κύκλοι τέμνωσιν ἀλλήλους, οὐκ ἔσται αὐτῶν τὸ αὐτὸ κέντρον

[15] Both τις and τυχόν are the kind of words it is wrong to try to translate. Doing wrong, then: τις may often be rendered by 'some . . .', while τυχόν may often be rendered by 'some chance . . .'. While these are the most important quantifiers, they are not the most common ones (these are the bare noun and the definite article, both used where generality is not an issue). Nor are they the only quantifiers implying generality: others include words such as ὁποιοσῦν, 'whatever' (e.g. *Elements* V.4, 14.14). But such variations are rare (ὁποιοσῦν in its various forms occurs 13 times in the entire *Elements* – only once per book), and therefore I will not discuss them separately. More common, but much less interesting, is πᾶς, 'all, every'. This is relatively common – 17 times in *Elements* I – but it almost always occurs in a single situation. This is in general enunciations of the form 'every triangle . . .' (i.e. the relevant feature studied by the proposition is not the having of a certain geometrical property, but simply being of a certain geometrical shape). This will be clarified in subsection 2.2 below.

[16] 176.5–6.

'If two circles cut each other, they will not have the same centre'.

The quantifier is 'two'; it is not suggested that these should be 'any two'.

More complex enunciations, however, may often include the *tis* as characterising at least a secondary object. I mean expressions such as:[17]

ἐὰν κύκλου ληφθῇ τι σημεῖον ἐντός

'If some point be taken inside a circle'.

The 'circle' is a bare noun;[18] the point – which is logically secondary, dependent upon that circle – is then qualified by *ti*. The same may happen, for instance, with tangents to a circle (e.g. *Elements* III.18). Another interesting case is *Elements* II.6. This is a case of so-called 'geometrical algebra', the theorem proving the 'equivalent' of $b(2a+b)+a^2=(a+b)^2$. The 'a' part (more exactly, '$2a$') is represented by a bisected line, the 'b' by a line juxtaposed to it:[19]

ἐὰν εὐθεῖα γραμμὴ τμηθῇ δίχα, προστέθῃ δέ τις αὐτῇ εὐθεῖα

'If a straight line is bisected [so far the 'a' part], and some line is added to it [the 'b' part]'.

Remarkably, 'a' and 'b' seem to occupy two different logical planes: 'a' is in the bare noun category, 'b' is in the *tis* category.

The first result, then: *an inclination to avoid tis with the primary noun of the enunciation.* This is not universal, but it is a very strong tendency.[20]

A further question is the place of *tis* in the proposition as a whole. The general enunciation is mirrored later by a particular setting-out, which is later followed by a construction, and all of these may use the *tis*.

Counting cases of *tis* and/or *tuchon* in Apollonius' *Conics* I, I have found 73 occurrences: 21 in enunciations, 30 in setting-outs, 1 in the definition of goal, 14 in constructions, and 7 in proofs. The occurrences in the proof are derivative, and reflect enunciation or setting-out in some earlier proposition which is then formulaically invoked in the process of the proof.[21] Quantifiers occur primarily, therefore, where

[17] Euclid's *Elements* III.9 190.12. [18] Greek does not have the indefinite article.

[19] 132.7–8. [20] I.14 is the only exception in the first three books of the *Elements*.

[21] E.g. in proposition 17, the *proof* contains the words (68.9–10) ἐπεὶ οὖν ἐν κώνου τομῇ εἴληπται τὐχὸν σημεῖον τὸ Γ. This is a formulaic reference to proposition 7, where the *setting-out* contained the words (24.29–26.1) καὶ εἰλήφθω τι σημεῖον ἐπὶ τῆς ΔΖΕ τομῆς τὸ Θ. As often happens elsewhere, a construction formula (εἰλήφθω τι σημεῖον) transforms into a predicate formula (εἴληπται τὐχὸν σημεῖον), in the context of an argumentation formula. Significantly, τι and τυχόν prove to be interchangeable.

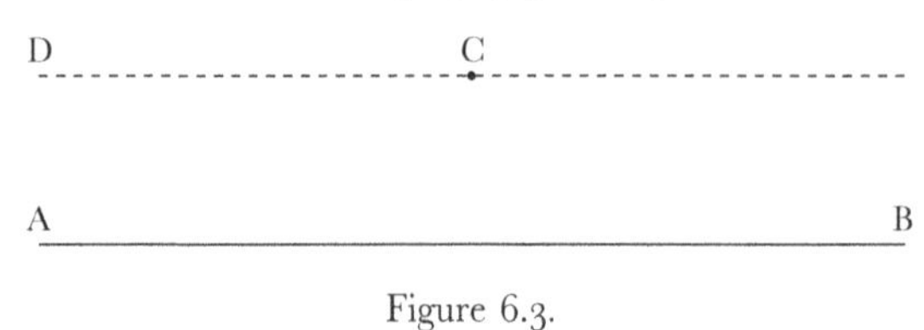

Figure 6.3.

objects are introduced (with construction formulae, then), and the introduction of objects takes place in these three contexts: enunciation, setting-out and construction. So let us concentrate on this triad: 21 in enunciations, 30 in setting-outs, 14 in constructions.

To make more progress, the concept of 'degrees of freedom' is required.

Suppose you draw a line *AB*: this is stage 1. Then you pick a point outside that line, *C*: this is stage 2. Now you draw a parallel to *AB* through *C*, say *CD*: stage 3 (fig. 6.3). In this example, stage 3 is unlike stage 2, because it includes no real choice. Once stages 1 and 2 were over, the parallel *CD* was given. So stage 3 includes no 'degree of freedom'. There are exactly two degrees of freedom in the situation above, namely that of stage 1 (the line *AB* has a freedom: it may be any line), and that of stage 2 (*C* has a freedom: it may be any point outside *AB*). *CD* involves no degree of freedom.

We have already seen one rule: that the object to be introduced – and therefore the first degree of freedom – is almost always without a *tis*.[22] Now we can introduce the second result: *where there is no degree of freedom, there is usually no tis or tuchon either.* (Most often the bare noun is used, though sometimes the definite article is used instead).[23] So the combination of these two results is simple: *tis* and *tuchon* are hardly used, unless in degrees of freedom, and then only in degrees of freedom which are not the first.

What about the remaining cases? Degrees of freedom which are not the first? Let us return to our main quantitative result: in *Conics* I, 21 cases in the enunciation, 30 cases in the setting-out, and 14 in the

[22] The setting-out starts from scratch in the particular domain, and has a first degree of freedom, as well, which is unmarked by *tis/tuchon*, just as in the enunciation.

[23] There are exceptions: Apollonius, for instance, may say that a line is in square '*ti chōrion*', 'a certain area' (e.g. *Conics* I.50, 150.1) – while this area has no degree of freedom. The truth is that *tis* has a wide semantic range, and may be as weak as the English indefinite article. In such cases, *tis* serves a certain stylistic function (especially given the absence of the indefinite article in Greek), but no logical one. Such cases occur from time to time (Archimedes: e.g. *SC* I.18 76.25, I.19 80.17, I.20 84.9), but in Euclid and Apollonius they are a negligible minority. We begin to see why the search for precise rules is difficult.

construction. This of course must be set against what we identified as the relevant background: the number of second or later degrees of freedom. Naturally, there are relatively few degrees of freedom in the *construction*. By the time the construction is under way, the parameters have already been given by the setting-out.

I have sampled propositions 11–20 in *Conics* I. There are 18 first degrees of freedom in the enunciations, the same number in setting-outs, and only 5 degrees of freedom in all the constructions. Of course this has the usual statistical limitations, but it is representative of the large-scale proportions. 18/18/5: how does that compare with 21/30/14 above? An interesting relation, which we can introduce as the third result: *the proportion of degrees of freedom which have 'tis' is higher in the setting-out than it is in the enunciation, and is (much) higher still in the construction.*

We should concentrate now on the *construction*. Do all degrees of freedom there get a *tis* or a *tuchon*?

I have checked this for *Elements* books I, III, VI, IX. The survey is large, but not as large as it seems: many propositions have no construction, and many constructions have no degrees of freedom. There are 35 relevant cases, and 19 of these get a *tis/tuchon*. I have not counted the number of cases in the *Conics*, but the behaviour there is very similar.

In both Euclid and Apollonius, the identity of the exceptions is significant. Most often, they belong to one of these two classes:

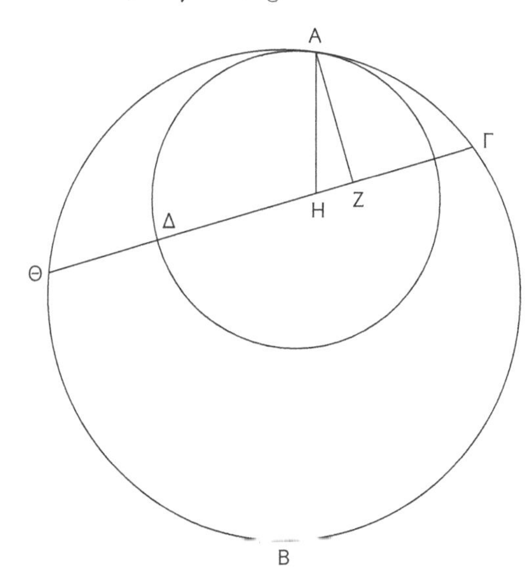

Figure 6.4. Euclid's *Elments* III.11.

(a) *Reductio* constructions, e.g.:

εἰ δύνατον, πιπτέτω ὡς ἡ ΖΗΘ[24]

'if possible, let it fall as the <line> ΖΗΘ'
(instead of: 'if possible, let it fall as *some* <line> ΖΗΘ' – see fig. 6.4).

(b) A line produced, e.g.:

ἐκβεβλήσθω . . . ἡ ΒΔ . . . ἐπὶ τὰ Θ, Λ σημεῖα[25]

'let the <line> ΒΔ be produced to the <points> Θ, Λ'

(instead of 'let the <line> ΒΔ be produced to *some* <points> Θ, Λ' – see fig. 6.5).

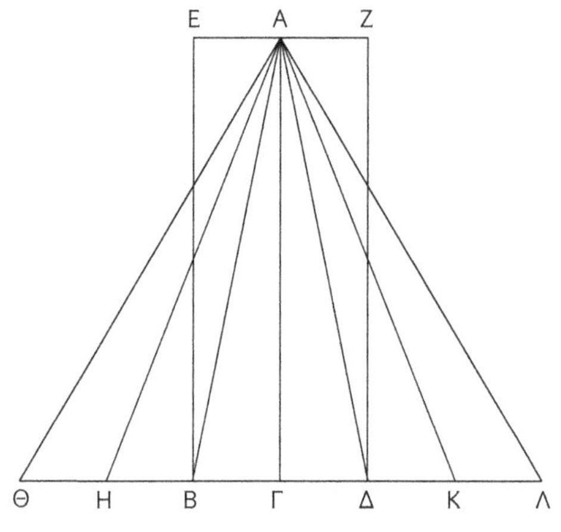

Figure 6.5. Euclid's *Elments* VI.1.

These two types of exceptions are differently motivated, but it can be seen that the exceptions prove the rule.[26]

To conclude, then, result four: *in the constructions of Euclid and Apollonius, whenever there is an active choice, and whenever the Greek allows this, the tis/tuchon is inserted.*

[24] *Elements* III.11, 196.4. [25] *Elements* VI.1, 70.9–10.

[26] The *reductio* is not properly a construction. The mathematician does not construct an object (after all, the object is impossible). The mathematician *contemplates* a hypothetical object. There is no real activity – hence, of necessity, no active choice on the part of the mathematician.

In the case of producing lines, there is an active choice – where to end producing the line. But this is awkward for the Greek. In the Greek, producing a line is an action *starting* at the line itself, which is prolonged until it hits a certain point (producing lines is viewed from the perspective of the start-line). The start-line, however, just cannot be quantified by *tis*, since it is after all a perfectly given line. Thus *tis* becomes inapplicable in such cases.

'Active choice' – this principle explains both why the *tis/tuchon* demands degrees of freedom (= 'choice') and why it is more common in the setting-out than in the enunciation, and most common in the construction (= 'active'). It also explains why the first degree of freedom eschews the *tis/tuchon*. This is because the first degree of freedom is unlike later degrees of freedom: it does not involve a real choice. When a proof starts with a 'let there be an ellipse', it is not as if you are confronted with an infinite number of ellipses from which you choose one. You simply draw a single one, over what was earlier a blank wax-tablet or a blank papyrus. There is no act of choice here. On the other hand, when, in the second degree of freedom, you are asked to pick a point on this ellipse, this involves a real choice between an actually present multitude of points.

We have travelled a long way by now, but we finally begin to see the emerging pattern: the presence of *tis/tuchon* is related to the presence of what is seen as an active choice.

What is the content of this quantifier? The use of *tuchon* may be helpful. This occurs 11 times in Apollonius' *Conics* book 1. Often it comes as an appendage to *tis*, e.g. as in proposition 11:[27]

εἰλήφθω τι σημεῖον ἐπὶ τῆς τομῆς τυχὸν τὸ K[28]

'Let some chance point K be taken on the section'.

Sometimes, however, *tuchon* replaces the *tis*, as in proposition 45:[29]

καὶ εἰλήφθω ἐπὶ τῆς τομῆς τυχὸν σημεῖον τὸ B[30]

'And let a chance point B be taken on the section'.

The *tuchon* variation is very popular in Euclid, especially in the form of an adverbial relative clause such as *hōs etuchen*. There are about 90 occurrences of related uses in the *Elements*. Many replace, rather than qualify, the *tis*; *tis* and *tuchon*, then, are interchangeable. This is not shocking news for ancient Greek, and LSJ quote this mathematical usage alongside other texts, e.g. Plato's, where a *tunchanō* derivative has the force of 'any' or 'just any' (i.e. 'nothing special'). Again, second-order language is somewhat less regimented than first-order language. The more emphatic *tuchon* may therefore serve to elucidate the content

[27] 38.25–6.

[28] This may be the place to note that the most typical place for a τις is the formula εἰλήφθω τι σημεῖον. I don't think this shows that τις has no independent significance; simply, points are very rarely in the first degree of freedom, and, more than other objects, they may often have a degree of freedom.

[29] 136.28–138.1. [30] Other references are: 68.10, 136.14, 140.18, 142.15.

of the *tis*. Both signal that the degree of freedom is indeed quite free, and it will not make any difference which object you may choose within the range of that degree of freedom – the object may be or should be chosen at random.[31]

So we begin to see how generality may be implied. It is implied by the invariance of the proof under the variability of the action. What the *tis/tuchon* signifies is that the active choices of the mathematician – what he *does* through the proof – are immaterial to the development of the argument. The proof is invariant to the active choices of the proposition. Earlier we established the principle 'repeatability of proof rather than generalisability of result'; now we add the principle 'invariance of the proof'. 'Repeatability', 'invariance' – how are these obtained in practice?

2 THE GREEK SOLUTION TO THE PROBLEM OF GENERALITY: A SUGGESTION

2.1 *The solution: an outline*

I have already described this chapter as a reflection upon Mueller (1981) 12–14, and what I am about to offer owes much to a suggestion there. I will therefore start with a quotation:[32]

It is natural to ask about the legitimacy of such a proof [namely] . . . How can one move from an argument based upon a particular example to a general conclusion, from an argument about the straight line *AB* to a conclusion about any straight line? I do not believe that the Greeks ever answered this question satisfactorily, but I suspect that the threefold repetition of what is to be proved reflects a sense of the complexity of the question. The *protasis* is formulated without letters to make the generality of what is being proved apparent. The *ekthesis* starts the proof, but, before the proof is continued, the *diorismos* insists that it is only necessary to establish something particular to establish the *protasis*. When the particular thing has been established, the *sumperasma* repeats what was insisted upon in the *diorismos*. Of course, insisting that the particular argument is sufficient to establish the general *protasis* is not a justification, but it does amount to laying down a rule of mathematical proof: to prove a particular case is to count as proving a general proposition.

[31] An important consideration is the variability of the practice. We have seen the variability of the expressions. We have also seen the non-obligatory nature of the expressions. It may be significant that they seem to be less and less used as we move from elementary works up to Archimedes: there are probably no more than 50 occurrences of *tis* in the relevant sense in the entire Archimedean corpus, and only very few cases of *tuchon*. As usual, Apollonius seems to occupy a position that is intermediate between Archimedes and Euclid.

[32] Mueller (1981) 13. From now on I will stick with Mueller's (and Proclus') terms, explained in the specimen of Greek mathematics (pp. 9–11).

Mueller must be right in viewing the structure of the proposition as the crucial feature. However, his description leaves the Greeks believing in what Mueller admits to be a logical absurdity – that of his last sentence. I therefore think we should reinterpret that structure of the proposition.

Before we embark on this project of reinterpretation, a word of warning is required. The structure of Greek propositions is not as rigid as Proclus seems to imply. There is much variability, which shows that this conventionalisation – like other conventions we have seen before – could not have been the result of an explicit codification. Furthermore, I think that the terms themselves are later impositions upon a pre-existing (shifting) pattern.[33] The structure of the proposition is a second-order formula, consisting of smaller formulae. Like other formulae, abbreviations and permutations may occur. But a stable kernel is noticeable, most clearly in Euclid's geometrical theorems. I will concentrate on this kernel in this subsection.

The first thing worthy of notice is that Mueller is very untypically wrong in the passage just quoted. He asserts that the *demonstrandum* is repeated three times in different forms; he also asserts that 'the *sumperasma* repeats what was insisted upon in the *diorismos*'. In fact, the *demonstrandum* is repeated *four* times, and the *sumperasma* repeats the *protasis*, not the *diorismos*.[34]

We have already seen the number four in fig. 4.1 in chapter 4. There, in the formulaic analysis of *Elements* II.2, we saw a fourfold sequence: construction formulae and then predicate formulae, repeated four times. We now look more closely at this rhythm of action and assertion, trying to find its logic. Taken in detail, the fourfold repetition is as follows.

First, the *protasis* sets out the *demonstrandum* of the proof, which is of course general. As will be explained in the next subsection, the typical structure of the general claim is a conditional, which I will represent by:

protasis: $C(x) \rightarrow P(x)$

To put this less anachronistically: 'if the situation so-and-so is made to exist, then so-and-so is true as well'.[35]

[33] I argue for this in detail in an article which I hope to publish separately.

[34] Mueller is echoing a difficulty Proclus had: both take proposition I.1 as their paradigmatic case, but this proposition is a problem, while the paradigmatic structure as described by Proclus applies to theorems only.

[35] I use 'x' to represent a general object ('all objects', 'any object'). Other lower-case letters represent particular objects ('a', 'b', etc. . . .). I use 'C' and 'P' as signs for kinds of formulae within which the objects may be fitted. 'C' represents construction formulae, and 'P' represents

Next comes a particular setting out of the situation, what may be described by:

ekthesis: $C(a)$.

The structure of the *ekthesis* is 'so-and-so is done in a particular case'.

Following this, Euclid adds *legō hoti* ('I say that' – I shall discuss the meaning of this below) and moves on to assert that the second part, the consequence of the conditional of the *protasis*, holds in the particular case, i.e. he asserts:

diorismos: $P(a)$

'so-and-so is true in the particular case'. This is the second appearance of the *demonstrandum*. The composite unit of the *ekthesis* and the *diorismos* is what we know as second-order formula 44.

Then follows a composite unit made of the two components construction (*kataskeuē*) and proof (*apodeixis*). (This composite unit starts usually with the particle *gar*, 'for', whose meaning I shall discuss below.) The *kataskeuē* is a construction-formulae structure, while the *apodeixis* is a predicate-formulae structure. Generally, C-formulae come before the P-formulae (though the two are sometimes intertwined). So we can represent what comes now as:

kataskeuē and *apodeixis*: $C(b), \ldots, C(n), P(b), \ldots, P(a)$

In other words, this is a sequence of construction formulae and predicate formulae, terminating when the predicate formula of the *diorismos* is obtained. The proof ends in reaching the *demonstrandum* asserted in the *diorismos*, i.e. the proof terminates when $P(a)$ has been established and stated explicitly. This ending of the proof is the third appearance of the *demonstrandum*.

Finally, Euclid repeats the *protasis*:

sumperasma: $C(x) \rightarrow P(x)$

The only difference is the addition of the particle *ara* – 'therefore' (I will immediately discuss its significance), and the concluding words, 'which it was required to prove' (what we know as second-order formula 43). The *sumperasma* is the fourth and final appearance of the *demonstrandum*.

To recapitulate, then, the structure is:

predicate formulae. $C(\ )$ is 'so-and-so being done', $P(\ )$ is 'so-and-so being true', (x) is 'in general', (a) is 'in the particular case a'. The structure of the *protasis* is 'so-and-so being done, so-and-so is true, in general': $C(x) \rightarrow P(x)$. This structure is what we know as the second-order formula 50.

('*protasis*') $\boldsymbol{C(x) \rightarrow P(x)}$.
('*ekthesis*') $\boldsymbol{C(a)}$. 'I say that' ('*diorismos*') $\boldsymbol{P(a)}$.
'for' ('*kataskeue* + *apodeixis*') $\boldsymbol{C(b), \ldots, C(n), P(b), \ldots, P(a)}$.
'therefore' ('*sumperasma*') $\boldsymbol{C(x) \rightarrow P(x)}$.

$P(a)$, at the *diorismos* and the end of the *kataskeuē* + *apodeixis*, is proved by the *kataskeuē* + *apodeixis*. The main point of the theory I am about to offer is this:

$P(a)$ is what the *kataskeuē* + *apodeixis* is taken to prove. The *kataskeuē* + *apodeixis* is *not*, in itself, taken to prove the *protasis*, that $C(x) \rightarrow P(x)$. This is taken to be proved by the combination of *ekthesis* and *diorismos*.

The *kataskeuē* + *apodeixis* proves $P(a)$, and in so doing immediately proves the provability of $P(a)$ on the basis of $C(a)$. For how can you better show the provability of $P(a)$ from $C(a)$, than by a proof of $P(a)$ from $C(a)$? The proof, then, does two things simultaneously: it proves $P(a)$ and, in its very existence, is a proof for the existence of a proof of $P(a)$ from $C(a)$, i.e. a proof of the provability of $P(a)$ from $C(a)$. This, the provability of $P(a)$ from $C(a)$, forms the ground for the claim $C(x) \rightarrow P(x)$.

I will immediately explain the grounds for my theory in the Greek practice. But it must first be said that it has one clear merit. It is, at least in part, true as a matter of logic, as pointed out by Mueller: there is a difference between proving that $P(a)$ and proving that $C(x) \rightarrow P(x)$. Proving a particular case is not proving a general statement.

To the arguments, then.

(a) The connector attached to the *kataskeuē* + *apodeixis*, the *gar*, 'for' at its start, shows what it does prove, and this is the *diorismos* which comes before it. The entire *kataskeuē* + *apodeixis* structure is a backward-looking argument. Had they started with something like *epei oun*, 'Now since . . .', leading onwards until the final *ara*, 'therefore', of the *sumperasma*, this would have meant that the proof led to the *sumperasma*. But the *gar*, 'for', is conspicuously put at the start of the *apodeixis* – conspicuously, given the tendency to avoid backward-looking justifications *during* the *apodeixis*, a feature described in chapter 5, section 3.[36] The first result, then: *the kataskeuē + apodeixis proves the diorismos, not the sumperasma.*

[36] This 'for' at the start of proofs is characteristic of much Greek mathematics, and not only of Euclid: e.g. in Apollonius (*Conics* I, 51) proofs start with a 'for'; only 7 fail to use it when it is appropriate. Euclid's *Elements*' first three books have 7 'for'-less proofs, 91 proofs with a 'for'. The *Data*, as far as I noticed, never omit the 'for'. The 'for' has become part of the formula at the start of *reductio* proofs, viz. 'for if possible . . .'. This is significant: surely the proof in the *reductio* – which establishes a *false* result – cannot be, in itself, the proof for the general claim. It merely proves the impossibility of the particular *reductio* hypothesis, a particular impossibility which is then translated into the general claim.

(b) Further: quite often, a *gar*, 'for', occurs at the very start of the *ekthesis*, i.e. immediately following the *protasis*.[37] This strengthens the suspicion that if the *sumperasma* is not supported by the *apodeixis*, it ought to be supported by the entire *ekthesis* to *apodeixis* sequence. The second result: *the protasis is proved by the sequence starting at the ekthesis.*

(c) The *legō hoti*, 'I say that', preceding the *diorismos* is ambiguous between asserting the result and anticipating it. It does not just say that something ought to be the case; it also says that it is. On the other hand, the subjectivity of the first person implies that the thing is not yet proved but merely asserted. It affirms an assertion and anticipates a proof of that assertion. The location of this affirmation and anticipation is significant. It immediately follows the *ekthesis*, which affirmed (as true by hypothesis) the truth of $C(a)$. The third result: *following the ekthesis, the provability of the protasis for the particular case is affirmed.*

The structure therefore should be interpreted as follows: $C(a)$ is affirmed (*ekthesis*). Immediately thereafter, $P(a)$ is affirmed (*diorismos*), the affirmation implying the (tentative) claim of the *provability* of $P(a)$. Next comes a proof (*kataskeuē* + *apodeixis*) – a parenthetical appendage to the affirmation of $P(a)$. The *kataskeuē* + *apodeixis* is a footnote to the *diorismos*, pointing out that the claim of the *diorismos* is in fact true (and, implicitly, that it is provable, indeed, on the basis of the *ekthesis* alone). This footnote established, the affirmation of $P(a)$ has no more tentativeness about it. The whole of this is then taken as a proof that $C(x) \rightarrow P(x)$.

What then validates this final move? The answer should by now be clear. The fact that the *diorismos* is provable on the assumption of the *ekthesis* – the fact that $P(a)$ may be necessarily inferred from $C(a)$ – is the proof for the general result, that $C(x) \rightarrow P(x)$. The crucial thing is that, assuming the *ekthesis* and nothing else, the *diorismos* is thereby necessarily true. The necessary nexus between $C(a)$ and $P(a)$ forms the ground for $C(x) \rightarrow P(x)$.

The essence of the structure of the proposition is that the combination of *ekthesis* and *diorismos* (the latter supported by *kataskeuē* + *apodeixis*) proves the *sumperasma*. This is valid, provided that the proof, that $P(a)$, was seen to be repeatable for any x; that one could go through the same motions from C to P. It must be seen that the proof – the sequence of *kataskeuē* + *apodeixis* – was indeed based on the *ekthesis* alone. The same proof must be repeatable for any other object as long as the

[37] See Euclid's *Elements* I.7, 12, 14, 15, 18, 20, 21, 27, 28, 29, 41, 48, II.2–10, III.5, 6, 10–13, 18, 19, 23, 24, 27, 32, 35–7.

same *ekthesis* applies to that object. And then, the repeatable provability of $P(a)$ on the assumption of $C(a)$ shows the generality of $C(x) \rightarrow P(x)$.

We see now how our two hints from the preceding section come together: repeatability of proof and invariance of proof.

The structure of the proof shows the validity of the general result. It shows this by pointing to the shadows of possible proofs, from $C(b)$ to $P(b)$, from $C(c)$ to $P(c)$, etc. This suggestion of extendability refers to the extendability of the particular proof rather than of the particular result. The suggestion is indeed a suggestion only, an implicit feature of the proof. And the key to it is that by varying C, by varying the specific content of the action, a new P, a new series of arguments, can be devised: there is an invariance of the extendability of P relative to the variability of the contents of C.

It is now possible to visualise the structure of the Euclidean geometrical theorem as offered in fig. 6.6:

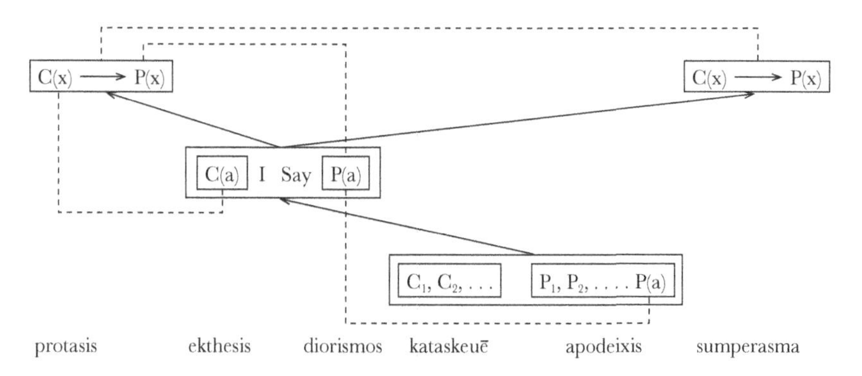

* The diagram gives the six main Proclean units, below their 'boxes'.

* Two types of relations are given in the diagram, three relations each:

* Relations of proof, represented by '⟶':

1. *ekthesis* + *diorismos* proves *protasis*
   (see *gar* at the start of the *ekthesis*)
2. *ekthesis* + *diorismos* proves *sumperasma*
   (see *ara* at the start of the *sumperasma*)
3. *kataskeuē* + *apodeixis* proves *diorismos*
   (see *gar* at the start of the *diorismos*)

* Relations of permutation, represented by '- - - -':

1. *protasis* permutes into *ekthesis* + *diorismos*
2. *protasis* permutes into *sumperasma*
3. *diorismos* permutes into the end of the *apodeixis*

Figure 6.6. The structure of the proposition: a diagrammatic survey.

I return to Mueller, one page ahead of my earlier quotation:

> In the case of Hilbert's theorem 3, the permissibility of such generalization is made clear by the fact that the conditional to which generalization is applied depends upon no assumptions in which the letters A and C occur as 'free' variables . . . Thus logic and the structural interpretation of mathematics make it possible to give a clear and reasonable account of ordinary mathematical reasoning. However, there is no reason to suppose the Greeks to have had anything like modern logic to represent actual mathematical argument, and the Euclidean style makes it look as though a proof is being carried out with respect to a particular object, but in a way assumed to be generalizable. In the absence of something like the rules of logic there is no uniform procedure for checking the correctness of this assumption in individual cases. Rather one must rely on general mathematical intelligence.

As you may see, the theory suggested above should be called Mueller's. But a few qualifications must be made. It should be noted that Hilbert actually does not use an explicit system of quantifiers. Mueller's point is not that Hilbert's theorems are quantified in a certain way, but that *it can be seen* that they *may* be quantified in a certain way. But surely this is a case of applying 'mathematical intelligence'. So Hilbert's actual generality, just like Euclid's, is not algorithmic but requires an understanding of the geometrical situation. Secondly, while no mechanical algorithm is present in the background of Euclid's generalisations, some algorithm is present in the background, namely, checking the proof and seeing that no step in it assumes anything particular about the object of proof, that no predicate formula depends on the specificity of the construction formulae. I have been trying to show that the structure of the proposition is a pointer towards that algorithm. Thus Hilbert is not completely algorithmic but rather points towards a possible algorithm, Euclid is not completely 'intelligent' but rather points – from a certain distance, perhaps even somewhat feebly, but still points – towards his own type of algorithm. It should be remembered that the written symbols of Hilbert are much more tempting from the point of view of explicit, mechanical algorithms than those oral, content-laden symbols used by Euclid. The solution of the problem of generality effected by Euclid is probably the best possible given the means of communication at hand, while the absence of a developed theory in support of this solution is yet another interesting case for the silence of Greek mathematicians in things philosophical.

2.2 *The structure of the* protasis

'[M]athematics is the class of all propositions of the form "*p* implies *q*", where [a certain qualification follows].'[38] Russell had his own reasons for defining mathematics in this way in the *Principles of Mathematics*. The insistence on implication as the form which mathematical propositions take owed a great deal to non-Euclidean geometries, whose different results would become non-exclusive once they were seen as asserting not the absolute truth of their conclusions but rather the truth of the implications from their axioms to their theorems.[39] The motivations for the insistence on implication as the natural format for mathematical propositions were thus different for Russell and for the Greeks; I will return at the end of this subsection to this difference.

Whatever the motivations for this feature, I argued that the logical form of the *protasis* was, as a rule, an implication. I now move on to show this conditional structure.[40] Very often this is the surface grammar of the Greek sentence. Excluding two propositions, 1 and 19, all the *protaseis* of theorems in Apollonius' *Conics* 1 are conditionals. All the theorems in the *Data* are of this form. The first nine books of Euclid's *Elements* contain about 200 theorems; of these, no more than 71 do *not* have this conditional form: a considerable minority, but still, the most natural way to state a *protasis* is by a conditional.

Problems do not have the surface grammar of a conditional, but they do not differ essentially from it. By far the most common way to state *protaseis* of problems is with the genitive absolute/infinitive construction, as in Euclid's *Elements* 1.3:[41]

> δύο δοθεισῶν εὐθειῶν ἀνίσων ἀπὸ τῆς μείζονος τῇ ἐλάσσονι ἴσην εὐθεῖαν ἀφελεῖν
>
> 'Given two unequal lines [so far genitive absolute], to cut from the greater a line equal to the smaller [infinitive construction]'.

The genitive absolute expresses the condition, the infinitive expresses the task under that condition. In a theorem, the existence of the suitable condition guarantees the existence of the result. In a problem, the

[38] This is a distorting out-of-context quotation from Russell (1903) 3. [39] Ibid. 5.

[40] I will not evolve any technical terminology for my own purposes here, and will speak loosely about 'implications', 'conditionals', 'derivations' or 'entailments'. The unit I refer to is a fuzzy linguistic structure, not any particular meta-mathematical interpretation of that structure.

[41] 14.17–18.

condition guarantees the *possibility* of the result, but a mathematician willing to bring about that result is still required for the result to come about. A conditional is therefore unsuitable for expressing the *protasis* of a problem: it is not the case that if two unequal lines are given, a straight line equal to the smaller is thereby cut off from the greater. It *may* be, but it does not have to. So a different surface grammar is needed, while the deep grammar of entailment remains.[42]

We may thus form a class consisting of theorems expressed by a conditional, on the one hand, and problems, on the other. This is a vast class, responsible for the great majority of Greek mathematics, and we see that the propositions in this class are, transparently, entailments. However, Greek mathematicians may sometimes use other grammatical forms. I do not see this as surprising. Once again, this is an example of the freedom of the second-order discourse. I would say, however, that the deep grammar, even in such cases, is that of an implication.

The most typical case is when the condition of the statement is the simple existence of an object having a more or less simple definition. Take Pythagoras' theorem. It may be expressed as a conditional stating that if a triangle is right-angled, the relation obtains. But it is very natural to conceptualise Pythagoras' theorem not as about triangles (in general) but as about right-angled triangles. So one wants something like 'if a right-angled triangle exists, etc.', which, however, would be unnatural. Euclid simply starts the proof by 'in right-angled triangles . . .'. The *ekthesis* then proceeds to assume the existence of a right-angled triangle, and to specify the identity of its right angle. Here, as usual, there are two elements, the background, on the one hand, and the substantive claim, on the other, the truth of the substantive claim being guaranteed by the existence of the background. Instead of expressing the background by a conditional clause, 1.47 uses an adverbial phrase, 'in a triangle', which delimits the extension of the verb.

Other variations are the *pas*, 'all' delimitation ('all so-and-so are such-and-such': the so-and-so being the background, the such-and-such being the claim)[43] and the simple definite article ('the so-and-so are such-and-such').[44] These classes, together with a few rarer subspecies, account, as stated above, for a minority of Euclid's (and

[42] Indeed, Archimedes is fond of stating problems in a form such as: κύκλων δοθέντων ὁποσωνοῦν τῷ πλήθει δύνατόν ἐστιν εὐθεῖαν λαβεῖν . . . – 'given . . . , it is possible to . . .' (*SL* 3 16.20–2. This is the consistent form of problems in *SL* and *CS*).

[43] See, e.g. Archimedes' *CS* 4.

[44] See, e.g. Euclid's *Elements* 1.35.

Archimedes') formulations. However, the deep grammar is always similar, and is on the whole not too deep, i.e. it is easy to separate background from claim. The main practice, that of conditional in theorems or genitive absolute in problems, is very commonly used, so it is naturally seen as the kernel for the variations surrounding it.

I have explained above that Russell's limitation of mathematical propositions to implications was due to non-Euclidean geometries. The conditionality of Greek propositions could not stem from such a motivation. I will, however, say that both conditionalities, that of Russell and that of the Greeks, resulted, at bottom, from the fact, common to Greek and modern mathematics, that mathematics is about proof from premisses. Russell had in mind, as premisses, mainly axioms. The Greeks had in mind, as premisses, the *ad hoc* hypotheses intrinsic to a given proof – its construction formulae. This is typical of the difference between ancient and modern mathematics.

2.3 *The permutations inside the proposition*

In three instances, elements in the ideal Euclidean geometrical theorem permute into each other:

(a) *protasis* permutes into *ekthesis* + *diorismos*
(b) *protasis* permutes into *sumperasma*
(c) *diorismos* permutes into the end of the *apodeixis*.

The second is the least interesting: confined to Euclid, it is a matter of simple, perfect copying.[45] Certainly no one ever read the *sumperasma*.[46] Its only importance lies in showing just how deeply Euclid cared for the structure described in subsection 2.1 above.

This leaves us with two transformations. How are they effected? In principle, both correspondences could be made explicit. The *diorismos* and the end of the *apodeixis* are both written in the same idiom (particularised, diagrammatic lettered objects as elements of predicate

[45] It should be said that at least part of this exact correspondence *may* be attributed to scribes and editors. In Euclid's *Elements* I.41, for instance, line 7, in the *protasis*, has ἔσται in most of the manuscripts, line 24, in the *sumperasma*, has ἐστι; Heiberg – based, *inter alia*, on the good authority of Proclus (who may, however, be more pedantic than his own copies?) – corrects the *protasis* to fit the *sumperasma*. Further, I suppose many scribes would find it easy to copy the *sumperasma* from their own already copied *protasis* rather than from the old and less legible *sumperasma* in their source. But it is clear that an assumption of a very close verbal fit between *protasis* and *sumperasma* must be there for such a process to make a start.

[46] And therefore Heath was justified in saving space by replacing the *sumperasma* by 'therefore etc.', the 'etc.' standing for the *protasis*.

formulae). The transformation from *diorismos* to the end of *apodeixis* could in principle be that of identity. As for *protasis/ekthesis* + *diorismos*, these could never be identical (the *protasis* is general, the *ekthesis/diorismos* is particular), but their relation could be very close.

But even Euclid's *Elements*, the most explicitly pedagogic Greek mathematical text, is usually much less explicit than that. For a typical case, consider the following three excerpts from Euclid's *Elements* I.6:[47]

protasis:

> 'ἐὰν τριγώνου αἱ δύο γωνίαι ἴσαι ἀλλήλαις ὦσιν, καὶ αἱ ὑπὸ τὰς ἴσας γωνίας ὑποτείνουσαι πλευραὶ ἴσαι ἀλλήλαις ἔσονται'
>
> 'If the two angles of a triangle are equal to each other, the sides subtending the equal angles will also be equal to each other'.

ekthesis + *diorismos*:

> 'ἔστω τρίγωνον τὸ ΑΒΓ ἴσην ἔχον τὴν ὑπὸ ΑΒΓ γωνίαν τῇ ὑπὸ ΑΓΒ γωνίᾳ. λέγω ὅτι καὶ πλευρὰ ἡ ΑΒ τῇ ΑΓ ἐστιν ἴση'
>
> 'Let there be a triangle ΑΒΓ having the angle <contained> by ΑΒΓ equal to the angle <contained> by ΑΓΒ; I say, that the side ΑΒ is also equal to the side ΑΓ'.

apodeixis:

> '. . . οὐκ ἄρα ἀνίσος ἐστιν ἡ ΑΒ τῇ ΑΓ. ἴση ἄρα'
>
> '. . . therefore ΑΒ is not unequal to ΑΓ; therefore <it is> equal'.

The correspondence between *protasis* and *ekthesis* + *diorismos* is fairly close, but is not absolute. For instance, it would have been easy to insert clauses stating that the relevant angles are indeed subtended by the relevant lines. Euclid, however, prefers to let this be clear from the diagram, instead. Similarly, there is no verbal correspondence between the *diorismos* and the end of the *apodeixis*. The logical content is identical, but nothing else.

The correspondences, then, are on the logical level of contents, and are not explicit. I have surveyed these correspondences for another 'elementary' work, this time not written by Euclid, namely Archimedes' *PE* I. It has the merit of being short: only 10 propositions count for our purposes. My results there show that *Elements* I.6 is typical: the fit between the permuted elements is very often at the level of logical contents only. Verbal identity is very rare. What is most common are

[47] 22.19–24; 24.6–7.

transformations consisting in changing the order of the constituents and in substituting equivalent descriptions for the same objects.[48] This is to be expected, following chapter 4 above. The main tool which Greek mathematics takes over from natural Greek language is the awareness of form. It is at this level of form that the relation between the permuted elements is perceived.

Another aspect of the same phenomenon is that representations of entities as either falling under a given description or as diagrammatic and lettered objects are perceived as equivalent. This of course is to be expected following chapter 1 above. The identity of the object is secured through the diagrammatic representation; this diagrammatic representation can therefore be used instead of any other description.

We have seen that the relations between the permuted elements in the proposition are important for the construction of generality. I have also explained that this structure is most carefully kept with Euclid. The structure of the proposition becomes looser as one moves on to more specialised works, as if the more expert mathematician needs less and less to establish the generality of the results. The case of the permutations described in this chapter is a good example of this process of maturation. The more expert a mathematician is, the more immediately he becomes aware of relations of form, and the more readily he reads off the information in the diagram. Thus, less and less is required, explicitly, to establish the necessary identities, and the relations between the various parts may become more and more fuzzy. In this way, we begin to see the role of 'mathematical intelligence' for the construction of generality.

2.4 *Pointers to the cognitive background*

The mechanism of subsection 2.1 demands:

(a) that the internal relationships inside the structure of the proposition be followed by the reader, and
(b) that the assertions (the predicate formulae) be inspectable, so that the reader could judge that they are invariant under a change in the specific contents of the actions (the construction formulae).

[48] In 7 out of the 10 cases, the relation *protasis/ekthesis* + *diorismos* is identity up to permutation of order or substitution of logically equivalent descriptions. In 5 out of the 10 cases, the relation *diorismos*/end of *apodeixis* is logical identity only. I suspect the difference results not from any deep logical reasons, but from the greater proximity of the *protasis* to the *ekthesis/diorismos*.

I have described in the preceding subsection how condition I is met. The remaining question – what is at the core of the riddle – is how individual assertions are seen to be independent of their particular content. We think of an assertion about an object, and of an operation on that object. The operation is: varying the object within the degree of freedom which it has. The question is: does the assertion still hold? In the simplest case, this can be seen as the following: if we have another object – another diagram – do we need to have different words? In more complex cases, the diagram does not capture all the information there is about the object (this is mainly true where some metric features are also relevant). Then the question is: is the assertion independent of the information about that object which is not present in the diagram?

There are, then, two types of variability. One is diagrammatic: the point 'taken at random' may have been taken elsewhere – would the statements about this point still be true? The other is undiagrammatic: the particular line in the diagram is given just as 'greater than *A*'. We do not look too closely at how large the line in the diagram actually is. But would the proof apply equally well for a line twice *A*, three times *A*, or four times *A*?

First, the diagrammatic variability. In subsection 1.1 we saw how in *Elements* III.1 only a few different cases had to be surveyed. There was no need to check whether the proof would apply for two different points in the same area of the diagram – for such a variation is not only immaterial, it is also unnoticeable. In the terms of perception itself, shapes are individuated by discrete *cases*, not by the infinite variations of precise measurements within cases. If a diagram includes a bisected line, and then a point selected arbitrarily on this line, then it makes a difference whether you choose the point on the bisection point or on the sides, and then it might make a difference which side you choose; but the exact position on a given side is not merely uninteresting. More strongly, such a variation does not yield a different diagram at all.[49] The extendability one is asked to ascertain is therefore an extendability over what is in effect a finite number of different cases. As explained in chapter 1, subsection 2.1.4, this limitation on the relevant variability of geometrical situations is a result of the reconceptualisation of the diagram via letters.

[49] I assume, of course, that no other relevant topological features interfere.

As suggested by the example above, very often the number of relevant cases may be reduced by simple symmetry considerations. This is sometimes made explicit: Aristotle, following a certain choice of a plane, reassures his audience that διοίσει γὰρ οὐδὲν ἂν ὁποιονοῦν, 'for it would make no difference which . . .'.[50] This indeed may be seen by visualising the symmetry between the different possible planes. Very similar remarks appear twice in Archimedes,[51] so – very rarely – the issue of generality was tackled head-on, and a certain invariance was stated explicitly. The explicit statement is, however, less important than the fact: a qualitative diagram, understood in terms of relations between lettered points (rather than as a metric icon), yields, naturally, a small number of qualitatively different and non-symmetrical configurations. This is an enormous simplification – nothing less than reducing infinity to a small, manageable number.

Second, non-diagrammatic variability. What is at issue here? We have the assertion; we want to know whether it still holds given a change in the construction. In other words, what we need to know is whether the grounds for necessity will be removed by a change in the construction. In *Elements* II.6, Euclid begins the *apodeixis* by saying

ἐπεὶ οὖν ἴση εστὶν ἡ ΑΓ τῇ ΓΒ[52]

'So since ΑΓ is equal to ΓΒ'.

What are the grounds for the necessity of this? This is a hypothesis starting-point, based on the *ekthesis*, a few lines earlier:

εὐθεῖα γάρ τις ἡ ΑΒ τετμήσθω δίχα κατὰ τὸ Γ σημεῖον[53]

'For let some line ΑΒ be bisected at the point Γ'.[54]

There is a degree of freedom involved – notice the 'some'. ΑΒ may be greater or smaller, and its position may vary. Everything about ΑΒ may vary, except the one thing which the hypothesis starting-point assumes, namely that it is bisected at Γ. So what is it that allows us to conclude that this hypothesis starting-point is invariant to changes in the construction – the only invariant thing about that line?

[50] *Meteor.* 375b33–4. [51] *PE* II.10, 206.5–6; *CS* 27, 392.27.
[52] *Elements* II.6 132.25. [53] *Elements* II.6 132.14–15.
[54] Notice that the texts in the *ekthesis* and in the *apodeixis* are not identical. The *ekthesis* has a 'bisection'; the *apodeixis* – 'equality'. The *ekthesis*, itself, does not even mention the objects ΑΓ, ΓΒ. This is typical, and similar to what we saw in subsection 2.3 above. Relations between statements are not mediated just in the surface level of text. The relevant object is not the text alone but the whole, composed of text and diagram.

First, there is a contribution from the tool-box. *Elements* I.9 shows, in a general way, how to bisect straight lines, so we know that the operation is repeatable with other lines. First step: whenever we are given a line (no matter which), we can bisect it.

The next question is: now that we have bisected that other line, would its two segments still be equal? '*How could they be otherwise?*' We are tempted to reply immediately: '*How can you bisect a line without its two segments being equal? What is bisection if not that?*' – and so it is the necessary nexus which provides the grounds for the generality.

We need not rerun the survey of possible grounds of necessity given in the previous chapter. But concentrate for the moment on this: the range of possible, qualitatively different expressions is limited. It is constricted by the possibilities of the mathematical language. What can the two sections be if not 'equal'? Not-equal, nothing else. In the mathematical world there are no shades of meaning. And this, the all-or-nothing nature of the mathematical predicates, is what makes the generality so obvious. If the sections are unequal, we should not call it a bisection. There are no nuances to bisection, no 'more or less bisected'. Just as the diagram is after all finite – a set of possible *cases* – so the language is finite and manageable. The language is both small and well defined. Variability is strictly limited and thus easily checked. Well may Socrates argue that the art of medicine does not study its own interest, but the needs of the body, and therefore in general every art seeks, not its own advantage, but the interest of the subject on which it is exercised.[55] Yes, we tend to respond, this has some truth in it. But how much? How general? How well could the statement be repeated, with other arts substituted for medicine? Checking a few cases (as Socrates does) is helpful, but does not solve our problem. We just cannot foresee how the terms may stretch, because the borders and the very constituents of the conceptual universe they inhabit are vaguely marked. The simplicity of the mathematical lexicon, on the other hand, makes it inspectable. We know not only what the text asserts, but what are the available options were we to try to manipulate it, to stretch it.

In short, then, the simplification of the universe, both in terms of the qualitative diagram and in terms of the small and well-regulated language, makes inspection of the entire universe possible. Hence generality is made possible.

[55] Cornford's translation/interpretation (for one must interpret, the concepts cannot be spelled out in any certainty) of *Republic* I 342C1–6.

2.5 *Conclusion: the generality of arithmetic*

A practice is often best understood through its nearest variations. Euclid's arithmetic is clearly related to his geometry, but is already significantly different. How and why is it different?

Most conspicuously, the full Euclidean theorem-*sumperasma* does not occur, as a rule, in Euclid's arithmetical books.[56] An arithmetical theorem tends to end with the conclusion of the *apodeixis*, i.e. with a particular conclusion.

Another difference relates to diagrams. Arithmetic diagrams represent numbers by lines. While the relative sizes of lines may have been taken as representing, in some way, the relative sizes of numbers,[57] it is clear that the relation between diagram and object represented by the diagram is much less iconic in arithmetic than it is in geometry. Both arithmetic and geometry involve a certain make-believe concerning the diagram, but the arithmetical make-believe is much more radical.

Another important difference has to do with the contents of arithmetical propositions. Often, these involve a two-dimensional degree of freedom. This happens when propositions are about sets of numbers, such as in *Elements* VII.33: 'Given any number of numbers, to find the least number which they measure.' (Such a proposition will be proved for a specific number or specific numbers, and the two degrees of freedom collapse into a single choice of the set of numbers with which the particular proof would proceed.) This sort of logical two-dimensionality is not so typically encountered in geometry.

Further, I mentioned the great simplification introduced into geometrical situations, where all one needs to survey is a finite number of *qualitatively* distinguishable cases, 'to the right of, on, to the left of', and the like. Such simplifications may happen in arithmetic: proofs may proceed along cases – 'greater than, equal, smaller' and the like or, more usefully, 'prime', 'not-prime' and the like – the arithmetical space divides according to non-contiguous, theory-laden divisions. But this sort of simplification is not so clear in the arithmetical case. The geometric space *must* break down into a finite number of qualitatively different areas. The space of integers is much smoother, and often the proof is about 'any integer', a quantity floating freely through the

[56] The exceptions are: VII.4, 31, 32, VIII.14, IX.35.

[57] To say whether this was the case we need to know much more about the original diagrams. From my acquaintance with the manuscripts of Archimedes, I tend to believe that no such iconicity was present in the original Greek diagrams.

entire space of integers, where it has no foothold, no barriers. Often, arithmetic deals with $\aleph_0$ qualitatively different cases. This may be why arithmetical diagrams represent numbers by lines and not by dots. A dot-representation implies a specific number, and therefore immediately gives rise to the problem of the generalisation from that particular to a general conclusion, from the finite to the infinite.

Greek mathematicians need, therefore, a representation of a number which would come close to the modern variable. This variable is not the letter *A* (which is a sign signifying the concrete line in the diagram); it is the line itself. The line functions as a variable because nothing is known about the real size of the number it represents. It is a black box of a number. What can be proved for the black box can be extended to any other number for, the black box being black, we did not know its contents and therefore did not argue on the basis of those contents. The radical aniconicity of the line-representation of numbers rules out any particular expectations based on the diagram.

This is not quite the modern symbolic mechanism of the variable, which is shown, finally, by the problematic status of propositions proving for sets of numbers. The Greeks cannot speak of 'A_1, A_2, . . . A_n'. What they must do is to use, effectively, something like a dot-representation: the general set of numbers is represented by a diagram consisting a of a *definite* number of lines. Here the generalisation procedure becomes very problematic, and I think the Greeks realised this. This is shown by their tendency to prove such propositions with a number of numbers above the required minimum.[58] This is an odd redundancy, untypical of Greek mathematical economy, and must represent what is after all a justified concern that the minimal case, being also a limiting case, might turn out to be unrepresentative. The fear is justified, but the case of $n = 3$ is only quantitatively different from the case of $n = 2$. The truth is that in these propositions Greeks actually prove for particular cases, the generalisation being no more than a guess; arithmeticians are prone to guess.[59]

[58] *Elements* VII.12: M(inimum)3, A(ctual)4; 14: M2, A3; 33: M2, A3; VIII.1: M3, A4; 3: M3, A4; 4: M2, A3; 6: M3, A5; 7: M3, A4; 13: M3, A3; IX.8: M6, A6; 9: M4, A6; 10: M5, A6; 11: M4, A5; 12: M3, A4; 13: M3, A4; 17: 17: M3, A4; 20: M2, A3; 35: M3, A4; 36: M3, A4.

[59] While this problem is essentially arithmetical, it may arise in geometry, as well, especially in proofs for 'any polygon'. Such a proof takes a doubly particular object: it has a particular *n*-gonality, and it is a specific *n*-gon. The particular *n*-gonality is irreducible and is similar to the (missing) subscript variables of the arithmetic case (see, e.g. Archimedes' *SC* I.1, where the choice of a *pentagon* as the token *n*-gon may represent a similar concern about limiting cases).

To sum up: in arithmetic, the generalisation is from a particular case to an infinite multitude of mathematically distinguishable cases. This must have exercised the Greeks. They came up with something of a solution for the case of a single infinity. The double infinity of sets of numbers left them defenceless. I suspect Euclid was aware of this, and thus did not consider his particular proofs as rigorous proofs for the general statement, hence the absence of the *sumperasma*. It is not that he had any doubt about the truth of the general conclusion, but he did feel the invalidity of the move to that conclusion.

The issue of mathematical induction belongs here.

Mathematical induction is a procedure similar to the one described in this chapter concerning Greek geometry. It is a procedure in which generality is sustained by repeatability. Here the similarity stops. The repeatability, in mathematical induction, is not left to be seen by the well-educated mathematical reader, but is proved.[60] Nothing in the practices of Greek geometry would suggest that a proof of repeatability is either possible or necessary. Everything in the practices of Greek geometry prepares one to accept the intuition of repeatability as a substitute for its proof. It is true that the result of this is that arithmetic is less tightly logically principled than geometry – reflecting the difference in their subject matters. Given the paradigmatic role of geometry in this mathematics, this need not surprise us.[61]

3 SUMMARY

Greek generality derives from repeatability (subsection 2.1). This is made possible because Greek propositions have the logical structure for which repetition is relevant, that of a move from one situation to another (subsection 2.2). Repeatability may be checked by Greeks by various cognitive means: the relevant permutations of general and particular are checked through the relations of diagrammatic content and formulae (subsection 2.3); most importantly, the scope for variability, in both diagram and text, is severely reduced, for reasons explained in the first four chapters (subsection 2.4). Arithmetic is in-

[60] Following Peano, we see that this involves a certain postulate, concerning the structuring principle of arithmetical cases. But we may for the moment be naive with Pascal (1954), and suppose that this structure is obviously true of integers.

[61] I therefore must side here with one of my friends, Unguru (1991), against another, Fowler (1994). Quasi-inductive intuitions are not cases of groping towards the principle of mathematical induction. They are the exact opposite of the essential mark of mathematical induction, i.e. its explicitness.

deed somewhat different, but the differences support the theory of this chapter (subsection 2.5).

The entire process is based upon the implicit repeatability of moves in the demonstration. That such repeatabilities are implied in the text was shown in section 1.

To sum up: generality is the repeatability of necessity. The awareness of repeatability rests upon the simplification of the mathematical universe, as explained in the first four chapters.

Part of my task has been completed. I have shown that the two aspects of deduction, that of necessity and that of generality, were shaped in Greek mathematics from a certain cognitive background (and not from some meta-mathematical theory). More specifically, they were shaped from two clusters of cognitive tools, the one organised around the lettered diagram and the other organised around the technical, formulaic language. It remains to describe the historical setting in which the Greek mathematical mode of communication emerged.

CHAPTER 7

The historical setting

INTRODUCTION AND PLAN OF THE CHAPTER

The question before us is 'What made Greek mathematicians write the way they did?'. This is related to the question 'What made Greek mathematicians *begin to* write the way they did?', but the two are not identical. Since what I study is not some verbalised 'discovery' (say, 'Mathematics is axiomatic!'), but a non-verbalised set of practices, explaining the emergence is not the same as explaining the persistence. Assertions, perhaps, simply stay put once they are propounded. The persistence of practices must represent some deeper stability in the context.

We are therefore obliged to adopt the perspective of the long duration. The question about the 'critical moment', the moment at which the cognitive mode started, is not 'What happened then?' but 'What is true of the entire period from then onwards, and is not true of the entire period before that?' – which does not rob the critical moment of its interest.[1] Something *has happened* there – and therefore I will raise the chronological question. When did Greek mathematics begin? When was the style described in this study fixed? Section 1 discusses such questions. (But I must warn the reader: I deliberately avoid the temptation of rich chronological detail, which can easily lead us astray. The chronological argument is briefly argued, and is more dogmatic – also for the reason explained above, that this is not after all the main issue.)

More central are the questions of the long-duration historical background. I ask two main questions: Who were the Greek mathematicians? And what cultural contexts are relevant for the cognitive mode? The first of these questions is what I call the demography of Greek mathematics. I discuss this in section 2. The second is the cultural

[1] Especially since part of the explanation for long-duration stability is not in the stability of the context but in the self-perpetuating mechanisms of the practice itself. Therefore, to explain why the practice got started is an important part of explaining why the practice continued.

background of Greek mathematics, discussed in section 3. Section 4 recapitulates the argument of the book in a few sentences.

1 THE CHRONOLOGY OF GREEK MATHEMATICS

1.1 The beginning of Greek mathematics

This is not the place to repeat the many discussions of this subject – which may all be seen as preliminaries to, or comments on, Burkert's *Lore and Science in Ancient Pythagoreanism*. Instead, I will survey the situation.

A few results are attributed by Proclus (fifth century AD), on the authority of Eudemus (fourth century BC), to Thales (sixth century BC).[2] As Proclus himself points out, Eudemus explicitly makes inferences on the basis of his partial knowledge of Thales. Few today would credit Thales with a book, hardly anyone with mathematical books.[3]

Pythagoras the mathematician perished finally AD 1962.[4]

In discussing the origins of Greek number theory, some prominence is given to Epicharmus' fr. 2. Knorr (1975) is twice reduced, after careful discussion, to concede that this fragment is the only passage in support of the presence of an arithmetical theory before very late fifth century.[5] The hard evidence is that Epicharmus, a comic author, refers to the fact that a number may be changed from odd to even and back by the addition of a single pebble (this as a simile for the fickleness of mankind).[6] All this is part of the pre-scientific background for mathematics. Of course, Epicharmus' audience may have included a few persons in possession of something like the theory of Euclid's *Elements* IX.24–36; but it is not to them that Epicharmus addresses himself, and he gives no evidence for their existence.[7]

So far, this evidence must be ruled out of court. Before we proceed to the real evidence, some general considerations are required.

[2] *In Eucl.* 157.10–11, 250.20–1, 299.3–4, 352.14–16. [3] See KRS 86–8.

[4] The original date of publication of Burkert (1972), a book which demolishes completely the idea of Pythagoras as a mathematician.

[5] 135–7, 145. In 136 the fragment represents 'theoretical arithmetic'; in 145 it '[documents] the technique of studying the structural properties of numbers'.

[6] The force of the simile is in the fact that the *minimal* change yields the *maximum* transformation, from one opposite to another. The simile became a topos – may indeed have been one in Epicharmus' time; see references in Lang (1957).

[7] Berk (1964) 88–93 sees here connections with various strands of pre-Socratic philosophy, including that of the Pythagoreans, concerning whom Berk is (unsurprisingly, given his date) probably over-sanguine. The main thrust of his work is to stress the dramatic nature of Epicharmus' work, which is in line with my interpretation.

The following, I find, is a useful metaphor. One is reminded here of the issue of gradualism versus catastrophism. Catastrophes are nowadays back in fashion in fields such as geology or biology.[8] What I suggest is that the early history of Greek mathematics was catastrophic, not gradual. The belief in gradualism in the early history of Greek mathematics amounts to the hypothesis that the earlier one goes, the more rudimentary are the levels of pre-Euclideanism one meets. This has two aspects. One is the continuous micro-accumulation. The other is that, at a certain stage, a certain super-rudimentary form must be imagined. We should imagine people gradually discovering Euclid's *Elements* book I: the first proposition would take a few years, say, and then new propositions would be discovered one by one, more or less yearly, until the entire book was discovered, thus laying the foundation for further developments. And this immediately shows that both aspects – the continuous micro-accumulation and the super-rudimentary form – are impossible. Whatever the first communication act containing Greek mathematical knowledge was, it included something worth communicating, something impressive and surprising; and whoever was capable of proving one Euclidean result was capable of proving most of them. I therefore suggest that the origin of Greek mathematics could have been a sudden explosion of knowledge. A relatively large number of interesting results would have been discovered practically simultaneously. No more than a generation would be required to find most of the elementary results of Euclid (though, perhaps, not yet to prove all of them equally rigorously).

As Knorr points out in the context of the early history of incommensurability, 'pre-Socratics' never allude to it.[9] They never allude to mathematics, in fact. Zeno's arguments look like mathematics only in Aristotle's reconstruction.[10] Milesian interest in precise shapes is interesting as an evidence for pre-scientific spatial thought, no more.[11] Empedocles is keen on mixtures, not on mathematical proportions.[12] The evidence for Anaxagoras' mathematical activities is feeble, and is

[8] Gould (1980), part 5.

[9] Knorr (1975) 36–40. 'Pre-Socratics' here means people who were born before Socrates.

[10] It is Aristotle who supplies the letters in his presentation of the moving rows argument, *Phys.* 239b33ff., as is clear from the οἷον in 240a4: Aristotle gives *his* example as an explanation of where the trouble with the argument is.

[11] I am referring to texts such as Anaximander, DK A25 (shape of Earth), A10 (shape of cosmos), and Anaximenes' much vaguer cosmology (see KRS 154–5). Herodotus IV.36, on circular maps, should be compared.

[12] Most important are DK B96, 98.

not reflected at all in his fragments (the *e silentio* being, in his case, a serious argument).[13]

Now we pass to the age of Socrates, so to speak, and a galaxy of figures confronts us, all of whom are relatively securely associated with real mathematical interests. These include Hippocrates of Chios, of course, as well as Oenopides,[14] and perhaps Euctemon[15] and Meton. Meton – best remembered as the air-measurer from Aristophanes' comedy *The Birds*[16] – reminds us of the important fact that in the time of another Aristophanes play, *The Clouds*, geometry was already part of the symbolism associated with Socrates' wisdom[17] (though it should be remembered that it is a symbol of an extreme avant-garde position). Other figures are Theodorus,[18] Hippias[19] and Antiphon;[20] one may add figures such as Hippasus[21] and, perhaps, Eurytus.[22] Philolaus, if not a practising mathematician, reflects an awareness of actual mathematics;[23] the same is true of Protagoras.[24] Democritus, certainly, was a mathematician.[25]

Up to and including the middle of the fifth century BC, not a single alleged reference to mathematics would bear scrutiny. From a little later on, many *sophoi* are associated with it in one way or another, and the impression is that most would have shown some awareness of mathematics had our evidence been complete.

Eudemus, setting out to write a history of geometry, no doubt tried to get a full collection of geometrical treatises written up to his day. He then wrote his history. The first book possibly included the introductory general remarks, the inevitable Egyptian speculations, and his reconstructions of Thales' and Pythagoras' mathematics. Then, in the second book, he wrote this: 'Also the quadratures of lunes, which seemed to be among the less superficial propositions by reason of their

[13] See KRS 356. The most likely place for mathematics would have been his theory of eclipse, DK A42; but this belongs to a very physical context.

[14] Proclus, *In Eucl.* 283, 333. There is no reason to dismiss Eudemus in this context, since, on other grounds, we know that in this time mathematics existed; and Eudemus, indeed, is not sceptical about Oenopides, as he was about Thales.

[15] Simplicius, *In de Cael.* 497.20. [16] Aristophanes, *Av.* 995ff. [17] *Nu.* 177–9, 202–4.

[18] *Theaetet.* 143dff., especially that curse of Greek mathematical historians, 147d.

[19] *Hip. Mai.* 285c; *Hip. Min.* 367d. [20] Simplicius, *In Phys.* 54.20–55.11.

[21] Iamblichus, *Vita Pyth.* 247. [22] Theophrastus, *Metaph.* 6a19ff.

[23] Of the greatest importance if genuine at all is testimony A7a (discussed in Huffman (1993) 193–9): 'Geometry is the source and mother city of . . .' The continuation is obviously Plutarch. This is the earliest reference to a mathematical field which makes sense only within a *scientific* interpretation (for no 'land measuring' – still very much the sense of 'geometry' in Aristophanes – could serve here).

[24] Aristotle, *Metaph.* 998a2. [25] Archimedes, *Meth.* 430. 1–9.

close relationship to the circle, were first proved by Hippocrates and also seemed to be set out methodically by him.'[26]

'Also' what? Clearly the most natural reading of the text is that this, like other things, was first proved (and methodically at that) by Hippocrates. This is echoed by the well-known reference in Proclus which may be interpreted as asserting that Hippocrates was the first to leave writings on Euclidean subject matter.[27] It is thus strongly probable that the earliest mathematical author known to Eudemus was Hippocrates, and, if so, he was probably the earliest mathematical author; and so, in an important sense, the first mathematician.

According to our evidence, mathematics appears suddenly, in full force. This is also what one should expect on *a priori* grounds. I therefore think mathematics, as a recognisable scientific activity, started somewhere after the middle of the fifth century BC. Dates are a useful tool: let us call this '440 BC'.

1.2 The emergence of Euclidean-style mathematics

The solid starting-point for Euclidean-style geometry is neither Euclid nor Autolycus, but Aristotle. His use of mathematics betrays an acquaintance with mathematics whose shape is only marginally different from that seen in Euclid; the marginal difference can, as a rule, be traced to the fact that Aristotle uses this mathematics for his own special purposes.[28] The natural interpretation of this is that the mathematics Aristotle was acquainted with – and he was bound to be acquainted with most – was largely of the shape we know from Euclid. This implies that by, say, *c.*360 BC, much of Greek mathematics was articulated in the Euclidean style. This is a first *ante quem*, then.

This may serve as the basis from which to approach the next step. Archimedes refers to Eudoxus as the first to prove something only asserted by Democritus.[29] Generally speaking, Eudoxus is such a central figure in pre-Euclidean mathematics that it would be difficult

[26] Second book: Simplicius, *In Phys.* 60.30–1. Reconstructions: Proclus, *In Eucl.* 379.2–4, 419.15–18, and references in n. 2 above. Quotation: Simplicius, ibid. 61.1–4: καὶ οἱ τῶν μηνίσκων δὲ τετραγωνισμοὶ δόξαντες εἶναι τῶν οὐκ ἐπιπολαίων διαγραμμάτων διὰ τὴν οἰκειότητα τὴν πρὸς τὸν κύκλον ὑφ' Ἱπποκράτους ἐγράφησάν τε πρώτου καὶ κατὰ τρόπον ἔδοξεν ἀποδοθῆναι. My translation is based on the Loeb as far as English vocabulary is concerned.

[27] Proclus, *In Eucl.* 66.7–8.

[28] See, for instance, Aristotle's use of letters, in chapter 1, subsection 2.3.1, and the logical structure of Aristotle's mathematical proofs, in chapter 5, subsection 3.3.

[29] See n. 25 above.

to imagine his style as strongly distinct from the Euclidean: if any pre-Euclidean could serve as a model for later mathematicians, surely this was Eudoxus. Given the Aristotelian starting-point, it is safe to assume that Eudoxus wrote his mathematics in the Euclidean style.

The dates of Eudoxus' activities are not among the safest of ancient dates.[30] Furthermore, while I think Archytas' solution to the problem of the duplication of the cube[31] may be genuine, nobody imagines we have Archytas' own words, so this fragment offers us nothing concerning early mathematical styles. Plato's works suggest that mathematicians already had a certain terminology – Plato does this by allowing his mathematical passages to be filled with what looks like jargon. This jargon is often different from the Euclidean one, but there is no reason to suppose Plato is trying to use the *correct* jargon.[32] Otherwise, Plato is strangely reticent about such aspects of mathematical practice as, for instance, the use of letters in diagrams. In general, his use of mathematics is done at a considerable distance from it, and does not allow us to see clearly what was the shape of the mathematics he knew.

I shall therefore return to the Archimedean passage, the introduction to *SC*: Democritus first knew what Eudoxus later proved properly. We do not know the exact dates of either Democritus or Eudoxus, let alone the exact dates of their mathematical works. We do not know what their mathematical works amounted to. But the two may serve as an emblematic pair. Democritus was the pioneer, one of a group of persons who had the luck to be there when mathematics first came to be studied as a field of theoretical enquiry. A generation or more later, Eudoxus was already a member of the very same well-defined tradition to which, later, Archimedes himself would contribute. The distance in time between the two, Democritus and Eudoxus, is not small. It is comparable with the distance between the so-called sophists and Plato. Can we say that Greek 'professional' mathematics passed through stages similar to those through which Greek 'professional' philosophy passed? For the time being I shall concentrate on two dates: 440 BC is my suggestion for the start of what may be called 'mathematics'; 360 is a firm *ante quem* for Euclidean-style mathematics.

This should be refined in the following way. What we discuss is not necessarily the emergence of a new object. Perhaps all the elements of the mathematical style were already there from 440. I doubt it, but this

[30] See Merlan (1960) 98–104, with the qualification of Waschkies (1977), chapter 2 (esp. 50), which makes the chronology even more obscure.

[31] Eutocius, *In SC* 84.12–88.2. [32] See Lloyd (1992).

is possible. The important question is not only when individual objects come into existence, but what the role of those objects is in a wider cultural structure. The process from 440 to 360 as I imagine it is not so much that of the introduction of new elements (this happens too, but is less important). The most important aspect of the development is a realignment. At the start, we have *sophoi*, many of them interested in mathematical questions. But the subject matter may not be recognised as such, and the boundaries may be blurred. Music has its own motivations; so has mechanics. Wider 'philosophical' issues may be the core of interest in any mathematical work. And then a new structure emerges: the mathematical subject matter becomes recognised, and is organised as a unity. This is what the evidence from Aristotle proves most clearly, especially as seen in the *Posterior Analytics*. This new structure, then, defines a unit, 'mathematics'; and its style may be said to have emerged.

But all this leads us beyond chronology to demography and wider contexts. This brief chronological introduction will have to suffice.

2 DEMOGRAPHY

The two questions I will raise here are who the Greek mathematicians were and how many they were. By 'who they were' I mean whether anything can be said about Greek mathematicians as a group. For both questions, some idea of what is meant by 'a Greek mathematician' must be given. I will not offer a definition, but I will try now to explain my guidelines. First, the term is not meant to be exclusive of other activities. Eratosthenes, for instance, certainly was not primarily a mathematician. But he did produce a mesolabion:[33] therefore I count him as a mathematician.

This is a relatively simple case: Eratosthenes is quoted by Eutocius in the context of a commentary on Archimedes, and he received letters from Archimedes. 'A Greek mathematician' is an indefinable concept, like the 'metre'. To see the mathematicianhood of a Greek, what you have to do is to measure him against others, and Archimedes is the best measure: he is the 'metre kept in Paris'. Apollonius and Euclid may be used as well, and by measuring against those three, many other authors, from Hippocrates of Chios to Eutocius himself, can

[33] Eutocius, *In SC* III.88.3ff. He did more mathematics, of course; my point is that even a single mathematical achievement, and even from a person who is predominantly a non-mathematician, would suffice to count here as a mathematician.

safely be called mathematicians. There is a great degree of continuity, in terms of subject matter discussed, on the one hand, and modes of presentation, on the other. A few borderline cases may occur. What does one have to say about numbers in order to count as a mathematician? How much theoretical astronomy must an astrologer do before he can enter my mathematical guild? Must musicians be Pythagorean to be called mathematicians? There can be no hard and fast rules. Take Aristoxenus. His style differs from that of the *sectio canonis* (the musical treatise in the Euclidean corpus), i.e. it is not as similar to my 'metre' as other musical works are. Aristoxenus' works have stylistic characteristics which remind one of mathematical works, but there are no diagrams, hence no lettered objects, and in general the presentation is closer to ordinary philosophical discursive Greek than to the mathematical system described in this book. Where to place Aristoxenus, then? Probably, the best course is to place him where he evidently wished to be placed, away from anyone else.[34] The liminality of Aristoxenus was voluntary – he is therefore the exception which proves the rule. As a rule, even idiosyncratic authors such as Diophantus, let alone Hero in his more strictly mathematical works, approach the 'Euclidean' style, and can easily be classified as mathematicians.

Astrology and number-lore present a different kind of problem. Putting them apart from mathematicians in the 'Euclidean' sense would not necessarily reflect ancient views. *Mathematicus* came to mean in Latin 'astrologer'.[35] As for number-lore, Plato's lecture had already led directly from the mathematical sciences to the conclusion that the good is unity.[36] But I will not count astrologers and numerologists as mathematicians. This is not cultural snobbery. Rather, I am concentrating on the issues which are important for my study, which are cognitive. Astrology and numerology, in themselves, do not impose the same cognitive regime as 'Euclidean' mathematical sciences do. A person who has devoted himself to astrology alone, with no element of astronomy, while he may have, say, calculated a great deal, has never proved anything, and he therefore cannot be seen as a mathematician from the persepctive adopted in this study.

Based on such criteria, I have made a list of all recorded Greek mathematicians (including those known only through anonymous works).[37] I now make a few comments on this list.

[34] For the character of Aristoxenus, see e.g. Barker (1989), vol. II.119: 'To judge by what [Aristoxenus] says, all previous writers on harmonics were incompetents or charlatans.'

[35] s.v. Lewis & Short, 2. 'post-Aug.'. [36] Aristoxenus, *El. Harm.* II.30.

[37] I intend to publish this list in a separate article.

2.1 *Class*

We consider ourselves lucky if we know for certain the century to which a Greek mathematician belonged. Typically, our evidence is in the form of a stretch of mathematics or a testimony on a mathematical work. The mathematics is stylised and impersonal, and the personal background is lost. Naturally, Greek mathematicians appear to us in fuller personal colours according to the extent to which they are exceptional, which in itself implies high status. A few do, which is noteworthy. Archytas was the leading citizen of Tarentum, the most important Italian Greek city of his time.[38] Eudoxus may have been a leading citizen, if not the leading citizen, at Cnidus.[39] Theaetetus' father was a well-reputed and, especially, a very rich citizen.[40] Meton may have been expected to furnish a trireme: an important citizen, therefore.[41] Hippias was a trusted ambassador of Elis; he was also very rich (though there is some indication that his wealth was self-made).[42] We have moved down, but we remain in the upper echelons of society.

Archimedes is more difficult, however. What little evidence[43] there is is contradictory, and it seems to express the authors' fancy rather than any specific knowledge. Neither Cicero nor Plutarch had access to Archimedes' bank statements: they had his writings, just as we do. He does converse with royalty, without excessive deference:[44] is this a mark of nobility or of proud poverty? Eratosthenes, again, writes to a king. His background is far better understood. As the librarian, he should be thought of as an important courtier:[45] High status, no doubt, but not quite cutting the same impressive figure as Archytas. Aristotle figures in my list as a mathematician, and he comes to mind as another famous royal functionary, this time a tutor. Vitruvius addressed Augustus;[46] Thrasyllus served Tiberius.[47] Philonides' role in the Seleucid court is comparable.[48] These last-mentioned persons – certainly if we exclude Vitruvius – no longer display the spirit of citizenship of Hippias, Eudoxus, Archytas, Theaetetus and Archimedes. They live away from their home towns. They may make their living from wages – no doubt very high wages, but still wages. Their class may approximate an

[38] DK A1.
[39] D.L. VIII.88, on the authority of Hermippus; doubtful, therefore, but inherently plausible.
[40] *Theaetet.* 144c5–8. [41] Sommerstein (1987), n. *ad* l.997, p. 264.
[42] *Hip. Mai.* 281a3–b1 (ambassador); 282d7 (wealth); 283a4–6: a self-made man?
[43] See Dijksterhuis (1938) 10.
[44] Archimedes simply uses the expression βασιλεῦ Γέλων twice, at the beginning and towards the end of the *Arenarius*: 216.2, 258.5.
[45] See Fraser (1972) 322–3. [46] Vitruvius, Introduction to book I.
[47] Tacitus, *Ann.* VI.20–1. [48] See, e.g. Fraser (1972) 416 and n. 322, vol. II, 602.

'intelligentsia' rather than the nobility though, again, they by no means live in damp attics on coffee and cigarettes.

Hippocrates of Chios may have been an *emporos*, 'merchant', a word in itself saying little about precise economic status. The evidence comes from Aristotle and from Philoponus, and one trusts Aristotle more. Both make him lose money through some misfortune; in Aristotle's version this is a huge sum. To lose a huge sum of money is a sign of some wealth. Philoponus – but not Aristotle – makes him live away from his home town.[49] Theodorus, a little later, probably did teach young students, as he is shown doing in the dialogue *Theaetetus*, presumably not without pay – though Plato does not mention money arrangements. Autolycus taught Arcesilaus,[50] though here the sense of 'taught' is as vague as it could be. One might as well learn that Plato taught Aristotle, and think of him as a paid teacher. Not much more illuminating is the reference to Carpus as a *mechanicus*.[51] He may have been an engineer; or he may have written on mechanics or even just been accidentally nicknamed. Setting aside Archimedes' war-engines, the only solid evidence for a Greek mathematician getting things moving in a big way[52] is Anthemius' of Tralles commission for building the Hagia Sophia in the sixth century AD[53] – very late evidence indeed.

I would like to make a detour, looking at the questions of age and gender. First of all, age. Very roughly speaking, Greek culture tended to appreciate old age more, and youth less, than some strands of modern culture.[54] Certainly the Greek world had many prodigies of achievement in old age.[55] Against this background, the slight evidence for mathematical achievement in youth becomes more noteworthy. Theaetetus is pictured as re-proving at least a result, as a mere boy.[56] The evidence is more than this mere attribution. Theaetetus, the Platonic character of the later dialogues, is a vivid figure, the bright youth, a charming contrast to the sagacity of most leading characters in the dialogues. Is it an accident that this embodiment of young brilliance

[49] The evidence is in DK A2. [50] D.L. IV.29. [51] Proclus, *In Eucl.* 241.19.

[52] I ignore toys, such as Ctesibius' organs; these are best seen as cultural objects. As far as class implications are concerned, to construct a toy is not necessarily more banausic than to write a book. Everything depends on the details of the construction and the dedication, which are generally unknown. I shall return this in the next section.

[53] Procopius, *de Aedific.* 1.1.24.

[54] Kleijwegt (1991) argues for the absence of adolescence-culture in the ancient world. Less controversially, he points out the value of adulthood in antiquity, e.g. 58ff., 188ff.

[55] The table of the longevity of philosophers, based on Diogenes Laertius, in Minois (1989) 55–6, is instructive (see also the wider discussion there, 53–7). The youngest death there is Eudoxus'.

[56] *Theaetet.* 147e–148a.

should be a mathematician? A literary topos is thus struck: in the *Amatores*, boys discuss mathematics.[57] And the popular image exists independently of Plato. Isocrates may be interpreted as associating an interest in mathematics with youth.[58] Aristotle's evidence is striking: he asserts that the young fare better at mathematics.[59] Is this his own impression, a cliché – or just a response to the *Theaetetus*?

Popular impressions aside, at what age did mathematicians produce their works? Very little is known, of course. Pythocles was known for his achievements when he was not yet eighteen, and was apparently interested especially in mathematics, or at least astronomy.[60] Eudoxus – if the tradition can be trusted[61] – died aged 52; is this young, given his achievement? On the other hand, we now know that Apollonius must have been quite old when he was producing the *Conics*,[62] but this is just the tip of his mathematical iceberg. Hipparchus' observations date from 161 to 126, at least 35 years of activity, perhaps considerably more: what does this make of his age at his début? Ptolemy's observations in the *Almagest* range from 125 to 141, Timocharis' from 292 to 279.[63] Only for Hipparchus is a career of some decades suggested, and it may equally have started by the age of 20 or 30. The starting-point is vague; what seems probable is that Greek mathematicians did not burn out in the way modern ones allegedly do. Again, this is typical of Greek culture in general. But no more hard evidence is available.[64]

The gender issue can be easily settled. Two women mathematicians are known: Hypatia and Pandrosion.[65] How many is two? Given the obvious obstacles, I think two is actually a lot.[66] And this is the general

[57] *Amat.* 132ab.

[58] *Antidosis* 261ff., where, unfortunately, the issue is complicated by the conjunction of mathematics with philosophy.

[59] *EN* 1141a11–13. [60] See Sedley (1976) 43–6. [61] See n. 30 above.

[62] See Fraser (1972) 415–16, or the article 'Apollonius' in the *Dictionary of Scientific Biography*, written by Toomer.

[63] In general for observations known through the *Almagest*, see Pedersen (1974), appendix A. Ptolemy, of course, could have been in mid-career or even younger when finishing the *Almagest* (and certainly when finishing its observational basis).

[64] The solid evidence gathered by Kleijwegt (1991), chapters 5–6, on the early achievement of physicians and lawyers, is important background. Compared to this, it would seem that mathematicians, if anything, were late maturers. However, the *e silentio* is meaningless: we simply have no comparable data for mathematicians. Rather than showing that young mathematicians were relatively rare, the absence of mathematicians from Kleijwegt's survey (largely based on inscriptions) is yet another argument for the absolute rarity of mathematicians in antiquity; for which see the next subsection.

[65] For Pandrosion, see Pappus, introduction to book III.

[66] Both come from late antiquity, a period when, in some contexts, exceptional women stood some chance; see Brown (1988) 145ff.

remark with which I would like to conclude this subsection. Access to education was extremely restricted in antiquity, with illiteracy, and the physical difficulty of obtaining books in a printless culture, practically barring entrance for anyone from outside the privileged classes. Further, generally speaking, ancient society was also strongly polarised. It is doubtful whether the concept of a 'middle class' has any meaning at all in antiquity. The rich could be more or less rich, the poor could be more or less poor, but the rich were rich and the poor were poor.[67] Thus, Greek mathematicians, by and large, should be assumed to have led a privileged life. However, money won't buy you mathematics. Mathematics, perhaps more than other disciplines, calls for specialised cognitive skills. They may be culturally developed – as I have insisted throughout this book – but a residue remains, of highly variable individual capacities. It is such capacities and inclinations, not wealth and status, which determined whether an individual would become a mathematician. Thus, some mathematicians could have been young; some could have been, perhaps, from a relatively modest background (though always within the privileged class); a few, indeed, could even be female.

2.2 *Numbers*

I have listed 144 individuals about whom we can make a *guess* that they may have been mathematicians. These are not authors for whom we have fragments, but a much wider group, including anyone for whom we have the slightest evidence – including anonymous authors. This number, 144, is therefore the minimum for our discussion.

A certain probabilistic argument may allow us to make a very speculative first guess concerning the absolute number of Greek mathematicians active in antiquity.[68] On the basis of this argument, I argue that

[67] de Ste Croix (1981).

[68] The argument is this. Suppose you make independent choices of balls from an urn. In the first choice you get out n balls, you return them, and then you draw out k balls. You are allowed to know that r balls came out both times. Since the assumption is that there is nothing special about the balls you took out in either choice, it must be assumed that the proportion of balls which have been taken twice (r) among balls taken out once (e.g. in the first choice, i.e. n) is the same as the proportion of all the balls taken in the other choice (i.e. with the same example, k) among the entire number of balls in the urn – let us call it N. In other words:

$$r/n = k/N$$

or

$$N = (n*k)/r$$

The total number of balls is most probably around the number of balls taken the first time multiplied by the number taken the second and divided by the number of balls taken twice.

When Proclus refers to earlier mathematicians, he is very much like a person picking balls

the absolute number of mathematicians whose names were at all accessible to anyone in late antiquity was around 300 (this is not merely the number of mathematicians active in late antiquity but includes all mathematicians, from Democritus onwards, known, in some way, to someone in late antiquity). How can we pass from what was known to late antiquity to the real number, then? This involves a second leap. But bear in mind that we talk not about the infinitely difficult (and well-known) survival of *works* in the manuscript tradition, but of the survival of mere *names*. This is not so difficult – a single manuscript surviving from Alexandria with 20 names on it will suffice to keep all those 20 in the list. It need not be a manuscript which Pappus consulted himself: suffice it that he *could*, theoretically, consult such a manuscript. What I claim is that the class of such names in late antiquity included no more than about 300 names. The number 300 thus cannot be a tiny fraction of the total number of ancient mathematicians. It is a sizeable proportion. Again, I will assume it represents a minority. I will therefore take the convenient number 1000.

Housman estimated that literary critics are rarer than the appearance of Halley's comet.[69] The heavenly body appropriate for measuring the appearances of Greek mathematicians is the sun. We cannot be far mistaken in assuming that, on average, no more than one or two Greek mathematicians were born each year. Probably the average was even lower.

My reader must be impatient. I know the limitations of such arguments, and soon I shall give more substantial ones. But before that, a chronological interlude is required, so please bear with me for a while.

Of course, Greek mathematicians were not as regular as the sun. Some years, especially in the Hellenistic period, must have been more plentiful than others. Each year around the third century BC may have

out of the urn of 'mathematicians he can possibly refer to' – mathematicians whose names are at all known around his time.

There are many difficulties with the application of the argument. The worst complication is that several mathematicians just *have* to be referred to (authors such as Euclid). There is nothing random about picking them. The rest, however, are much more arbitrary. I have isolated seven such names and struck them off the list. With the remaining names, I have proceeded to make the calculations as described above, identitying four acts of 'choice from the urn': Pappus, Proclus and Eutocius, and the most important choice, that of the manuscript tradition. This gives us six pairs, and the results, applying the equation above, are: 130, 104, 135, 101, 303, 89. All except one cohere around the same number but, as a safer guess, I will take the higher 300 (Proclus/Manuscripts).

[69] Housman (1988) 302.

seen the birth of two or three mathematicians, but no more than this, since otherwise the Hellenistic period would have exhausted the stock of mathematicians allotted to antiquity. The point is that while the Hellenistic period was relatively plentiful, nothing suggests it was not more or less *consistently* plentiful, which means that the number of mathematicians per year could not have risen dramatically at any given continuous stretch.

Nor were the downward fluctuations complete. Mathematics was never completely extinguished. There are a few gaps in the chronology. For instance, I can find no mathematical activity in the period between Thrasyllus' death (AD 36) and Nicomachus' career (late first century AD?) – neither, it should be said, a particularly inspiring mathematician. This is a wilderness between two deserts. But this period is in general a low point for the documentation of Greek culture, and it is probable that some mathematical activity went on throughout the period. That the birth-rate of mathematicians could have been as low as one per decade is possible, though even that seems extreme. We can imagine mathematically inspired magi doing their rounds, adoring every born mathematician – they are never pressed for time as they get from one birthplace to the next, but on the other hand they are never out of work for whole generations.[70]

So this is the fantasy. Now the evidence.

The essence of the probabilistic argument is the close prosopographic repetition between our various sources. No matter through what angle we look at Greek mathematics, we always see the same faces, and this leads to the conclusion that there were not so many faces there to begin with. The trouble with this is that we do not actually look at Greek mathematics from radically different angles. The commentators, by and large, shared the same set of interests and had similar access to the past; both these interests and this access are reflected by the selection of the manuscript tradition. It is thus vital to get a completely different hold, to try and view Greek mathematics from as independent a viewpoint as possible.

First, the perspective from which the mathematicians themselves viewed mathematics. Indeed, both Archimedes and Apollonius alone testify to the existence of a group of mathematicians: Dositheus, Pheidias and Zeuxippus in the case of Archimedes; Attalus, Eudemus, Philonides,

[70] And their travels take them around all the eastern Mediterranean – returning time and again to Alexandria, but still travelling widely. I have identified 51 mathematical sites – a very large number for 144 individuals (for many of whom no location can be assigned).

Thrasydaeus and Naucrates in the case of Apollonius. These are the near-contemporaries of Archimedes and of Apollonius, some of whom probably belonged to the passive audience of mathematics, no more. As for predecessors, Apollonius refers to Euclid alone,[71] Archimedes to Democritus, Eudoxus, Aristarchus and (perhaps) Euclid.[72] Furthermore, one of the contemporaries to whom Archimedes writes is Eratosthenes, who is well known as a mathematician from other sources, and another, Conon, is referred to by Apollonius as well.[73] In other words: of the five contemporaries referred to by Archimedes, two are known independently, which prima facie suggests, according to the probabilistic argument, 12.5 as a minimum approximation to the number of contemporaries.

This should be considered in all seriousness. We are witnessing here the heyday of Greek mathematics, a period where I am willing to imagine the birth of up to three mathematicians a year, i.e. the total number of contemporaries may be as large as 100.[74] The brief glance Apollonius and Archimedes allow us of their world can not be compatible with anything more than this. When Archimedes first approaches Dositheus, following the death of Conon, he is as desperate as any veteran 'lonely hearts' column correspondent.[75] Apollonius, approaching Attalus in the introduction to *Conics* IV, is more direct, but the picture is similar. The main feature is that contact is made on strictly individual grounds. No 'school of mathematics' is ever hinted at by the mathematicians themselves, and the death of a single person seriously disrupts the network. The vision of the mathematical past, as available to the mathematicians, is narrow, and does not add to what we know from elsewhere.

[71] *Conics* I.4.14. The persons alluded to at IV are best seen as contemporaries, though no doubt Conon was dead by that date, and so presumably was Thrasydaeus.

[72] Democritus: II.430.7; Eudoxus: I.4.5 and elsewhere; Aristarchus: II.218.7 and elsewhere. Euclid: I.12.3 – which may be a gloss.

[73] *Conics* IV.2.15.

[74] This involves the important question of longevity. The number of contemporary mathematicians is the number of mathematicians born per year multiplied by the average *mathematical* longevity. Mathematical longevity is the period in which you are mathematically active. To put it at 30 is extremely generous given ancient life expectancies and the fact that many mathematicians (1) may have become mathematicians relatively late in life or (2) may have become bored with mathematics after a while (a non-exclusive disjunction).

[75] *QP* 262.2–8 (Heath's version): 'Archimedes to Dositheus greeting. When I heard that Conon, who was my friend in his life-time, was dead, but that you were acquainted with Conon and withal versed in geometry, while I grieved for the loss not only of a friend but of an admirable mathematician, I set myself the task of communicating to you, as I had intended to send to Conon . . .'

Archimedes hopes (in vain) that the method of the *Method* will be picked up by others.[76] Look at what he writes at *SL* 2.18–21:

> ... πολλῶν ἐτέων ἐπιγεγενημένων οὐδ' ὑφ' ἑτὸς οὐδὲν τῶν προβλημάτων αἰσθανόμεθα κεκινημένον
>
> 'Though many years have elapsed ... I do not find that any one of the problems has been stirred by a single person'.[77]

The natural translation of κινέω here is not 'advance' but, as Heath correctly translates, 'stir', i.e., Archimedes complains that for many years he has not heard about a single person even *trying* to solve the open problems he had offered to the world via Conon. No wonder he was so desperate following Conon's death. Archimedes, in the *Method* and elsewhere, gives a sense of boundless intellectual energy, crying out for some collaboration; the world did not collaborate.[78]

Another set of evidence – another angle on the mathematical world – is offered by the documentary evidence from antiquity: inscriptions and papyri. This is especially valuable, since this is the closest we come anywhere in classical studies to a random sample. In particular, the papyri from Egypt present a more or less precise picture of the frequency of different kinds of literature in the drier parts of Egypt, especially in the Roman period. Neither the period nor the location is perfect for our purposes, but Alexandria, after all, is not far away.

To say nothing of pure literature, papyri have given us the *Hellenica Oxyrhynchia*, the *Constitution of Athens*, the *Anonymus Londinensis* medical treatise. The closest parallel in mathematics is the brief and semi-mathematical *Ars Eudoxi*. We have seen that the literary evidence for Greek mathematics repeats itself. The papyrological evidence is almost non-existent. The few bits which do exist repeat material known from Euclid.[79] It is symptomatic that the most extensive and serious piece of 'papyrus' mathematics is on a series of ostraka.[80] This should be

[76] 430.15–18. [77] I use Heath's translation, Heath (1897) 151.

[78] Before setting aside this sort of evidence, the Eratosthenes' fragment preserved by Eutocius should be mentioned (together with Knorr (1989), I take it to be genuine). Eratosthenes was extraordinarily placed as an antiquarian. He does offer us here a fragment of an unknown tragedy, and a precious morsel of information on Hippocrates of Chios. In general the interest of part of the fragment is decidedly historical. All this boils down to is a mention of Hippocrates of Chios, Archytas, Eudoxus and Menaechmus – all well known from elsewhere.

[79] For a full discussion, see Fowler (1987), section 6.2, supplemented by to Brashear (1994) 29, according to whom there are altogether six pieces of literary papyri relating to Euclidean material. Five of these, it should be pointed out, relate to book 1. I do not know of any other Euclidean-style mathematical papyri.

[80] Mau and Mueller (1962).

compared to the rich papyrological evidence on very elementary numeracy and 'geometry'.[81]

Furthermore, the documentary evidence, both in papyri and in inscriptions, amply testifies to the existence of philosophers, not to mention grammarians, rhetoricians and, of course, doctors. Very often (relatively speaking), such professions are mentioned in decrees, on funerary steles, and in everyday correspondence.[82] On the other hand, the noun μαθηματικός occurs only once in the Duke papyrological collection.[83] Invariably, the use of γεωμετρία and its cognate forms – exceedingly common in documentary papyri – refers there to land measurement for tax purposes, a point to which we shall return below. *IGChEg.* 246.p1, a Cyrenaic inscription of unknown date, laments the death of Isaac the geometer – a fine name for a mathematician – aged 33.[84] *IG* VII.22. col. B.7, from Megara, sets up the salaries of various teachers, including a geometer: he is paid '50 per child', as much as a παιδαγωγος, χαμαιδιδασκαλος or γραμματικος Ελληνικος ητοι Ρομαικος is paid; in other words, the 'geometer' in question is an elementary schoolteacher, the 'geometry' means elementary measurements, and any active mathematician would be overqualified – not that he would need the job, as explained in the preceding subsection.[85] Another inscription from Megara has . . . κυκλει μαθη . . . , which very probably encodes some reference to mathematics.[86] The series of inventory inscriptions from Delos, *c.*150 BC,[87] listing a complex

[81] See especially Fowler (1987), 7.5 (a) 271–9. Worth quoting in this connection is Beck (1975) 16: 'there is amongst all the extant vases only one which unequivocally alludes to the teaching of mathematics'. (This is Louvre G 318, reproduced ibid. fig. 84. Very low-powered 'mathematics' – no more than someone drawing something on the sand for children – so was it mathematics? The vase is dated 500–475!) Given the obvious visual possibilities of geometry, this is a useful piece of evidence.

[82] I have made a CD-ROM survey of the following four sequences: ιατρ, φιλοσοφ, ρητωρ, γραμματικ, in the Attica inscriptions and the *POxy*. I have ascertained that the usages are relevant.

Attica: ιατρ – 103, φιλοσοφ – 35, ρητωρ – 36, γραμματικ – 8.
POxy.: ιατρ – 67, φιλοσοφ – 12, ρητωρ – 10, γραμματικ – 8.

It is a statistician's joy to see how a certain reliability emerges from the figures; but naturally Attica had many philosophers and rhetoricians.

Neither of these two sources had any γεωμετρ, αστρονομ, or αστρολογ, and the single μαθημ in *POxy.* (10.1296.r.6) is irrelevant.

[83] *PCair.Zen.* 5.59853 rp.r.3: the text is . . . εστιν δε και μαθη . . . which possibly refers to the presence of a mathematician, though this is far from certain.

[84] A further slight piece of evidence for youthful mathematical achievement? Unfortunately, this may equally or more probably be an indication of youthful land-taxing achievement.

[85] The inscription should be compared with Diocletian's Edict on Prices, 7.70.

[86] *IG* VII.114.3. The inscription is undatable.

[87] *ID* 3.1426 face B col. II.52; 1442 face B.42; 1443 face B col II.109.

astronomical diagram as one of the dedications to the temple, has been discussed in chapter 1 above. This list mentions a name: Eudoxus, yet another repetition of a name known from elsewhere. Astronomers or astrologers are referred to in two inscriptions: 146 BC, at Delphi,[88] and an unknown date (in Christian times) at Nikaia.[89] The same nouns are also mentioned in three sets of papyrological sources: first century BC, Thebaid;[90] second century AD, Philadelphia;[91] fourth century AD, Hermopolis.[92] This sums up our documentary evidence for the relevant words as presented in the current CD-ROM of documentary sources. I repeat: the important piece of evidence is not just the absolute number, but the fact that similar 'search all' checks for words such as ἰατρός (doctor), γραμματικός (grammarian), ῥήτωρ (rhetor) or φιλόσοφος (philosopher) break down under the sheer number of references.

A similar comparison can be made through Diogenes Laertius' list, based on Demetrius, of persons having the same name. This list is not a random sample. It was edited with a view to intellectual pursuits, with a surprising degree of interest in the visual arts. Of the 188 persons in the list, 34 are poets, the same number are historians (in a wide sense), 28 are rhetoricians, the same number are philosophers (this does not include the original philosophers, whose biographies Diogenes gives), 16 are visual artists, 13 are physicians and 11 are grammarians (all these categories may overlap). Mathematicians are: II.103, the Theodorus known from Plato's dialogues; IV.23, a certain Crates who wrote a geometrical work; IV.58, a Bion, a mathematical astronomer; IV.94, a Heraclides – not of course the subject of Diogenes Laertius' biography, to which the list of 'having the same name' is appended – who may have been an astronomer (but could easily be an astrologer); the Protagoras of IX.56, who was probably a real astronomer (though he need not have been a mathematical astronomer). Between three and five mathematicians is slightly more than what I would expect in such a list, but the numbers are so small that fluctuations may easily occur.

Finally, the *Digest*:[93] various exemptions from civic duties are accorded to doctors (5 to 10 per city, depending on the size of the city), sophists or rhetors (3 to 5; 'sophist' and 'rhetor' are interchangeable here) and grammarians (3 to 5, again). The text goes on to explain that[94] 'the number of philosophers has not been laid down, since there

[88] *FD* 3;1.578.col. I.18. [89] *IK* Nikaia, 575.4–5. [90] *Ostras.* 787.5.
[91] *BGU*.7.1674.r.8. [92] *PHermLandl.* I rp.17.245; II rp.XXII.460.
[93] 27.1.6 (dating from Antoninus Pius' reign). [94] Ibid. I.7.

are so few philosophers'. The numbers mentioned in this text are no more than thresholds, something like tenured positions for the various professions, though one doubts whether all small cities filled their quota, or, if they did, what sort of sophists, say, they came up with.[95] The numbers, then, give little indication of absolute numbers. As usual, however, a sense of the relative numbers is made clear: perhaps as many as half of the professional intellectuals are doctors, the rest being mainly teachers of skills related to language. Mathematicians are not even mentioned.

To sum up all the evidence above: the quadrivium is a myth. Very few bothered at all in antiquity with mathematics, let alone became creative mathematicians.

I have surveyed the occurrences of the sequence γεωμετρ in all ancient Greek literary sources – within the current limitations of the *TLG* CD-ROM. I have found 108 authors, responsible for 2618 occurences.[96] Only 19 authors have 30 or more occurrences, and, of these, almost all are either Aristotle and his commentators (about a half of the entire set of occurrences) or Plato and his commentators (with another fifth). Only two are from outside this Platonic–Aristotelian group: Galen and Sextus Empiricus – and are they really outside it? Other references are more or less accidental. Historians refer to the literary topos of the origin of geometry in Egypt – or to the story of Archimedes. A few mathematicians did manage somehow to filter through the anti-mathematical zeal of the *TLG*. Following Galen, some reference to mathematics becomes common among physicians. Xenophon, Demosthenes and especially Isocrates refer fleetingly and dismissively to the Platonic tradition at its inception. The corpora of the Fathers of the Church would include any sequence of Greek letters. There are a few useful surprises: a comical fragment by Nicomachus, a detail on Oppian's education. Perhaps in the class of surprises should be put the relatively large number of uses by Chrysippus (this belongs to the wider issue of the relation between mathematics and philosophy, to which I shall return below). But on the whole, Greek culture, excluding the Platonic–Aristotelian tradition, knew no mathematics.

[95] In general, for the significance of this evidence, see the discussion by Duncan-Jones – to whom I owe the reference – in Duncan-Jones (1990) 161.

[96] This includes only 'real' references, i.e. I checked all and got rid of spurious uses of the sequence (e.g. in the sense 'land measurement'). Such spurious uses, however, are rare in the main literary tradition (they occur in the Septuagint, for instance).

The evidence accumulated so far is, I hope, sufficient confirmation for my basic deflationary estimate of the number of Greek mathematicians. The discussion will be amplified when we come to the position of mathematics within the wider culture. But the main consideration concerning the relative unpopularity of mathematics is quite simple. Mathematics is difficult. So is medicine, no doubt: but medicine has structural forces demanding its spread. Philosophy, it may be argued, did not have similar forces working for it, at least not always. And philosophy is difficult, indeed I will say Greek philosophy at its most difficult is more difficult than Greek mathematics at its most difficult.[97] But mathematics has a major disadvantage peculiar to it. This is its all-or-nothing nature. The most rugged Roman general can spend some time with Greek philosophers, apparently to find some satisfaction in his dim understanding of their utterances on Truth and the Good Life. But what satisfaction is there for him in Euclid? Only the frustration of the feeling of inferiority, so well known to anyone who has passed through our educational system. We invest enormous social and economic capital in forcing children over this hurdle, and still most fail to make it. Lacking these forces, the ancients did not try.

To be more precise: we all know the fate of a book which suddenly becomes a bestseller after being turned into a film – in the version 'according to the film'. This process originated in south Italy in the late fifth century BC, but it was Plato who turned 'Mathematics: The Movie' into a compelling vision. This vision remained to haunt western culture, sending it back again and again to 'The Book according to the Film' – the numerology associated with Pythagoreanism and Neoplatonism. A few people, especially in the Aristotelian tradition, went back to the original, until, emerging from the last Platonic revival of the Renaissance, mathematics exploded in the sixteenth century and left Platonism behind it with the rest of philosophy and the humanities. We now take this centrality of mathematics for granted; we should not project it into the past.

[97] And yet, there were more philosophers than mathematicians. The number of ancient philosophers whose names are known and who can be assumed to have written anything – a considerably smaller class than that of 'people who can be assumed to have been at least part-time philosophers' – is given as at least 316 in Runia (1989) (note also that this excludes Christians). Adjusting for the difference in definitions, this is considerably more than twice my number for mathematicians. I also assume that a smaller group will by definition survive better proportionally (the 'stars' are more numerous in a smaller group), so the number of philosophers in antiquity was at least three or four times that of mathematicians (and I suspect it was much more than this, with many semi-philosophers; I shall go on to explain this).

I conclude, therefore: in antiquity, each year saw the birth of a single mathematician on average, perhaps less. A handful of people interested passively in mathematics may have been born as well, but not more than a handful and, possibly, their numbers were quite negligible. In every generation, then, a few dozens at most of active mathematicians had to discover each other and to reach for their tiny audience. They were thinly spread across the eastern Mediterranean (excluding, to some extent, the more densely populated centre in Alexandria). They were thus doubly isolated, in time and in space. Here is the point for which the subsection was a preparation. Physical continuities were the exception in Greek mathematics. Hippocrates of Chios taught Aeschylus; did his pupil become a mathematician?[98] Theodorus taught Theaetetus (if Platonic dialogues can be read so literally). The Proclean summary contains two micro-traditions: Neoclides taught Leon; Eudoxus taught Menaechmus, whose brother was Dinostratus.[99] This may have substance to it. The relation Oenopides–Zenodotus–Andron is far less clear.[100] Timocharis taught Aristyllus. We know of Archimedes' father. Aristarchus *may* have taught Conon – dates and places allow this. But Arcesilaus and Philonides did not pursue, as far as we can tell, their mathematical studies. Pappus' picture of Apollonius studying with the pupils of Euclid must be seen as his interpretation. But Alexandria may indeed be the exception to my rule[101] – an exception not be overestimated, since there were never more than a handful of Alexandrian mathematicians. They formed a literary tradition, not a school.

The general rule is best seen through the list of observations used in the *Almagest*.[102] This is a set composed of intermittent explosions. No site of observation was kept for more than a few consecutive decades – excluding the Babylonian set, which was crucially different. Here we have found the important pattern. Greek mathematics, unlike Babylonian mathematics, was not a guild, a 'Scribes *enuma anu enlil*'.[103] It was an enterprise pursued by *ad hoc* networks of amateurish autodidacts – networks for which the written form is essential; constantly emerging and disappearing, hardly ever obtaining any institutional

[98] Aristotle, *Meteor.* 343a1. [99] *In Eucl.* 66–7. [100] Ibid. 80.

[101] Another exception, in time rather than space, is late antiquity, when Neoplatonists took mathematics so seriously to create traditions of mathematics, parasitic as it were upon the traditions of philosophy: Theon and his daughter Hypatia, the chain leading from Proclus to Eutocius and further.

[102] Pedersen (1974), appendix A. [103] See Neugebauer (1955) vol. I.13ff.

foothold. The engine does not glide forward evenly and smoothly: it jolts and jerks, ever starting and restarting.[104] Our expectations of a 'scientific discipline' should be forgotten. An 'intellectual game' will be a closer approximation. But this already leads into the question of the position of mathematics within the wider cultural setting.

3 MATHEMATICS WITHIN GREEK CULTURE

In this section, I set out to locate Greek mathematics within Greek culture. Spatiality is a metaphor for relations between Greek mathematics and other aspects of the culture – and a useful metaphor. We will look at one *intersection* within which Greek mathematics was located, and at two *borders* separating Greek mathematics from other activities.

3.1 The social-political background: an intersection

There is one type of historical setting suggested for the development of Greek science in general, by authors such as Vernant, Vidal-Naquet and Detienne.[105] It is now very well understood thanks to the work of Lloyd (1979 and later studies). I will therefore say relatively little on this. A résumé of that work, however, is in place.

The historical setting discussed in these works is the role of the public domain in Greek culture. Lloyd stresses the role of *debate* in Greek culture – the way in which debate was essentially *open* to participants and audience, and the way in which it was *radical* in its willingness to challenge everything.[106] It is this polemical background which explains the role of forms of persuasion in Greek culture. One should stress also the *orality* of this setting. By 'orality' it should be understood not that the political life of the Greek *polis* was uninfluenced by literacy but that the characteristic mode of political debate – which is the background most important in this context – was oral. And indeed, as Lloyd shows, there are many intellectual domains where presentation is heavily influenced by the form of an *epideixis*, a public, oral presentation, akin in a sense to a political speech.[107]

[104] We should not imagine a smooth tradition in disciplines other than mathematics; Glucker (1978) shows the intermittent reality behind the smooth facade of the Platonic Academy; which, *a fortiori*, strengthens my argument.

[105] Vernant (1957, 1962), Detienne (1967), Vidal-Naquet (1967).

[106] See esp. Lloyd (1979) 246ff.

[107] See esp. ibid., 86ff.

Lloyd noted the significance of this background for the development of mathematics, especially in 1990, chapter 3. Thus my arguing for the role of an oral background for the structure of Greek proofs should come as no surprise. Following the discussion of chapter 5 (especially section 3), we may now say that the mathematical *apodeixis* is, partly, a development of the rhetorical *epideixis*.

This background is certainly a feature of the classical *polis* throughout its history; it did not disappear in the Hellenistic or even the Roman *polis*, where the public domain, though transformed, did not disappear, certainly not universally.[108] The tradition of the free *polis* remained important in the background of the culture. However, if taken as a single factor, the picture presented by this background is especially valid for the fifth century BC, especially in the development of pre-Socratic philosophy and Hippocratic medicine, both mainly fifth-century creations. This background is thus of extreme importance. However, the formative stage for the emergence of mathematical form is the very late fifth century BC and, especially, the first half of the fourth century, a period already different in several ways from the fifth century. The background of the political city is still an immensely influential background, but it no longer has quite the same significance for the group of elite members in whom we are interested.

I shall concentrate now on two studies, Carter (1986) and Herman (1987). Carter's argument is that one possible reaction to the existence of the public domain was abstention from public life.[109] This abstention could reflect a critical attitude towards the nature of the public domain, certainly in a city such as Athens, a democracy (Carter's study deals exclusively with Athens). Another motive for abstaining may have been the growing dangerousness of the public domain. This is an Athenian feature, no doubt, but the impression is that political life became more and more dangerous throughout the Greek world at the time of the Peloponnesian war, reaching a new plateau then.[110]

It is important to follow Carter in realising that 'quietness' would have a new meaning within a 'non-quiet' society. '*apragmōsynē* grew out

[108] Rhodes, for instance, an important intellectual centre, led a highly politicised life well into the second century BC. As for traces of the political in cities in general, see, e.g. Bulloch et al. (1993), part 4; or even Jones (1940), chapter 20.

[109] Some of the reviews of Carter have been quite critical, especially concerning detail; it is often noted that he is stronger on the fifth century. For some of the references for the later period, see Campbell (1984). For another highly scholarly discussion of the same topic, see Nestle (1926).

[110] At least, stasis seems to become a more and more common feature of Greek political life; see Fuks (1984).

of the Athenian democracy – as a product of it and a reaction against it. Peasant farmers had been quietly working their farms for generations . . . But once the radical democracy took its final shape the character of their lives changed – in so far as they did not respond to the new regime they become *apragmōnes*'.[111] Carter discusses at length the impact of quietism on intellectual life in Athens. The interest for Carter is especially in the *contents* of intellectual activity, leading to Plato's insistence on contemplation. This is useful in showing that important intellectuals were quietists. The issue of the influence of quietism on the *form* in which intellectual life took place is not touched upon by Carter.[112] It is a natural question to raise, starting from Lloyd's analysis of the role of the public domain in the forms of Greek intellectual life. If indeed, as Lloyd shows, the public domain is an important background for the emergence of scientific activities, what are we to make of the fact that elite members were actively abstaining from that public domain?

I now move on to Herman (1987). This study describes 'ritualised friendship', a feature of Greek society – 'Greek' understood here to cover at least the period from Homer to Roman times.[113] The fact that this is 'ritualised' is important, for my purpose, especially in underlining the fact that the system of friendships described by Herman cannot be reduced to a mere sum of chance encounters between individuals. Ritualised friendship was an actor-concept, an institution of Greek society. It is defined by Herman as 'a bond of solidarity manifesting itself in an exchange of goods and services between individuals originating from separate social units'.[114] The components of the definition important for our purposes are the 'bond of solidarity', the 'individuals' and the 'separate social units'. The key point is that important social relations in the Greek world were not located inside the *polis* and in the social level of the *polis*, but in the inter-*polis* domain, and in the social level of individuals. While nothing in the definition demands this, the realities of the situation made sure that the relation of ritualised friendship applied to the top strata of Greek society alone. That some of their members belonged to this international network

[111] Carter (1986) 187. *apragmosynē*, *apragmones* may be translated for our purposes as 'quietness', 'quiet'.

[112] Not a criticism of Carter, whose concerns are different.

[113] Indeed, this form exists in other cultures besides the Greek, though the Greek form had its idiosyncratic features. See Herman (1987) 31ff.

[114] Ibid. 10.

was thus a characteristic feature of these top strata. Whereas the social world of the lower strata was limited to their immediate social unit, the social world of the upper strata was international. In Herman's metaphor, adapted from Gellner (1983), the structure of the social life of the ruling class was 'horizontal'; below it, the mass of producers were bound into a 'vertical' system of relations. This metaphor was first offered by Gellner in the context of a far more general theory. The uniqueness of the Greek case in the classical period, as argued by Herman, was the emergence of the *polis* – the vertical dimension – as another political factor, with which the ruling class had to reckon. Thus, this ruling class was enmeshed in a double system of significant relations.[115]

I do not claim that Greek mathematicians formed a network of ritualised friendship. Ritualised friendship is a technical concept, demanding precise acts, which cannot be proved for any relationship between mathematicians – not that this would matter. What is important for our purposes is the existence of such a system in the background of all civic life, and the international horizons of the ruling class which such a system implies, independently of any individual ties.

It should not be thought that international involvement was especially connected with quietism. The prototypical case taken by Herman, of an individual Greek whose political life was influenced by the network of ritualised friendship, is that of Alcibiades,[116] anything but a Quiet Athenian. Quietism and ritualised friendship are similar, however, in that they are practices which were given a new set of meanings in the new context of the intense public domain of the classical age. Both antedated the new significance of the *polis*, yet neither could fail to be a reaction to it. The period in which we are most interested, that from the Peloponnesian war onwards, is the period in which this reaction took definite shape.

Both quietism and ritualised friendship highlight the fact that Greek society, however democratic at times and places, was an aristocratic society. It was this aristocracy which was responsible for the emergence of Greek science. It is thus necessary to qualify any understanding of the forms taken by Greek science which is made through the characteristic features of the public domain alone. Not, of course, that such an understanding is by any means flawed, since the public domain is a constant and essential feature of Greek culture. In fact, any

[115] Ibid. 162–5. [116] Ibid. 116ff.

attempt to understand Greek intellectual life apart from the public domain, just by taking into account the role of quietism and international involvement, will be seriously misleading. It is only by taking into account both features of Greek culture – by understanding the *polis* as exercising both centrifugal and centripetal forces – that Greek intellectual life can be understood. As stressed by Herman, it is the *duality* which is significant, not just one of the two terms.

How this duality may influence the form of intellectual life is perhaps best seen through Plato's works, which define the period of most importance for us. The role of quietism in Plato has been stressed by Carter. As for his international involvement, I will not lay too much stress on his possible ritualised friendship, in the technical sense, with Dion,[117] nor on the fact that, given his family background, he was certain to inherit a network of ritualised friendships.[118] My main points are that (a) Plato's political life happened abroad; (b) in the later dialogues, divorced from Socrates' historical life,[119] the discussion is often done indoors, with the essential aid of foreign participants.[120] The culmination of this is what may be seen as the metamorphosis of the Athenian speaker – the descendant of Socrates – into an Athenian *stranger*, in the *Laws*. The horizons of the intellectual world envisaged by the older Plato are international.

The implication of this is obvious: an international intellectual world cannot be approached orally. If indeed some of the Platonic letters are genuine, this would be emblematic, for this would then be among the first significant uses of the letter form, later to dominate so much of intellectual life in antiquity. Plato is an author to whom the written form is essential. The hidden structures of his work demand reading and rereading rather than performance. The elaborate prose style used by Plato is an alternative to the *poetic* – and therefore more aural, if not oral – form of much previous philosophy. On the other hand, Plato is unimaginable without orality either. After all, he writes dialogues.

[117] This is suggested by the seventh letter, 328c–329b, where Plato (?) describes the motives for his second Sicilian visit as arising from the obligations of friendship, indeed the obligation towards Ζεὺς ξένιος (329b4). This is in line with Herman (1987) 118ff. Apparently the author of the seventh letter at least wishes to create the impression that a formal ritualised friendship existed between Plato and Dion.

[118] None is recorded for Plato's family in ibid., appendix C, but the evidence is extremely fragmentary.

[119] As noted by Carter (1986) 186, Socrates was very unquiet; and his lack of internationality is well known.

[120] Who thus should be distinguished from figures such as Meno and Hippias, who do not contribute positively to the philosophical content of discussions.

The image of discussion as the vehicle of intellectual life is central to Plato, as, of course, he says explicitly in the *Phaedrus*.[121] What Plato did was to adapt a form whose origin was set firmly in the public domain of democratic Athens – that of the historical Socratic discussion – into a written form. This form then participates in the international, horizontal dimension; it no longer participates in the local, vertical dimension, since literacy, in antiquity, was limited effectively to the elite.[122] An object belonging to the horizontal dimension shows clear marks of an earlier stage in the vertical dimension: such is the duality characteristic of Greek culture from Plato's times onwards.[123]

Not all Greek mathematicians were 'quiet'. Two of the most important early mathematicians, Archytas and Eudoxus, were not.[124] But whatever their civic life, Greek mathematicians had to lead their *mathematical* life outside the public domain of their cities, if for no other reason than because of sheer numbers. Indeed, it is curious to see how persistently cosmopolitan mathematical relations are. Archimedes (Syracuse) and Conon (Samos), Dositheus (Pelusion) and Eratosthenes (Alexandria); Apollonius (Alexandria) and Attalus and Eudemus (Pergamon); Theodorus (Cyrene) and Theaetetus (Athens); the observations of Hipparchus, conducted both at Rhodes and at Alexandria:[125] as long as the Mediterranean was divided into political units, no matter how large they became, mathematical relations crossed those boundaries.

Greek mathematics reflects the importance of persuasion. It reflects the role of orality, in the use of formulae, in the structures of proofs, and in its reference to an immediately present visual object. But this orality is regimented into a written form, where vocabulary is limited, presentations follow a relatively rigid pattern, and the immediate object

[121] *Phaedr.* 274e–275e. That there is a certain tension in this passage is of course part of my thesis.

[122] See Harris (1989). The claim is sweeping, but if 'literacy' is taken to mean something like participation in literate culture, this claim is clearly true.

[123] I will thus argue that the 'literate revolution' described by Havelock (1982), inasmuch as it corresponds to any reality, does not reflect any technological breakthrough or even greater accessibility to a technology; it is a classic case where means of production change their significance within new relations of production.

[124] Meton, however, was 'quiet', apparently, if indeed his political life amounted to not furnishing a trireme on a single occasion, for which see n. 41 above; the implication of the *Theaetetus* is slightly in favour of his civic career being limited to good soldiering. On the other hand, Hippias may have been a leading citizen of Elis, for which see n. 42 above. As always, the projection of the image of Thales is significant. Heraclides, in the fourth century, made him say in a dialogue that he lived alone, as an ἰδιαστής (D.L. 1.26).

[125] Pedersen (1974), appendix A; of course, the indication of localities is not safe biographical evidence.

is transformed into the written diagram – doubly written, for it is now inscribed with letters, so that even the visual object of mathematics becomes incomprehensible for one's less privileged compatriots. It is at once oral and written, a feature we have stressed many times so far in the book. We can now begin to see that this intersection represents an even more basic one – in simple terms, the intersection of democracy and aristocracy.

3.3 A border: mathematics and 'the material'

There are two ways in which the relation between mathematics and 'the material' may be understood. One is a mathematical reflection upon the material world: for instance, musical theory may be mainly a reflection upon the empirical data of musical activity. This is one kind of materiality Socrates objects to in the *Republic* (and correspondingly, he objects to viewing astronomy as a reflection upon the actual stars. There 'the material' becomes less tangible, but the problematic is similar). Another way in which mathematics may be related to the material world is when mathematics is 'applied', i.e. mathematical knowledge is used in areas such as economic or military activity. Socrates is no friend to this either, but he concedes the possible usefulness of mathematics from this point of view, insisting however (at least, in the context of the curriculum offered in the *Republic*) that this is of minor significance.[126]

There are therefore two problems. One is whether Greek mathematicians were interested in the physical world. Another is whether Greek mathematics was 'applied'. The second, in a sense, implies the first – if your mathematics is related to the productive or the military spheres, then it is bound to be related to the physical, material world in which production and wars take place. The opposite does not hold, and this is why I insist upon the distinction. Looking at the stars is not an action in the productive sphere. Constructing a toy is no more an action in the productive sphere than writing a book is. No isolated act can belong *per se* to the productive sphere. It is by entering the system of economic relations that an act is endowed with an economic significance. So the second question is much more complicated than the first one.

[126] The practical role of arithmetic: 522c. Qualified: a long discussion summed up at 525bc. A quick repetition of the argument with geometry: 526c–527c. Practical role of astronomy: 527d. This is dismissed as irrelevant. Then the philosophical role, in some sense detached from the material world, is described: esp. 530c. No practical value is claimed for music (however, this is perhaps obvious), and its philosophical role is explained in 531bc.

To begin with this first question, then, I would say that the very formulation, 'whether Greek mathematicians were interested in the physical world', shows the unreal nature of this sort of discussion. Of course they were, as humans in general and scientists in particular had always been. This *a priori* claim can be easily sustained, to begin with, by Plato's words themselves, which explicitly blame 'them', in a very general way, for doing music and astronomy with great attention to physical instantiations.[127] Mechanics was part of mathematics, in Aristotle's judgement and in Archimedes' practice, not to mention lesser mathematicians.[128] Astronomy was, *pace* Plato, the theory of the actual sky,[129] in several cases involving the materiality of planetaria and, later, special astronomical devices.[130] I have argued in chapter 1 that the assertions of geometry were understood to hold, at least partly, for concrete objects – which, as may be argued following Burnyeat (1987), is not a proposition Plato himself would have disagreed with. In short, we should let go of the appearances of the *Republic* and assume, as is obvious, that Greek mathematicians had their curiosities about stars, musical instruments and spatial objects in general. I therefore now move on to the problem of applied mathematics in antiquity.

As I have just said, Plato conceded the existence of applied mathematics. It is necessary to understand what this amounted to. The most explicit description of 'applied mathematics' is τὸ ἕν τε καὶ τὰ δύο καὶ τὰ τρία διαγιγνώσκειν, 'to know well one, two and three'. This is meant to be necessary for knowing how many feet one has. The general description of this sort of knowledge is λογίζεσθαί τε καὶ ἀριθμεῖν, 'to calculate and to count'.[131] We would call this basic numeracy skills. The case for applied geometry is less clear. The setting up of camps is mentioned, for instance, by Glaucon, but it is not explicitly said how geometry is meant to aid such activities. Rather than try to guess this, we would do better to heed Socrates' answer to Glaucon: for such purposes 'a tiny fraction of arithmetic and geometry will do'.[132] In all, the impression is that the kind of mathematics envisaged as useful by Plato and Plato's readers is the most basic mathematical knowledge, knowledge which undoubtedly preceded

[127] See esp. 530e7–531a3. Glaucon, significantly, answers that he knows this well.
[128] Aristotle, *APo.* 1076a24. For Archimedes, alongside his well-known works on mechanics in a strictly mathematical sense, the σφαιροποιία (see Archimedes vol. II, 551ff.) should be mentioned.
[129] See Lloyd (1978). [130] See chapter 1 above, subsection 3.2.2.
[131] 522c5–6; d6; e2. The relevant Greek is set in such a rich philosophical context that one hardly dares to translate.
[132] 526de, esp. d7–8.

mathematics itself: numeracy and the basic qualitative facts concerning shapes.[133] The Spartans did well, in military terms, with the barest minimum of numeracy;[134] it is difficult to see how a thorough understanding of Euclid's books VII–IX would have helped them in the Peloponnesian wars. 'Why learn mathematics?' asked Brecht ironically: 'that two loaves of bread are more than one you'll learn to know anyway'.[135] Brechtian mathematics would have been sufficient in Greek economics, with some very qualified exceptions.

Let us go through the curriculum. Greek practical numerical systems showed no sign of any scientific rationalisation until Babylonian science cast its influence upon Greek *science* and science alone: everyday calculations never became positional. More important still, Greek calculations used a system of fractions which is extremely difficult to rationalise. Greek arithmetic and proportion theory show no trace of any effort to rationalise that system.[136]

The word *geōmetria* is an interesting one, meaning quite different things in literary sources ('geometry') and documentary sources such as papyri ('land measurement'). That the land measurement of the documentary evidence has nothing to do with Euclid is clear from the algorithms used: size was assumed to be proportionate to circumference (*sic*),[137] or it was measured by multiplication of the averages of opposite sides.[138] Both systems are mathematically invalid, though the second at least involves a certain expertise in calculation. Babylonian clay tablets are high-tech in comparison.

Stereometry is in principle relevant to engineering. Most notably, the scaling of machines may be conceptualised as a difficult stereometrical problem known as the duplication of the cube. Philo of Byzantium, or perhaps Ctesibius before him, may have been responsible for this conceptualisation.[139] The ancient mechanical author to whom real practical interests can most plausibly be ascribed, Vitruvius, is aware of this conceptualisation, but offers a recipe of numerical

[133] That this level of mathematics was an essential part of Greek education is amply shown by papyrological evidence, for which see n. 81 above.

[134] *Hip. Mai.* 285c3–5. [135] *1940*, ll. 1–3.

[136] Fowler (1987), chapter 7; see especially how little the slaveboy and the accountant have in common at their meeting (268–70). The slaveboy's wonder at the end of this dialogue – whether anthyphaeretic ratio could be related to the system of fractions used in antiquity – is, significantly, Fowler's wonder, not that of any Greek mathematician.

[137] See the very early example discussed in Finley (1951) 3. As explained at 58 there, there are very few examples for land-measurement from Athens, especially from the period we are interested in (five from the entire period 500–200 BC) – in itself significant.

[138] Fowler (1987) 230–4. [139] See Knorr (1989), chapter 3.

relations instead of the geometrical approach, and the recipe is so 'bad', geometrically speaking, that one suspects that the physical contamination by friction, etc., could have made trial and error numerical recipes better than the geometrical approach.[140]

None of the many Greek civic calendars of antiquity paid the least attention to astronomical science.[141] The most 'scientific' calendar of antiquity, the Julian, was a product of Egyptian common sense rather than of Greek science.[142] Again, this should be compared to Mesopotamia. Astronomy has some relevance for navigation – i.e. if you cross the Atlantic. Greek ships, hopping by day from island to island or along the coast, would need to tell their position only in the most extreme weather, when astronomy would be of little help anyway. The relatively late mastering of the Red Sea was a continuation of ancient traffic, and if the Greeks introduced to it the use of astronomy, then the *Periplous of the Red Sea* is silent about this. In very general terms, the stars have always been, until recent times, part of the system by which humans situate themselves in time and space, and this is ubiquitous as early as Hesiod. But this is not 'applied mathematics'.

Musical theory in the mathematical sense may have influenced practising musicians. At any rate it seems highly probable that at least some mathematical music was done with a view to actual implementation.[143] However, Aristoxenus has no time for mathematisations, and *a fortiori* we may assume that most practising musicians did not tune their lyres according to mathematical manuals.

We should now move on to disciplines which are not covered by the curriculum. Optics is comparable in a sense to music. Euclid's *Optics* 36 is a unique proposition, asserting that a *wheel* would not look circular from a given angle. Catoptrics (the theory of mirrors) is a mathematical discipline defined by an artefact. All of which does not show that mirrors were produced by mathematicians (there's no evidence, impossible legends about Archimedes apart) or that Greek mathematicians actually developed a theory of perspective designed for the use of painters. Greek painters developed a perspectival system of sorts,[144] and there is some obscure evidence showing that a corresponding mathematical theory may have existed; the evidence does not allow us to tell the relation between theory and practice in this case.[145]

[140] Vitruvius x.11.3. [141] Bickerman (1968) 29ff. [142] Ibid. 27ff.

[143] The introduction to the *sectio canonis* and Archytas B1 are remarkable cases of the introduction of physical considerations into Greek mathematics.

[144] See chapter 1, subsection 1.2. [145] Vitruvius vii.11.

And finally we reach mechanics itself. Here one must tread very carefully.[146] Nowhere is the distinction between the two senses of the problem, as described above, as important. That Greek mathematicians were often interested in the field called *mechanica* is clear. Part of this activity was purely theoretical, as in Archimedes' statics and hydrostatics. Another part involved the passive contemplation of machines which the mathematician neither built nor apparently had any wish to build. Such, for example, is the pseudo-Aristotelian *Mechanics*. The first mechanism described in that tract[147] *may* have been built according to some theoretical principles, i.e. we would then have a case where a theoretician built a machine. This machine, however, is a miniature toy set up as a dedication in a temple. This brings us to another class, then: a mathematician constructing a 'cultural', rather than an 'economic' object.[148] That this procedure was common is clear from the evidence on *sphairopoiia*, discussed in chapter 1 above, as well as from some of the evidence on Archytas,[149] Ctesibius and, of course, the *locus classicus* for that sort of activity, which is the bulk of Hero's *Pneumatics* and his *Automata*.

Finally, there are cases where mathematical authors appear to have contributed to machines of real economic value. The evidence for this is uneven. There's the Cochlias attributed to Archimedes; which attribution, however, is very doubtful.[150] Then there's Ctesibius' water-pump; and not much more.[151]

How to gauge this evidence? First, the large class of mechanical tracts which have no economical value is significant. Even if it does not exhaust the field of mechanics, contemplative mechanics was probably

[146] As guides, one should take Lloyd (1973), chapter 7, Pleket (1973) and Landels (1980). White, in Green (1993), is eager to downplay the forces working against technology in antiquity, which is useful as an antidote to over-pessimistic interpretations of the evidence; but this eagerness does lead him to ignore the reality of these forces, and the result is on the whole over-optimistic. Finley (1965) should still be the starting-point for any discussion.

[147] Ps.-Aristotle's *Mech.* 848a20.

[148] My sense, I hope, is clear: the artefact has an aesthetic or an intellectual, rather than a productive, value. If it fetches a price (which in general we should not assume), this price reflects an appreciation of the skill of the artificer, not a valuation of the commercial value of the artefact. It is comparable indeed to another kind of artefact: the *book*. Before print, before mass-production, the tracts describing a machine and the toy specimen of a machine are very similar indeed.

[149] *DK* A10.

[150] See Dijksterhuis (1938) 21–2. The presence of spirals within the machine would, in itself, be an excellent ground for misattribution.

[151] Vitruvius x.7. As for Hero's 'serious' artefacts (e.g. in the *Mechanics* and the *Dioptra*), these seem to belong to the category of a mathematician *describing* machines already existent, an exercise with a mainly literary interest.

the most common type of mechanics in antiquity. Greek mathematicians were strangely content to understand machines without bothering to change the world with the aid of these machines. Secondly, how does mechanics become an economic force? What are the social forces which transform an intellectual product into an economic product? Now the one ancient author about whose economic activities there is no doubt is Vitruvius, and this fact leads to the following argument:[152] Vitruvius was a practical man, a person in the economic sphere. It was from his position in the economic sphere that he used Greek intellectual products, both as aids for action and as sources of prestige.

This ties together with the preceding point, on the general passivity of mathematicians. As a rule – and such generalisations cannot be more than a guide – the (few and feeble) forces connecting mathematics and economics in antiquity come from economy, not from mathematics. Mathematicians do not reach out to try and implement their work in the economic sphere; some practical men reach out to the mathematical literature to see what this has to offer. Thus the story of ancient technology, even to the limited degree that it is indebted to mathematics, is not part of the story of ancient mathematics. The one great exception is, of course, war-engines, which I have ignored so far. The tradition ascribing them to Archimedes is very sound, going as far back as Polybius.[153] Philo of Byzantium probably meant business. In general, as the case of Archimedes himself shows, war is a necessity which asserts itself. Otherwise, what is clear is the estrangement between the theoretical and the practical, and undoubtedly this estrangement is on the whole due to what may be called the banausic anxiety of the ancient upper classes – to whom, as was argued above, the mathematicians belonged.[154] Plutarch is mildly hysterical when it comes to the question of whether Archimedes was a practical man.[155] In general, we constantly see how praises of mathematics are qualified. *'Do it, but don't overdo it'*: this is the suggestion of Isocrates,[156] Xenophon[157] and Polybius.[158] *'Do it, but only in a certain, limited way'*: this is the suggestion of Plato[159] and Plutarch.[160] One is reminded of Aristotle on the

[152] The evidence for Vitruvius' activities is mainly in V.1.6ff. [153] Polybius VIII.5ff.

[154] As de Ste Croix insists in (1981) 274–5, there is no reason to suppose that this anxiety was felt anywhere but by (some) members of the ruling class. This stresses the role of this anxiety in the self-definition of the ruling class.

[155] *Vita Marc.* 17.4; but then Plutarch can always be counted upon for cultural snobbery, for which see reference in n. 154 above.

[156] *Antidosis* 261ff., esp. 268. [157] *Mem.* IV.7.2–3. [158] IX.20.5–6.

[159] This, after all, is the main result of the curriculum passage in the *Republic*.

[160] *Vita Marc.* 17.3–7.

study of music[161] (never very far away from mathematics!) where both strands are combined in a very explicit statement: *'Yes, study, in a certain way, and don't overdo it – lest you become banausic.'*

It needs no stressing that the thrust of the Platonic project in the curriculum is the separation between the world of action and the world of contemplation. I do not need to argue here that something like the banausic complex was a feature of the attitudes of the Greek elite; and within this prevailing cultural attitude mathematics occupied an uneasy position, being liminal.[162] It was essentially related to the material world: it was often known just as *geōmetria*, in itself a term of great economic significance. It was related to practices such as the drawing of diagrams, not to mention the construction of mechanical gadgets, which bordered on the banausic. On the other hand, it was extremely theoretical in the sense that most of it had no obvious connection to a person's life (this could not be said, for instance, of ancient philosophy.)[163] The impractical mathematician is a figure known to antiquity. Thales gazed at the skies and fell into a well; Hippocrates of Chios was swindled.[164] On the other hand, Thales could also make a fortune through his astronomy;[165] the worry over the possible banausic attachment of the mathematician is a theme of the Curriculum passage in the *Republic* – and may be a theme of Epicurus' argument against mathematicians.[166] There is a duality here, best symbolised by the legend of Archimedes. An old man, he could resist the entire

[161] *Politics* VIII.6.

[162] I use the term 'liminal' in the technical anthropological sense (for which see Turner 1969); my reasons for attributing such a liminality to Greek mathematics will become apparent in the following. As Turner points out (125), the concept covers phenomena as diverse as 'neophytes in the liminal phase of ritual, subjugated autochthones, court jesters, holy mendicants, good Samaritans, millenarian movements, "dharma bums", matrilaterality in patrilineal systems, patrilaterality in matrilineal systems, and monastic orders'. Surely adding Greek mathematicians should not be impossible.

[163] Burnyeat (1982) 25ff. [164] Thales, *Theaetet.* 174a; Hippocrates, DK A2.

[165] See the evidence at KRS 73, 80.

[166] I am referring to Epicurus' *On Nature* XI, col. IIa, ll. 7–11 and its parallel in the *Letter to Pythocles*, 93. In the first passage (Sedley 1976: 32), Epicurus mentions the ἀνδραποδείας . . . ὑπὸ τῶν διδασκαλιῶν, which is parallel to ἀνδραποδώδεις ἀστρολόγων τεχνιτείας in the second passage. The context must in both cases be the use of specific astronomical instruments. Sedley (ibid., 39) suggests that the notion of 'enslavement' has to do with a doctrinaire training, which is of course possible; another possibility, I suggest, is a reference to the low status attributed to the use of instruments. The word διδάσκαλος was quoted by Epicurus as a term of abuse, used against him by Nausiphanes, and implying low status (D.L. x.8); Nausiphanes actually merely replied to Epicurus who (in the *On Nature* itself?) mentioned Nausiphanes as being, *inter alia*, καθάπερ καὶ ἄλλοι πολλοὶ τῶν ἀνδραπόδων (ibid., x.7). My suggestion is that the passage from *On Nature* XI quoted above is a combination of such terms of abuse implying low status.

Roman army – and then be killed because he was too preoccupied with a diagram.[167] Both too practical and too ethereal, the Greek mathematician occupied a difficult, liminal position in a society where the borderline between the practical and the non-practical was particularly meaningful.

So what is the significance of all this? The obvious answer is that the existence of this border zone explains the ontologically neutral tone of Greek mathematics, the avoidance of clear references to the material object of mathematics. This is true, but it is just an approximation. This is because what we have said so far is too general. Some form of rejection of the material is a universal of human culture, seen in many forms of religion, philosophy or ideology. The material may be shunned because it is the location for death, pain, taboos, exploitation, war and cruelty. The desire to escape from the material can be no more than a sign of human sensibility. But this desire may take on many various forms, and we should look for cultural specifics. Bourdieu (1993), for instance, describes a special kind of economy: the anti-economy of modern literature, based on rejecting commercial success. Here then is one kind of rejection of the material where the specific form of 'the material' rejected is 'commercialism'. The authors try to put on display a claim that their work is untouched by the impact of market forces. This is the specific form of the rejection of the material, in western European literature, from about the middle of the nineteenth century onwards. And it is possible to see the specific background for this rejection in the sociological position of this literature – a task performed by Bourdieu.

I cannot match Bourdieu, but a few preliminary remarks can be made. Greek mathematics is the product of Greek elite members addressing other elite members. Commercialism is not an issue, of course – this was a pre-print culture, and writing in general was not predominantly a commodity. The possible taint is a different matter. What the ancient author had to put on display was that he was part of *culture*, that his writings were specifically meant for highly educated, elite members, and were not just technical compilations (and thus of possible interest to uneducated people). And the mathematician was under a special pressure to demonstrate this, just because of the liminal position described above. The Greek mathematician had to prove that he

[167] See Dijksterhuis (1938) 30–1. Of course, there is no reason to suppose this tradition is historical, but it is a good reflection of the mathematician's image.

did not act within the 'vertical' context of offering technical guidance for his social inferiors (which would immediately arouse the suspicion that he himself might be one of them!), but within the 'horizontal' context of writings addressed to a few sympathetic elite members. He had to prove that his writings were a form of literature in their own right – that they belonged to a *genre*. Here then is the fundamental importance of the specific form of the rejection of the material in Greek culture. First, Greek mathematics has a style, and this sense of style overrides, as we saw, any other consideration. The mode of presentation is the constant, contents are the variable (which is seen, e.g. in the use of geometrical diagrams for arithmetic). Greek mathematics is an area where style dominates content. We can now see why: if it were not so, Greek mathematics would stand in danger of being viewed not as literary product (characterised by its style) but as technical manual (characterised by its content). Secondly, Greek mathematics assumes a readership of initiates. And again the purpose of that is clear: the need to demarcate the possible readership is pressing and real, representing the need to keep a sense that this is a product aimed at a few elite members and no one else. So perhaps some of the blame for the small number of ancient mathematicians lay with the ancient mathematicians themselves.

3.4 *A border: mathematics and other disciplines*

The subject can be understood in two ways: first, as a question about *persons*. This involves issues such as the relations between mathematicians and other intellectuals. Another approach is to view the subject as a question concerning *activities*. Thus, even if the same person was both a philosopher and a mathematician, the questions can still be raised: how were these two activities related; did they influence each other, or were they pursued in isolation?

In general, Greek intellectual life was a field with very fuzzy borders. It must be remembered that institutionalised faculties were not a feature of the ancient world.[168] And in fact, from Empedocles to Sextus Empiricus, philosophers claimed to heal; from the first to the latest 'sophists', philosophers taught rhetoric; and they have dealt with any

[168] The institutionalisation of the Hellenistic schools took differing *dogmas* for boundary lines, rather than differing subject matters. The library in Alexandria catered for the omnivorous bibliophile. But these are exceptions to the rule of absence of institutions.

subject under the sun (and above it). Mathematics takes part in this scheme of things. Democritus did everything. Oenopides, Hippocrates of Chios, Euctemon and Meton can all with varying degrees of security be assumed to be interested in astronomy in a more philosophical sense.[169] The label 'sophist', misleading as it is, suits Hippias and Antiphon much better than the label 'mathematician'. Hippasus and Eurytus, little-known figures, may have been interested in mathematics as part of a wider Pythagorean scheme; this is probably true for Archytas. Eudoxus studied pleasure and chronology, to name two subjects.[170] In later periods, a few persons can be named who approach mathematics from the outside, and thus of course combine it with other activities: certainly Aristotle and Eudemus, perhaps Eratosthenes. This brings us into the Hellenistic period: up to that period most mathematicians about whose activities anything at all is known are not exclusively mathematicians. Why not, after all? The *a priori* argument is overwhelming: lacking institutional constraints, a lifetime of leisure dedicated to a single activity is humanly impossible. Human beings do not possess the relentless *energeia* of the stars. On the other hand, it is only natural that some of the greatest names in the catalogue of mathematicians should also be those most intent on mathematics: names like Euclid, Archimedes, Apollonius and (the more philosophical) Ptolemy. Some degree of single-mindedness may help to produce intellectual work of such extensive nature and quality – yet Eudoxus, perhaps the greatest of them all, was one of the most colourful figures. This line of thought is inevitably complicated by the variety of individual traits.

What is more widely applicable is that mathematics tends to be associated with some disciplines more than with others. It is probably right to consider Hippocrates of Chios' astronomy as an extra-mathematical activity. It seems more like pre-Socratic philosophy than like mathematics. However, there is a certain affinity of interests between geometry and astronomy, even in this non-mathematical sense of astronomy. Many of such extra-mathematical interests are within the Archytean–Platonic system: fields like astronomy and music. At the other end of the scale stands medicine, with its imprecision and irreducible empirical involvement. And it is curious to see how few of the

[169] For Hippocrates of Chios there's the DK testimony, A5–6, which in turn sheds a new light on Oenopides' and Meton's gnomons and Euctemon's and Meton's calendars.

[170] Pleasure: Aristotle, *EN* 1101b27, 1172b9. Chronology: Pliny, *Nat. Hist.* XXX.3.

mathematicians were physicians.[171] There were not many literary persons in the strict sense, though Antiphon and Hippias gave speeches and Eratosthenes was a glaring counter-example. Even Eudoxus was interested, after all, in *chronology*: as an accomplished astronomer, he would have been in a position to realise its difficulties and interests. Thus a certain order does emerge from the chaotic fuzziness. An affinity between a group of disciplines led to the notion of the mathematical sciences; some mathematicians strayed to other fields, but mostly within 'the Platonic curriculum' or, more generally, 'philosophy'.

Here we should return to the evidence mentioned in the preceding section: the *TLG* CD-ROM survey of words related to '*geōmetria*'. This in fact gives us the mirror image of what we have seen so far. Just as we have seen that the extracurricular activities of mathematicians belong mainly to philosophy, so the non-mathematicians whose writings mention mathematics are philosophers, in fact those of a certain tradition, the Platonic–Aristotelian. The *e silentio* evidence surrounding those islands of interest in mathematics is eloquent. I have already noted in chapter 1 how the most conspicuous element of Greek mathematical style – the lettered diagram – had left no traces on Greek culture as a whole (e.g., nothing like our everyday use of *X* ever emerged). And then there are individual cases of qualitative significance. Thucydides is unsurpassed in sheer intelligence. His description of the plague may show him to be indebted to Hippocratic medicine. The Melian dialogue is an important text for understanding what is known as the Sophistic movement. Yet there is not the slightest hint in his work that anything like mathematics was at all known to him. He estimates the size of Sicily by its circumference.[172] Similarly, Isocrates' references to mathematics are dismissive and betray no mathematical knowledge.[173] There's no need to go on retracing ground already covered in the preceding section. Against the general Greek background of interrelated intellectual activities, mathematics emerges as a well-bounded field.

We have already moved to the problem of the relation between activities. The relation between mathematics and other disciplines is seen to consist mainly in the relation between mathematics and

[171] Though Diogenes Laertius (VIII.88) gives, as one of Eudoxus' professions, ἰατρός (but this is very unconvincing). A certain fascination with mathematical precision is noticeable in the physicians themselves, first with the interest in numbers shown in some Hippocratic texts, for which see e.g. references in Longrigg (1993: 98) (but this is a very diluted sense of mathematics). Later, of course, there's the Galenic programme, whose propaganda, at least, makes great use of mathematics.

[172] VI.1. [173] See especially *Antidosis* 261ff.

philosophy, e.g. philosophy of mathematics or astronomy understood in a more philosophical sense. It is the relation between mathematics and the more philosophical aspect of mathematics: mathematics and meta-mathematics (reminding ourselves that we do not at all mean the modern technical sense of meta-mathematics but simply philosophy with a mathematical content).[174] How can we account for this? Part of the answer must lie in numbers. Small groups would have a strong sense of identity. Surrounded by a world of doctors, sophists and rhetoricians, the person who has seriously engaged in mathematical studies would inevitably feel closer to the few other eccentrics known to him who pursued the same interests. This can produce quite an aggressive stance, as shown by Cuomo (1994) concerning Pappus. Here is one possible background to – relative – isolationism. Another possible contributing cause may have been the liminal position of mathematics. Close association between mathematics and meta-mathematics would imply, for instance, that a musical theorist combines physics and mathematics, that speculation about the nature of the stars is indistinguishable from theorems, i.e., mathematics would be seen as far more physical – which the mathematicians tried to avoid. But still, there is something radical about the isolationism of Greek mathematics, compared with the general background. By bringing in this general background, the *explanandum* becomes very remarkable. And I think that the main explanation for this isolationism must be found in this very background.

The following is a reformulation of the argument of Lloyd (1990), chapter 3: the development of rigorous arguments in both philosophy and mathematics must be seen against the background of rhetoric, with its own notion of proof. It was the obvious shortcomings of rhetoric which led to the bid for incontrovertibility, for a proof which goes beyond mere persuasion. For the sake of my development of the argument, I will put this in the following terms. The rigorous arguments of philosophy and mathematics, when understood as part of a larger structure, consisting, *inter alia*, of rhetoric, acquire a meaning different from that which they have in isolation. They are not only 'compelling' arguments *per se*. They are also 'more compelling' arguments, arguments which stand in a certain relation to the arguments offered by,

[174] Cf. here the result of chapter 3: the abrupt terminological break between introductions to mathematical treatises and the mathematical treatises themselves. Even when engaging in the inevitable philosophical (in a very wide sense) component of their study, mathematicians consistently set this component apart from the main work, and cast it in a language different from that of the main work.

e.g. rhetoricians. The various kinds of argument thus form a structure. Within Greek culture, with its stress on public debate, this structure has a special significance. So let us look at this structure in action.

Parmenides' arguments certainly are rigorous in a sense. But are they incontrovertible? They were controversial in fact, and this is the major point. In a crucial passage of Isocrates, where he wishes to dismiss *both* mathematics and philosophy in the more speculative sense, he clinches his argument by pointing out that some ancient thinkers thought there were infinitely many beings, Empedocles four, Ion no more than three, Alcmaeon two, Parmenides and Melissus one and Gorgias none.[175] The controversial, polemic nature of philosophy shows its controvertibility. But in a sense this cannot be done for mathematics, *and in fact this was not done by Isocrates.* He gives no examples for current controversies in mathematics. And this because it would be self-defeating for him: it would immediately stress the fact that mathematics, in a sense, is beyond controversy. The squaring of the circle may be an area for debate, but there are considerable areas where mathematics simply does not allow this kind of polemical exchange.

Within Greek polemical culture, this feature of mathematics acquired a meaning which it did not possess in China or in Mesopotamia. For the Greeks, mathematics was radically different in this respect from other disciplines and therefore mathematicians pursued their studies with a degree of isolationism. Babylonian mathematics is not all that much different in form from Babylonian omen literature;[176] and I ascribe this to the absence of a radically critical attitude in their culture. It is such a critical attitude which makes mathematics stand out from other fields. The harsh light of a critical attitude separates clearly those fields which are by their nature less open to criticism, from others.

So there is, first, an activity of great prestige for the Greeks: that of making compelling arguments. And there is one type of argument which is more compelling than others, which leaves less room for controversy than others. This is mathematics. Viewed from the point of view of its *form*, it has a special advantage, which the Greek mathematicians would take care to stress, to put apart from other, competing forms of argument. On the other hand, viewed from the point of view of its *content*, it is suspect: fascinating, no doubt, but still suspect, in its connection with the material. As explained above, in its subject matter Greek mathematics is liminal, dangerous.

[175] *Antidosis* 268. [176] I owe this observation to J. Ritter.

Little wonder, then, that Greek mathematics stresses form. Throughout the book, I have stressed form rather than content, partly as a method of getting at the cognitive reality behind texts, but partly – and this is the fundamental justification of my approach – because this is the place where stress should be placed, if we are to be sensitive to the historical context of Greek mathematics. Greek mathematics, to put it briefly, was a cultural practice in which the dominant was the form.

4 SUMMARY

Greek mathematical form emerged in the period roughly corresponding to Plato's lifetime (section 1). It was a form used by a small international group of members of the literate elite (section 2). It reflected in part the inevitable background of their life, the Greek *polis* as influenced by the spectacular rise of Athenian democracy; it also reflected the reaction of some of them to this background. While critical, argumentative, looking for convincing persuasive speech, they were also committed to writing and to a certain kind of professionalisation (in the sense not of disciplinary distinctions but of expertise). They thus used a form oral in its essence, most importantly in its insistence upon persuasion, but modified by the possibilities of the written form, most importantly, perhaps, in the introduction of letters to the diagram. Their language was regimented and formulaic, partly because of the attempt to isolate their field, and partly in a way reflecting the double origin of their style, oral and written (subsection 3.1).

They had their curiosity about the material world around them, but also their reservations about productive activity in that world. They occupied a liminal position. Trying to maintain a balance there, their form became neutral as regards interpretation. Most importantly, they preferred to see themselves as 'authors', not as 'practitioners'. Hence a stress on the *presentation*, the tendency to produce mathematics within a recognisable *genre* (subsection 3.2).

Against the polemical background of their culture, the incontrovertibility of a certain aspect of their enterprise became striking, thus encouraging further the neutral, isolated format. This formal incontrovertability dominated their self-perception, and thus their *genre* centred around forms of presentation rather than contents (subsection 3.3).

It is possible to go through the main aspects of the practice described so far and to see the mark of the historical context. Chapters 1 and 2 described the dual nature of the diagram – a material object tamed by the use of letters, the mark of literate culture. Chapter 3, on the lexicon, stressed the role of isolationism on the one hand (the tendency to separate first- from second-order discourse) and the role of 'professionalism' in some limited sense. Chapter 4, on formulae, described how an essentially pre-written practice survives in a highly literate environment. Chapter 5, on necessity, showed how these aspects of the practice aided deduction (subsections 5.1–5.2). It then added two kinds of practices. First, it described the structure of arguments as an idealised, written version of oral argument (subsection 5.3), and again stressed the double oral and written nature of the tool-box, as well as the professionalism assumed by it (subsection 5.4). The material prepared in chapters 1–5 was then seen to be sufficient for generality, the subject of chapter 6. Historical location explains the emergence of the form; the form explains the emergence of deduction.

Partly because some aspects of the relevant background remained in force throughout antiquity, but mainly because of the various self-perpetuating mechanisms of Greek mathematics described in the work, this form survived to remain the distinctive feature of mathematics throughout antiquity. Indeed – albeit with considerable modifications – this form is a distinctive feature of modern science, as well. The tension between the proclaimed universal openness, on the one hand, and the insistence on professionalisation, on the other hand – at bottom, a reflection of the tension between democracy and oligarchy – is a feature of science inherited from the Greeks; possibly, their most lasting legacy.

APPENDIX

The main Greek mathematicians cited in the book

APOLLONIUS

Active at around the end of the third century BC, Apollonius produced a great *œuvre*, ranging from the theory of irrational numbers to astronomy. Only some parts of one work are extant in Greek. This work, however, is the *Conics*, an outstanding presentation and development of the main advanced mathematical theory of pre-modern mathematics:

Heiberg, J. L. (1891–) *Opera Omnia* (2 vols.). Leipzig.

Books I–III are in the first volume. Book IV, together with Eutocius' commentaries, is in the second volume (Eutocius, a sixth-century AD commentator, was an intelligent scholar and mathematician).

ARCHIMEDES

Traditionally given the dates 287–212 BC, Archimedes is one of the greatest mathematicians of all times. A relatively large number of his works has survived. The works are usually dedicated to a single set of problems. Archimedes gradually develops the tools for dealing with those problems, and then, in remarkable *tours-de-force*, the tools are brought to bear on the problems. Measurement of curvilinear objects is a constant theme, but for that purpose he uses a very wide range of techniques, from almost abstract proportion theory (in the *SL*, *CS*) to mechanical considerations (*QP*, *Meth.*). Mechanics, especially, was an Archimedean interest, and he made a seminal contribution to the mathematisation of the physical world, in his *PE*, *CF*. The edition used here is:

Heiberg, J. L. (1910–) *Opera Omnia* (3 vols.). Leipzig. I use the following abbreviations for Archimedean treatises:

Vol. I:
SC On Sphere and Cylinder
DC Measurement of Circle
CS On Conoids and Spheroids
Vol. II:
SL Spiral Lines
PE Plane Equilibria
QP Quadrature of the Parabola

Arenarius (unabbreviated)
CF On Floating Bodies
Meth. The Method
Vol. III contains Eutocius' commentary to *SC*, *DC* and *PE*.

ARISTARCHUS

Aristarchus was active at around the beginning of the third century BC. He has a certain romantic aura, due to the fact that he suggested (we do not know in what detail) the heliocentric hypothesis. Clearly he was mainly an astronomer, and a brilliant mathematical astronomer, as can be seen from his only surviving work, *On the Sizes and Distances of the Sun and the Moon*, which I refer to through the edition included in:

Heath, T. L. (1913/1981) *Aristarchus of Samos*. New York.

(Incidentally, this surviving treatise is logically neutral as regards geocentricity or heliocentricity, nor does it offer any explicit comments related to this question.)

AUTOLYCUS

Autolycus was almost certainly active in the second half of the fourth century BC. Our evidence for Euclid's chronology is very vague, so we cannot tell the exact relation in time between Euclid and Autolycus. But it is probable that, among extant mathematical authors, Autolycus is the earliest. This is his only distinction. His treatises give uninspired proofs of very elementary, boring theorems in mathematical astronomy. The edition used in this work is:

Aujac, G. (1979) *Autolycus de Pitane*. Paris.

EUCLID

We know almost nothing of Euclid. It is probably safest to stick to the established view, which makes him active around the year 300 BC. His (?) extant works are all presentations of elementary mathematics, but it should be borne in mind that this 'elementary' may often become quite difficult. He may have been no more than a compilator; certainly he was intelligent and accurate. I use in this book especially the following works, through the following editions:

Heiberg, J. L. (1883–) *Elementa* (4 vols.). Leipzig.
(1895) *Optica*, etc. Leipzig.
Menge, H. (1896) *Data*. Leipzig.

HERO

Some Hero was probably active in the second half of the first century AD. The works in the Heronian corpus are 'practical', in some sense. When they deal with geometry, they seem to be interested in actual measurements of material objects. They are certainly interested in mechanical gadgets. But while the corpus of extant works projects a clear persona, it is also a difficult corpus in textual terms, and it is never certain who wrote what and when: 'Hero' is generally a better appellation than Hero. To 'Hero', then, I refer mainly through the following:

Schoene, H. (1903) *Metrica*, etc. (vol. III of series). Leipzig.
Heiberg, J. L. (1912) *Definitiones*, etc. (vol. IV of series). Leipzig.
(1914) *Stereometrica*, etc. (vol. V of series). Leipzig.

HIPPOCRATES OF CHIOS

Hippocrates of Chios was active in the second half of the fifth century BC. He is thus the first mathematician about whom anything substantial is known. Of course, none of his works is extant, but we possess an extensive fragment (which generates, however, enormous philological complications), which I refer to through either:

Diels, H. (1882) Simplicius, *In Phys.* Berlin,

or, more often, through the best philological reconstruction, Becker (1936b).

PAPPUS

Pappus was active at the beginning of the fourth century AD. He is known mainly through 'The Collection' – an anthology of mathematical treatises, occupying a mid-position between a commentary on earlier works and an original study. The extant text represents most, but not all, of the original. He is, mathematically, the most competent among ancient mathematical commentators. I refer to him through:

Hultsch, F. (1875/1965) *Collectio*. Amsterdam.

PROCLUS

Active in the fifth century AD, Proclus was an important Neoplatonist philosopher, head of the Neoplatonist Academy in Athens. Of his large *œuvre*, we are interested in one work, a commentary on Euclid's *Elements* book I. This is the most substantial work in the philosophy of mathematics surviving from antiquity, and also contains much historical and mathematical detail. I refer to it through:

Friedlein, G. (1873) *In Euclidem*. Leipzig.

PTOLEMY

Active in the second century AD, Ptolemy is a great astronomer by any standards, and certainly the most important astronomer whose works survive from antiquity – besides producing further works of mathematical interest (and other works, interesting in other respects). I refer mainly to the following:

Duering, I. (1980) *Die Harmonielehre des Klaudios Ptolemaios*. New York.
Heiberg, J. L. (1898–) *Syntaxis* (2 vols.). Leipzig.

Bibliography

Ascher, M. (1981) *Code of the Quipu*. Ann Arbor, MI.

Aujac, G. (1984) 'Le Langage formulaire dans la géométrie grecque', *Revue d'histoire des Sciences* 37: 97–109.

Barker, A. (1989) *Greek Musical Writings* (2 vols.). Cambridge.

Beard, M. (1991) 'Adopting an Approach', in *Looking at Greek Vases*, ed. T. Rasmussen and N. Spivey. Cambridge.

Beck, F. A. G. (1975) *Album of Greek Education*. Sydney.

Becker, O. (1936a) 'Lehre vom Geraden und Ungeraden im Neunten Buch der Euklidischen Elemente', *Quellen und Studien zur Geschichte der Mathematik, Astronomie und Physik* 3: 533–53.

(1936b) 'Zur Textgestellung des Eudemischen Berichts über die Quadratur der Möndchen durch Hippokrates von Chios', *Quellen und Studien zur Geschichte der Mathematik, Astronomie und Physik* 3: 411–19.

Berk, L. (1964) *Epicharmus*. Groningen (in Dutch).

Besthorn, R. O. and Heiberg, J. L. (eds.) (1900) *Codex Leidenensis 399,1 vol. II*. Copenhagen.

Betz, H. D. (1992) *The Greek Magical Papyri* (2nd edn). Chicago.

Bickerman, E. (1968) *Chronology of the Ancient World*. London.

Blass, F. (1887) *Eudoxi Ars Astronomica*. Kiel.

Blomquist, J. (1969) *Greek Particles in Hellenistic Greek*. Lund.

Bourdieu, P. (1993) *The Field of Cultural Production*. Cambridge.

Brashear, W. (1994) 'Vier neue Texte zum antiken Bildungswesen', *Archiv für Papyrusforschung* 40: 29–35.

Brown, P. (1988) *The Body and Society*. New York.

Bulloch, A. et al. (eds.) (1993) *Images and Ideologies*. Berkeley, CA.

Burkert, W. (1962/1972) *Lore and Science in Ancient Pythagoreanism*. Cambridge, MA.

Burnyeat, M. F. (1982) 'Idealism in Greek Philosophy: What Descartes Saw and Berkeley Missed', *Philosophical Review* 91: 3–40.

(1987) 'Platonism and Mathematics: A Prelude to Discussion', in *Mathematics and Metaphysics in Aristotle*, ed. A. Graeser, 213–40. Berne.

Campbell, B. (1984) 'Thought and Political Action in Athenian Tradition: The Emergence of the "Alienated" Intellectual', *History of Political Thought* 5: 17–59.

Carter, L. B. (1986) *The Quiet Athenian*. Oxford.

Chemla, K. (1994) 'Essai sur la signification mathématique des marqueurs de couleur chez Liu Hui (3ème siècle)', *Cahiers de Linguistique Asie Orientale* 23: 61–76.
Coulton, J. J. (1977) *Greek Architects at Work.* London.
Cuomo, S. (1994) 'The Ghost of Mathematicians Past', Ph.D. thesis, Cambridge.
Dantzig, T. (1967) *Number: The Language of Science.* New York.
Dehaene, S. (1992) 'Varieties of Numerical Abilities', in *Numerical Cognition* (a special issue of *Cognition*), ed. S. Dehaene, 1–42.
Descartes, R. (1937/1954) *The Geometry.* New York.
Detienne, M. (1967) *Les Maîtres de vérité dans la Grèce archaïque.* Paris.
Dijksterhuis, E. J. (1938/1956) *Archimedes.* Princeton, NJ.
Dilke, O. A. W. (1985) *Greek and Roman Maps.* London.
Dornseiff, F. (1925) *Das Alphabet in Mystik und Magie* (2nd edn). Leipzig.
Duncan-Jones, R. (1990) *Structure and Scale in the Roman Economy.* Cambridge.
Einarson, B. (1936) 'On Certain Mathematical Terms in Aristotle's Logic', *American Journal of Philology* 57: 33–54, 151–72.
Eisenstein, E. L. (1979) *The Printing Press as an Agent of Change.* Cambridge.
Federspiel, M. (1992) 'Sur la locution ἐφ' οὗ / ἐφ' ᾧ servant à désigner des êtres géométriques par des lettres', in *Mathématiques dans l'antiquité*, ed. J. Y. Guillaumin, 9–25. Saint-Etienne.
Finnegan, R. (1977) *Oral Poetry.* Cambridge.
Finley, M. (1951/1985) *Studies in Land and Credit in Ancient Athens, 500–200 BC.* New Brunswick.
(1965) 'Technical Innovation and Economic Progress in the Ancient World', *Economic History Review* 18: 29–45.
Fodor, J. (1983) *The Modularity of Mind.* Cambridge, MA.
Fowler, D. (1987) *The Mathematics of Plato's Academy.* Oxford.
(1994) 'Could the Greeks Have Used Mathematical Induction? Did They Use It?', *Physis* 31: 253–65.
(1995) 'Further Arithmetical Tables', *Zeitschrift für Papyrologie und Epigraphik* 105: 225–8.
Fraser, P. M. (1972) *Ptolemaic Alexandria.* Oxford.
Fuks, A. (1984) *Social Conflict in Ancient Greece.* Leiden.
Gardies, J. L. (1991) 'La Proposition 14 du livre v dans l'économie des *Eléments* d'Euclide', *Revue d'histoire des Sciences* 44: 457–67.
Gardthausen, V. (1911) *Griechische Palaeographie* (2nd edn). Leipzig.
Gellner, E. (1983) *Nations and Nationalism.* Oxford.
Gillies, D. (1992) *Revolutions in Mathematics.* Oxford.
Glucker, J. (1978) *Antiochus and the Late Academy.* Göttingen.
Gooding, D. (1990) *Experiment and the Making of Meaning.* Dordrecht.
Goody, J. (1977) *The Domestication of the Savage Mind.* Cambridge.
(1987) *The Interface between the Written and the Oral.* Cambridge.
Goody, J. and Watt, I. P. (1963) 'The Consequences of Literacy', *Comparative Studies in Society and History* 5: 304–45.

Gould, S. J. (1980) *The Panda's Thumb*. London.
Goulet, R. (ed.) (1989) *Dictionnaire des philosophes, vol. I*. Paris.
Green, P. (1993) *Hellenistic History and Culture*. Berkeley, CA.
Hainsworth, J. B. (1968) *The Flexibility of the Homeric Formula*. Oxford.
(1993) *The Iliad: A Commentary, Books 9–12*. Cambridge.
Harris, W. (1989) *Ancient Literacy*. Cambridge, MA.
Havelock, E. A. (1982) *The Literate Revolution in Greece and its Cultural Consequences*. Princeton, NJ.
Heath, T. L. (1896) *Apollonius*. Cambridge.
(1897) *The Works of Archimedes*. Cambridge.
(1921) *A History of Greek Mathematics*. Oxford.
(1926) *The Thirteen Books of Euclid's Elements* (2nd edn). Cambridge.
Heitz, E. (1865) *Die Verlorenen Schriften des Aristoteles*. Leipzig.
Herman, G. (1987) *Ritualized Friendship and the Greek City*. Cambridge.
Herreman, A. (1996) 'Eléments d'histoire sémiotique de l'homologie', Ph.D. thesis, Paris 7.
Hilbert, D. (1899/1902) *The Foundations of Geometry*. London.
Hintikka, J. and Remes, U. (1974) *The Method of Analysis*. Dordrecht.
Hoekstra, A. (1965) *Homeric Modifications of Formulaic Prototypes*. Amsterdam.
Housman, A. E. (1988) *Collected Poems and Selected Prose*. London.
Hoyrup, J. (1985) 'Varieties of Mathematical Discourse in Pre-modern Cultural Contexts: Mesopotamia, Greece, and the Latin Middle Ages', *Science and Society* 49: 4–41.
(1990a) 'Algebra and Naive Geometry. An Investigation of Some Basic Aspects of Old Babylonian Mathematical Thought', *Altorientalische Forschungen* 17: 27–69, 262–354.
(1990b) 'Dynamis, the Babylonians and Theaetetus 147c7–148d7', *Historia Mathematica* 17: 201–22.
Huffman, C. (1993) *Philolaus of Croton*. Cambridge.
Jackson, H. (1920) 'Aristotle's Lecture-Room and Lectures', *The Journal of Philology* 35: 191–200.
Janus, C. (1895–) *Musici Graeci* (2 vols.). Leipzig.
Johnson-Laird, P. (1983) *Mental Models*. Cambridge.
Johnston, A. E. M. (1967) 'The Earliest Preserved Greek Map: A New Ionian Coin-Type', *Journal of Hellenic Studies* 87: 86–94.
Jones, A. (1986) *Book 7 of the Collection/Pappus*. New York.
Jones, A. H. M. (1940) *The Greek City*. Oxford.
Jones, A. H. M. et al. (1971) *The Prosopography of the Later Roman Empire, vol. I*. Cambridge.
Kästner, E. (1959) *Pünktchen und Anton*, Gesammelte Schriften, vol. 6. Cologne.
Kiparsky, P. (1974) 'Oral Poetry: Some Linguistic and Typological Considerations', in *Oral Literature and the Formula*, ed. B. A. Stolz and R. S. Shannon, 73–106. Ann Arbor, MI.
Kleijwegt, M. (1991) *Ancient Youth*. Amsterdam.
Klein, F. (1925/1939) *Geometry*. New York.

Klein, J. (1934–6/1968) *Greek Mathematical Thought and the Origins of Algebra*. Cambridge, MA.
Knorr, W. (1975) *The Evolution of the Euclidean Elements*. Dordrecht.
(1981) 'On the Early History of Axiomatics: The Interaction of Mathematics and Philosophy in Greek Antiquity', in *Theory Change, Ancient Axiomatics, and Galileo's Methodology*: Proceedings of the 1978 Pisa Conference on the History and Philosophy of Science vol. 1, ed. J. Hintika, D. Gruender and E. Agazzi, 145–86. Dordrecht.
(1986) *The Ancient Tradition of Geometric Problems*. Boston.
(1989) *Textual Studies in Ancient and Medieval Geometry*. Boston.
(1992) 'When Circles Don't Look Like Circles: An Optical Theorem in Euclid and Pappus', *Archive for the History of Exact Sciences* 44: 287–329.
(1994) 'Pseudo-Euclidean Reflections in Ancient Optics: A Re-examination of Textual Issues Pertaining to the Euclidean Optica and Catoptrica', *Physis* 31: 1–45.
Kuhn, T. S. (1962/1970) *The Structure of Scientific Revolutions*. Chicago.
Lachterman, D. R. (1989) *The Ethics of Geometry*. New York.
Lackner, M. (1992) 'Argumentation par diagrammes: une architecture a base de mots', *Extrême-Orient – Extrême-Occident* 14: 131–68.
Landels, J. G. (1980) *Engineering in the Ancient World*. London.
Lang, M. (1957) 'Herodotus and the Abacus', *Hesperia* 26: 271–87.
Lear, J. (1982) 'Aristotle's Philosophy of Mathematics', *The Philosophical Review* 91: 161–92.
Ledger, G. (1989) *Re-counting Plato*. Oxford.
Lefevre, W. (1988) 'Rechensteine und Sprache', in *Rechenstein, Experiment, Sprache*, ed. P. Damerow, 115–69. Stuttgart.
Lloyd, G. E. R. (1966) *Polarity and Analogy*. Cambridge.
(1973) *Greek Science after Aristotle*. London.
(1978) 'Saving the Appearances', *Classical Quarterly* 28: 202–22.
(1979) *Magic, Reason and Experience*. Cambridge.
(1983) *Science, Folklore and Ideology*. Cambridge.
(1990) *Demystifying Mentalities*. Cambridge.
(1991) *Methods and Problems in Greek Science*. Cambridge.
(1992) 'The Meno and the Mysteries of Mathematics', *Phronesis* 37: 166–83.
Long, A. A. and Sedley, D. N. (1987) *The Hellenistic Philosophers*. Cambridge.
Longrigg, J. (1993) *Greek Rational Medicine*. London.
Lord, A. B. (1960) *The Singer of Tales*. Cambridge, MA.
Maclane, S. and Birkhoff, G. (1968) *Algebra*. New York.
Marrou, H. I. (1956) *A History of Education in Antiquity*. London.
Martindale, J. R. (1980) *The Prosopography of the Later Roman Empire, vol. II*. Cambridge.
Mau, J. and Mueller, W. (1962) 'Mathematische Ostraka aus der Berliner Sammlung', *Archiv für Papyrusforschung* 17: 1–10.
Merlan, P. (1960) *Studies in Epicurus and Aristotle*, Klassisch-Philologische Studien 22: 1–112.

Mill, J. S. (1973) *A System of Logic*, vols. VII–VIII of *Collected Works*. Toronto.
Milne, H. J. M. (1934) *Greek Shorthand Manuals*. Oxford.
Minois, G. (1989) *History of Old Age*. Cambridge.
Mogenet, J. (1950) *Autolycus de Pitane*. Louvain.
Mueller, I. (1981) *Philosophy of Mathematics and Deductive Structure in Euclid's Elements*. Cambridge, MA.
(1982) 'Geometry and Skepticism', in *Science and Speculation*, ed. J. Barnes, J. Brunschwig, M. Burnyeat and M. Schofield 69–95. Cambridge.
(1991) 'On the Notion of a Mathematical Starting Point in Plato, Aristotle and Euclid', in *Science and Philosophy in Classical Greece*, ed. A. C. Bowen, 59–97. New York.
Mugler, C. (1958) *Dictionnaire historique de la terminologie géométrique des Grecs*. Paris.
Nagler, M. N. (1967) 'Towards a Generative View of the Oral Formula', *Transactions of the American Philological Association* 98: 269–311.
(1974) *Spontaneity and Tradition*. Berkeley.
Nasr, S. H. (1968) *Science and Civilization in Islam*. Cambridge, MA.
Needham, J. (1954–) *Science and Civilization in China*. Cambridge.
Nestle, W. (1926) 'ΑΠΡΑΓΜΟΣΥΝΗ', *Philologus* 81: 129–40.
Netz, R. (forthcoming) 'The Limits of Text in Greek Mathematics', in *History of Science, History of Text*, ed. K. Chemla, Dordrecht.
Neugebauer, O. (1945) *Mathematical Cuneiform Texts*. New Haven.
(1955) *Astronomical Cuneiform Texts*. London.
Neugebauer, O. and Van Hösen, H. B. (1959) *Greek Horoscopes*. Philadelphia.
Noble, J. V. (1988) *The Techniques of Painted Attic Pottery*. London.
Olson, D. (1994) *The World on Paper*. Cambridge.
O'Meara, D. (1989) *Pythagoras Revived*. Oxford.
Parry, M. (1971) *The Making of Homeric Verse*. Oxford.
Pascal, B. (1954) *Œuvres complètes*. Paris.
Pedersen, O. (1974) *A Survey of the Almagest*. Odense.
Peermans, W. and Van't Dack, E. (1968) *Prosopographica Ptolemaica, vol. VI*. Louvain.
Peirce, C. S. (1932) *Elements of Logic* (vol. II of *Collected Papers*). Cambridge, MA.
Philip, J. A. (1951/1966) *Pythagoras and Early Pythagoreanism*. Toronto.
Pleket, H. W. (1973) 'Technology in the Greco-Roman World: A General Report', *Talanta* 5: 6–47.
Poincaré, H. (1963) *Mathematics and Science: Last Essays*. New York.
Price, D. (1974) 'Gears from the Greeks', *Transactions of the American Philosophical Society* 64(7): 1–70.
Riggsby, A. M. (1992) 'Homeric Speech Introductions and the Theory of Homeric Composition', *Transactions of the American Philological Society* 122: 99–114.
Rips, L. J. (1994) *The Psychology of Proof*. Cambridge, MA.
Runia, D. T. (1989) 'Aristotle and Theophrastus Conjoined in the Writings of Cicero', in *Cicero's Knowledge of the Peripatos*, ed. W. F. Fortenbaugh and P. Steinmetz, 23–38. New Brunswick.

Russell, B. (1903/1937) *The Principles of Mathematics*. London.
de Ste Croix, G. E. M. (1981) *The Class Struggle in the Ancient Greek World*. London.
Saito, K. (1985) 'Book II of Euclid's *Elements* in the Light of the Theory of Conic Sections', *Historia Scientiarum* 28: 31–60.
(1994) 'Proposition 14 of Book V of the *Elements* – a Proposition that Remained a Local Lemma', *Revue d'histoire des Sciences* 47: 273–82.
Schmandt-Besserat, D. (1992) *Before Writing*. Austin.
Scribner, S. and Cole, M. (1981) *The Psychology of Literacy*. Cambridge, MA.
Sedley, D. N. (1976) 'Epicurus and the Mathematicians of Cyzicus', *Cronache Ercolanesi* 6: 23–54.
(1989) 'Philosophical Allegiance in the Greco-Roman World', in *Philosophia Togata*, ed. M. Griffin and J. Barnes, 97–119. Oxford.
Shapin, S. (1994) *A Social History of Truth*. Chicago.
Sommerstein, A. H. (1987) *Birds/Aristophanes*. Warminster.
Szabo, A. (1969) *Anfänge der griechischen Mathematik*. Munich.
Toomer, G. J. (1976) *On Burning Mirrors/Diocles*. New York.
(1990) *Conics: Books V to VII / Apollonius: The Arabic Translation*. New York.
Turner, E. G. (1968) *Greek Papyri*. Oxford.
Turner, V. (1969) *The Ritual Process*. London.
Unguru, S. (1975) 'On the Need to Rewrite the History of Greek Mathematics', *Archive for the History of Exact Sciences* 15: 67–114.
(1979) 'History of Ancient Mathematics: Some Reflections on the State of the Art', *Isis* 70: 555–64.
(1991) 'Greek Mathematics and Mathematical Induction', *Physis* 28: 273–89.
Unguru, S. and Fried, M. (forthcoming) 'On the Synthetic Geometric Character of Apollonius' Conica', *Mathesis*.
Unguru, S. and Rowe, D. (1981–2) 'Does the Quadratic Equation Have Greek Roots?', *Libertas Mathematica* 1: 1–49, 2: 1–62.
Urmson, J. O. (1990) *The Greek Philosophical Vocabulary*. London.
Van der Waerden, B. L. (1954) *Science Awakening*. Groningen.
Vernant, J. P. (1957) 'La Formation de la pensée positive dans la Grèce archaïque', *Annales* 12: 183–206.
(1962) *Les Origines de la pensée grecque*. Paris.
Vidal-Naquet, P. (1967) 'La raison grecque et la cité', *Raison Présente* 2: 51–61.
Visser, E. (1987) *Homerische Versifikationstechnik*. Frankfurt.
Vita, V. (1982) 'Il punto nella terminologia matematica greca', *Archive for the History of Exact Sciences* 27: 101–14.
Waschkies, H. J. (1977) *Von Eudoxus zu Aristotles*. Amsterdam.
(1993) 'Mündliche, graphische und schriftliche Vermittlung von geometrischem Wissen im Alten Orient und bei den Griechen', in *Vermittlung und Tradierung von Wissen in der griechischen Kultur*, ed. W. Kullman and J. Althoff, 39–70. Tübingen.
Weitzman, K. (1971) *Studies in Classical and Byzantine Manuscript Illumination*. Chicago.

West, M. L. (1992) *Ancient Greek Music*. Oxford.
White, J. E. C. T. (1956) *Perspective in Ancient Drawing and Painting* (Society for the Promotion of Hellenistic Studies, Suppl. Paper 7).
Whittaker, J. (1990) *Alcinoos*. Paris.
Wos, L. et al. (1991) *Automated Reasoning*. New York.

Index

IDEAS IN CONTEXT

Edited by QUENTIN SKINNER (*General Editor*)
LORRAINE DASTON, WOLF LEPENIES,
J. B. SCHNEEWIND and JAMES TULLY

1 RICHARD RORTY, J. B. SCHNEEWIND and QUENTIN SKINNER (eds.)
Philosophy in History
*Essays in the historiography of philosophy**

2 J. G. A. POCOCK
Virtue, Commerce and History
*Essays on political thought and history, chiefly in the eighteenth century**

3 M. M. GOLDSMITH
Private Vices, Public Benefits
Bernard Mandeville's social and political thought

4 ANTHONY PAGDEN (ed.)
The Languages of Political Theory in Early Modern Europe*

5 DAVID SUMMERS
The Judgment of Sense
*Renaissance nationalism and the rise of aesthetics**

6 LAURENCE DICKEY
Hegel: Religion, Economics and the Politics of Spirit, 1770–1807*

7 MARGO TODD
Christian Humanism and the Puritan Social Order

8 LYNN SUMIDA JOY
Gassendi the Atomist
Advocate of history in an age of science

9 EDMUND LEITES (ed.)
Conscience and Casuistry in Early Modern Europe

10 WOLF LEPENIES
Between Literature and Science: The Rise of Sociology*

11 TERENCE BALL, JAMES FARR and RUSSELL L. HANSON (eds.)
Political Innovation and Conceptual Change*

12 GERD GIGERENZER *et al.*
The Empire of Chance
*How probability changed science and everyday life**

13 PETER NOVICK
That Noble Dream
*The 'objectivity question' and the American historical profession**

14 DAVID LIEBERMAN
The Province of Legislation Determined
Legal theory in eighteenth-century Britain

15 DANIEL PICK
Faces of Degeneration
*A European disorder, c.1848–c.1918**

16 KEITH BAKER
Approaching the French Revolution
*Essays on French political culture in the eighteenth century**

17 IAN HACKING
The Taming of Chance*

18 GISELA BOCK, QUENTIN SKINNER and MAURIZIO VIROLI (eds.)
Machiavelli and Republicanism*

19 DOROTHY ROSS
The Origins of American Social Science*

20 KLAUS CHRISTIAN KOHNKE
The Rise of Neo-Kantianism
German Academic Philosophy between Idealism and Positivism

21 IAN MACLEAN
Interpretation and Meaning in the Renaissance
The Case of Law

22 MAURIZIO VIROLI
From Politics to Reason of State
The Acquisition and Transformation of the Language of Politics 1250–1600

23 MARTIN VAN GELDEREN
The Political Thought of the Dutch Revolt 1555–1590

24 NICHOLAS PHILLIPSON and QUENTIN SKINNER (eds.)
Political Discourse in Early Modern Britain

25 JAMES TULLY
An Approach to Political Philosophy: Locke in Contexts*

26 RICHARD TUCK
Philosophy and Government 1572–1651*

27 RICHARD R. YEO
Defining Science
William Whewell, Natural Knowledge and Public Debate in Early Victorian Britain

28 MARTIN WARNKE
The Court Artist
The Ancestry of the Modern Artist

29 PETER N. MILLER
Defining the Common Good
Empire, Religion and Philosophy in Eighteenth-Century Britain

30 CHRISTOPHER J. BERRY
The Idea of Luxury
*A Conceptual and Historical Investigation**

31 E. J. HUNDERT
The Enlightenment's 'Fable'
Bernard Mandeville and the Discovery of Society

32 JULIA STAPLETON
Englishness and the Study of Politics
The Social and Political Thought of Ernest Barker

33 KEITH TRIBE
Strategies of Economic Order
German Economic Discourse, 1750–1950

34 SACHIKO KUSUKAWA
The Transformation of Natural Philosophy
The Case of Philip Melancthon

35 DAVID ARMITAGE, ARMAND HIMY and QUENTIN SKINNER (eds.)
Milton and Republicanism

36 MARKKU PELTONEN
Classical Humanism and Republicanism in English Political Thought 1570–1640

37 PHILIP IRONSIDE
The Social and Political Thought of Bertrand Russell
The Development of an Aristocratic Liberalism

38 NANCY CARTWRIGHT, JORDI CAT, LOLA FLECK and THOMAS E. UEBEL
Otto Neurath: Philosophy between Science and Politics

39 DONALD WINCH
Riches and Poverty
*An Intellectual History of Political Economy in Britain, 1750–1834**

40 JENNIFER PLATT
A History of Sociological Research Methods in America

41 KNUD HAAKONSSEN (ed.)
Enlightenment and Religion
Rational Dissent in Eighteenth-Century Britain

42 G. E. R. LLOYD
Adversaries and Authorities
*Investigations into Ancient Greek and Chinese Science**

43 ROLF LINDNER
The Reportage of Urban Culture
Robert Part and the Chicago School

44 ANNABEL BRETT
Liberty, Right and Nature
Individual Rights in Later Scholastic Thought

45 STEWART J. BROWN (ed.)
William Robertson and the Expansion of Empire

46 HELENA ROSENBLATT
Rousseau and Geneva
From the *First Discourse* to the *Social Contract*, 1749–1762

47 DAVID RUNCIMAN
Pluralism and the Personality of the State

48 ANNABEL PATTERSON
Early Modern Liberalism

49 DAVID WEINSTEIN
Equal Freedom and Utility
Herbert Spencer's Liberal Utilitarianism

50 YUN LEE TOO and NIALL LIVINGSTONE (eds.)
Pedagogy and Power
Rhetorics of Classical Learning

51 REVIEL NETZ
The Shaping of Deduction in Greek Mathematics
A Study in Cognitive History

Titles marked with an asterisk are also available in paperback

Printed in the United States
By Bookmasters